Fuzzy Control

Advances in Soft Computing

Editor-in-chief

Prof. Janusz Kacprzyk

Systems Research Institute

Polish Academy of Sciences

ul. Newelska 6

01-447 Warsaw, Poland

E-mail: kacprzyk@ibspan.waw.pl

http://www.springer.de/cgi-bin/search-bock.pl?series=4240

Esko Turunen
Mathematics Behind Fuzzy Logic
1999. ISBN 3-7908-1221-8

Robert Fullér
Introduction to Neuro-Fuzzy Systems
2000. ISBN 3-7908-1256-0

Robert John and Ralph Birkenhead (Eds.)
*Soft Computing Techniques
and Applications*
2000. ISBN 3-7908-1257-9

Mieczysław Kłopotek, Maciej Michalewicz
and Sławomir T. Wierzchon (Eds.)
Intelligent Information Systems
2000. ISBN 3-7908-1309-5

Peter Sinčák, Ján Vaščák, Vladimír Kvasnička,
and Radko Mesiar (Eds.)
*The State of the Art
in Computational Intelligence*
2000. ISBN 3-7908-1322-2

Rainer Hampel
Michael Wagenknecht
Nasredin Chaker (Eds.)

Fuzzy Control

Theory and Practice

With 207 Figures
and 33 Tables

Physica-Verlag

A Springer-Verlag Company

Prof. Dr. Rainer Hampel
Dr. Michael Wagenknecht
Dr. Nasredin Chaker
University of Applied Sciences Zittau/Görlitz
Institute of Process Technique, Automation
and Measuring Technique (IPM)
Theodor-Körner-Allee 16
02763 Zittau
Germany
E-mail: r.hampel@htw-zittau.de
       m.wagenknecht@htw-zittau.de
       n.chaker@htw-zittau.de

ISSN 1615-3871
ISBN 3-7908-1327-3 Physica-Verlag Heidelberg New York

Die Deutsche Bibliothek – CIP-Einheitsaufnahme
Fuzzy control: with 33 tables / Rainer Hampel, Michael Wagenknecht, Nasredin Chaker (eds.). – Heidelberg;
New York: Physica-Verl., 2000
  (Advances in soft computing)
  ISBN 3-7908-1327-3

Physica-Verlag Heidelberg New York
a member of BertelsmannSpringer Science+Business Media GmbH
© Physica-Verlag Heidelberg 2000
Printed in Germany

Softcover Design: Erich Kirchner, Heidelberg

SPIN 10776205   88/2202-5 4 3 2 1 0 – Printed on acid-free paper

Foreword

The present edited volume is of special importance, and for various reasons. First of all, it is one of the most comprehensive and multifaceted coverage of broadly perceived fuzzy control in the literature. The editors have succeeded to collect papers from leading scholars and researchers on various subjects related to the topic of the volume. What is relevant and original is that - as opposed to so many volumes on fuzzy control published by virtually all major publishing houses that are strongly technically oriented and covering a narrow spectrum of issues relevant to fuzzy control itself – the editors have adopted a more general and far sighted approach.

Basically, the perspective assumed in the volume is that though fuzzy control has reached such a level of maturity and implementability that it has become a part of industrial practice, science and academic research still have a relevant role to play in this area. One should however take into account that by their very nature, the role of science and academic research is very peculiar and going beyond straightforward applications, ad hoc solutions, "quick and dirty" tools and techniques, etc. that are usually effective and efficient for solving practical problems. This does not mean that aspects of practical implementations should not be accounted for by scholars and researchers.

The above perspective, maybe even a rationale, has been very well reflected in this edited volume. The editors have found a very good balance between, on the one hand, conceptual and theoretical aspects, and tools and techniques that are relevant for broadly perceived fuzzy control, and on the other hand, on more practical aspects. Moreover, they have attracted prominent contributors whose original works are highly regarded in the field.

The volume *starts* with an introductory part with more general papers by prominent scientists, Professors Zadeh, Peschel and the editors, that present remarks and visions on paradigms which may be of general relevance to broadly perceived fuzzy technology, notably fuzzy control. These include computing with words, relations between fuzzy and other logics, and a relevant and general issue of the use of fuzzy logic in knowledge representation that is a key issue for intelligent systems in general.

The *second* part contains papers on theoretical aspects of fuzzy sets and possibility theories, including aggregation operators, logical operators, fuzzy inference, fuzzy equations and their solution, fuzzy optimization, etc. These papers concern more general issues, maybe not directly related to narrowly meant fuzzy control, but whose general character is relevant from a broader perspective.

The *third* part contains papers directly related to fuzzy control including those on the theory of fuzzy control, rule bases and inference for fuzzy controllers, the use of Petri nets for the design of control systems, comparative analyses of fuzzy and non-fuzzy controllers, etc.

The *fourth* part is concerned with applications of fuzzy tools, notably of fuzzy control, and includes both technological application as, e.g., in process control, and non-technological ones as, e.g., in medical diagnosis and social sciences.

The *fifth* part discusses various aspects of hybrid approaches, including elements of fuzzy logic, neural networks and evolutionary computation. Both general issues related to, e.g., proper architectures of hybrid systems or inference schemes, and applied aspects are dealt with.

The *last* part of the volume contains various papers on the theory and applications of fuzzy tools and techniques in modeling, forecasting and prediction, monitoring and fault diagnosis, decision making and planning, etc. These areas will certainly gain more and more relevance in the future.

While trying to assess this volume one should also emphasize the following fact. The volume may be viewed as a natural consequence of a series of Fuzzy Colloquia that have been held in Zittau (Germany) for some years, with the eight to come in 2000. Initiated as a small local scientific gathering of scholars and researchers from the neighboring regions of Germany, Poland and the Czech Republic, they have soon become a larger, more global scientific event attracting more and more prominent people from all parts of the world. This is an unquestionable achievement of Professor Hampel, and his collaborators, notably the two other editors of this volume, Dr. Wagenknecht and Dr. Chaker.

The success of Fuzzy Colloquia is also no surprise to many people, including myself. One should bear in mind that long before the reunification of Germany, in the former German Democratic Republic, research in fuzzy sets theory and its applications was flourishing. Professor Manfred Peschel, then an influential official in the Academy of Sciences, had quite early discovered huge potentials of that new field. He was, and still is, a man of great knowledge, vision and enthusiasm, and had attracted many collaborators, including his younger disciples. These people had quickly become leading experts in both theory and applications. Notable examples were here results by, say, Professors Gottwald and Bandemer in more theoretical areas, on the one hand, and Professors Hampel, Bocklisch, Wagenknecht, Straube, Lippe, etc. in more applied areas; I hope that other friends whose names were here omitted would forgive me for this. From this perspective, it is no surprise that since most of these people are still very active, and now have new young and talented disciples, research in fuzzy sets and related areas (e.g., neural networks, evolutionary computation, etc.) in that part of Germany is now very intensive and successful too. And further, the success of the Zittau Colloquia is naturally implied by the fact that it is organized by Professor Hampel together with Dr. Wagenknecht and Dr. Chaker, and that a considerable part of participants consists of people from leading centers from the Eastern lands of Germany headed by people I have just mentioned above.

To summarize this short foreword, I would like to congratulate Professor Hampel, Dr. Wagenknecht and Dr. Chaker, the editors of this volume, for their outstanding editorial work that has resulted in a volume that should be on the shelf of all interested in new trends and developments in broadly perceived fuzzy control, in its basic and applied aspects.

Warsaw, April 2000

Janusz Kacprzyk
Editor in chief
"Advances in Soft Computing"

Preface

The first Zittau Fuzzy Colloquium took place in 1993 on initiative of Professor Peschel, the then director of the International University Institute in Zittau. This idea was picked up and continued by the staff of the *Institute of Process Technique, Process Automation and Measuring Technique (IPM)* at the University of Zittau/Görlitz. Nowadays we can resume with gratification that this meeting has already become a tradition in our academic life.

From the very beginning and first of all, the aim of our activities was to contribute to the dissemination of Fuzzy Logic bases and their potentials of application. Starting in 1988, initial works were carried out at the IPM dealing with fuzzy-supported control of combustion in a heating power plant with raw lignite firing. In consequence of the rapid development of computing technique and particularly, of the introduction of digital signal processing in process automation, the potentials of application of Fuzzy Logic were significantly increasing. Correspondingly, the Fuzzy Set Theory developed by Prof. Zadeh in 1965 became an independent scientific discipline within automation engineering and information processing.

Regarding the application of Fuzzy Logic in process automation, it seems to be useful to distinguish two fundamental classes:

1. applications with high degree of repetition (e.g. in consumer good engineering, automotive industry);
2. applications with lower degree of repetition (e.g. in energy and process engineering).

In both cases it is important that for open and closed loop control, monitoring, and diagnosis non-algorithmic expert knowledge can be exploited. While in the first case, the scientific expenditure for expert knowledge acquisition and resulting structuring, parameterization, and optimization of Fuzzy signal processing can be very high (without increasing the costs for the product), in the second case one has to count on effective costs because of problems complexity. From it the following conclusions can be drawn:

- For the second class mentioned above, tools for structuring, parameterization and optimization must be developed that can compete with classical design methods.

- Experiences from classical signal processing have to be implemented. They play the role of quality criteria.

That is why a large number of publications deal with the design of Fuzzy Controllers having PID-similar behaviour.

The present volume subdivides into the following chapters

Introductory Sections
Fuzzy Sets Theory
Fuzzy Control Theory
Fuzzy Control Application
Neuro Fuzzy Systems and Genetic Algorithms
Diagnosis, Monitoring, and Decision Support Systems

and the contributions therein are widely ranging in the above discussed problems.

Beside the already mentioned features and due to its particular status, the Zittau Fuzzy Colloquium has been engaging in the promotion of cooperation between universities, institutions, and industry in the areas of teaching, research, and development between Eastern and Western European countries; also basing on the "three-countries-corner" built by Poland, Czech Republic and Germany, or broader, between Eastern and Western countries. These ambitions have permanently been encouraged by Professor L. A. Zadeh and other outstanding scientific authorities.

At this place we want to remind of the great engagement and brilliant scientific contributions made by our colleague E. Czogala who died in a tragic accident in October 1998. So, we decided to include one of his earlier conference papers.

The editors would like to express their thanks to all authors, Professor Kacprzyk, and Dr. Bihn for benevolent promotion and support during publishing this volume. Finally, we gratefully acknowledge the generous support by the Saxon State Ministry of Science and Arts (SMWK), and the German Academic Exchange Service (DAAD) which has been one of the decisive pre-conditions for our scientific activities.

March, 2000

Rainer Hampel   University of Applied Sciences
Michael Wagenknecht   Zittau/Görlitz
Nasredin Chaker

Contents

Introductory Sections

Fuzzy Sets Theory

Fuzzy Control Theory

Fuzzy Control Applications

Neuro-Fuzzy Systems and Genetic Algorithms

Diagnosis, Monitoring, and Decision Support Systems

Introductory Sections

From Computing with Numbers to Computing with Words - From Manipulation of Measurements to Manipulation of Perceptions[1]

Lotfi. A. Zadeh

Computer Science Division
University of California Berkeley, CA
zadeh@cs.berkeley.edu

Abstract. Computing, in its usual sense, is centered on manipulation of numbers and symbols. In contrast, computing with words, or CW for short, is a methodology in which the objects of computation are words and propositopns drawn from a natural language, e.g., *small, large, far, heavy, not very likely, the price of gas is low and declining, Berkeley is near San Francisco, CA, it is very unlikely that there will be a significant increase in the price of oil in the near future*, etc. Computing with words (CW) is inspired by the remarkable human capability to perform a wide variety of physical and mental tasks without any measurements and any computations. Familiar examples of such tasks are parking a car, driving in heavy traffic, playing golf, riding a bicycle, understanding speech, and summarizing a story. Underlying this remarkable capability is the brain's crucial ability to manipulate perceptions - perceptions of distance, size, weight, color, speed, time, direction, force, number, truth, likelihood, and other characteristics of physical and mental objects. Manipulation of perceptions plays a key role in human recognition, decision and execution processes. As a methodology, computing with words provides a foundation for a computational theory of perceptions - a theory which may have an important bearing on how humans make - and machines might make - perception-based rational decisions in an environment of imprecision, uncertainty and partial truth.

A basic difference between perceptions and measurements is that, in general, measurements are crisp whereas perceptions are fuzzy. One of the fundamental aims of science has been and continues to be that of processing from perceptions to measurements. Pursuit of this aim has led to brilliant successes. We have sent men to the moon; we can build computers that are capable of performing billions of computations per second; we have constructed telescopes that can explore the far reaches of the universe; and we can date the age of rocks that are millions of years old. But alongside the brilliant successes stand conspicuous underachievements and outright failures. We cannot build robots which can move with the agility of animals or humans; we cannot automate driving in heavy traffic; we cannot translate from one language to another at the level of a human interpreter; we cannot create programs which can summarize

[1] (c) 1999 IEEE. Reprinted, with kind permission, from IEEE Transactions on Circuits and Systems - I: Fundamental Theory and Applications 45, 105-119 (1999).

nontrivial stories; our ability to model the behavior of economic systems leaves much to be desired; and we cannot build machines that can compete with children in the performance of a wide variety of physical and cognitive tasks.

It may be argued that underlying the underachievements and failures is the unavailability of a methodology for reasoning and computing with perceptions rather than measurements. An outline of such a methodology - referred to as a computational theory of perceptions is presented in this paper. The computational theory of perceptions, or CTP for short, is based on the methodology of CW. In CTP, words play the role of labels of perceptions and, more generally, perceptions are expressed as propositions in a natural language. CW-based techniques are employed to translate propositions expressed in a natural language into what is called the Generalized Constraint Language (GCL). In this language, the meaning of a proposition is expressed as a generalized constraint, X isr R, where X is the constrained variable, R is the constraining relation and isr is a variable copula in which r is a variable whose value defines the way in which R constrains X. Among the basic types of constraints are: possibilistic, veristic, probabilistic, random set, Pawlak set, fuzzy, graph and usuality. The wide variety of constraints in GCL makes GCL a much more expressive language than the language of predicate logic.

In CW, the initial and terminal data sets, IDS and TDS, are assumed to consist of propositions expresed in a natural language. These propositions are translated, respectively, into antecedent and consequent constraints. Consequent constraints are derived from antecedent constraints through the use of rules of constraint propagation. The principal constraint propagation rule is the generalized extension principle. The derived constraints are retranslated into a natural language, yielding the terminal data set (TDS). The rules of constraint propagation in CW coincide with the rules of inference in fuzzy logic. A basic problem in CW is that of explicitation of X, R, and r in a generalized constraint, X isr R, which represents the meaning of a proposition, p in a natural language.

There are two major imperatives for computing with words. First, computing with words is a necessity when the available information is too imprecise to justify the use of numbers; and second, when there is a tolerance for imprecision which can be exploited to achieve tractability, robustness, low solution cost and better rapport with reality. Exploitation of the tolerance for imprecision is an issue of central importance in CW and CTP. At this juncture, the computational theory of perceptions - which is based on CW - is in its initial stages of development. In time it may come to play an important role in the conception, design and utilization of information/intelligent systems. The role model for CW and CTP is the human mind.

1 Introduction

In the Fifties, and especially late Fifties, circuit theory was at the height of importance and visibility. It played a pivotal role in the conception and design of electronic circuits and was enriched by basic contributions of Darlington, Bode, McMillan, Guillemin, Carlin, Youla, Kuh, Desoer, Sandberg, and other pioneers.

However, what could be discerned at the time was that circuit theory was evolving into a more general theory - system theory - a theory in which the physical identity of the elements of a system is subordinated to a mathematical characterization of their input/output relations. This evolution was a step in the direction of greater generality, and, like most generalizations, it was driven by a quest for models which make it possible to reduce the distance between an object that is modeled - the modelizand - and its model in a specified class of systems.

In a paper published in 1961 entitled "From Circuit Theory to System Theory," [33] I discussed the evolution of circuit theory into system theory and observed that the high effectiveness of system theory in dealing with mechanistic systems stood in sharp contrast to its low effectiveness in the realm of humanistic systems - exemplified by economic systems, biological systems, social systems, political systems and, more generally, man-machine systems of various types. In more specific terms, I wrote:

> There is a fairly wide gap between what might be regarded as 'animate' systems theorists and 'inanimate' systems theorists at the present time, and it not at all certain that this gap will be narrowed, much less closed, in the near future. There are some who feel that this gap reflects the fundamental inadequacy of conventional mathematics - the mathematics of precisely-defined points, functions, sets, probability measures, etc. - for coping with the analysis of biological systems, and that to deal effectively with such systems, which are generally orders of magnitude more complex than man-made systems, we need a radically different kind of mathematics, the mathematics of fuzzy or cloudy quantities which are not describable in terms of probability distributions. Indeed, the need for such mathematics is becoming increasingly apparent even in the realm of inanimate systems, for in most practical cases the *a priory* data as well as the criteria by which the performance of a man-made system are judged are far from being precisely specified or having accurately-known probability distributions.

It was this observation that motivated my development of the theory of fuzzy sets starting with the 1965 paper "Fuzzy Sets" [34], which was published in *Information and Control*.

Subsequently, in apaper published in 1973, "Outline of a New Approach to the Analysis of Complex Systems and Decision Processes," [37] I introduced the concept of a linguistic variable whose values are words rather than numbers. The concept of a linguistic variable has played and is continuing to play a pivotal role in the development of fuzzy logic and its applications.

The initial reception of the concept of a linguistic variable was far from positive, largely because my advocacy of the use of words in systems and decision analysis clashed with the deepseated tradition of respect for numbers and disrespect for words. The essence of this tradition was succintly stated in 1883 by Lord Kelvin:

> In physical science the first essential step in the direction of learning any subject is to find principles of numerical reckoning and practicable methods for measuring some quality connected with it. I often say that when you can measure what you are speaking about and express it in numbers, you know something about it; but when you cannot measure it, when you cannot express it in numbers, your knowledge is of a meagre and unsatisfactory kind: it may be the

beginning of knowledge but you have scarcely, in your thoughts, advanced to the state of science, whatever the matter may be.

The depth of scientific tradition of respect for numbers and derision for words was reflected in the intensity of hostile reaction to my ideas by some of the prominent members of the scientific elite. In commenting on my first exposition of the concept of a linguistic variable in 1972, Rudolph Kalman had this to say:

I would like to comment briefly on Professor Zadeh's presentation. His proposals could be severely, ferociously, even brutally criticized from a technical point of view. This would be out of place here. But a blunt question remains: Is Professor Zadeh presenting important ideas or is he indulging in wishful thinking? No doubt Professor Zadeh's enthusiasm for fuzziness has been reinforced by the prevailing climate in the U.S. - one of unprecedented permissiveness; it tends to result in socially appealing slogans unaccompanied by the discipline of hard scientific work and patient observation.

In a similar vein, my esteemed colleague Professor William Kahan - a man with a brilliant mind - offered this assessment in 1975:

"Fuzzy theory is wrong, wrong, and pernicious." says William Kahan, a professor of computer sciences and mathematics at Cal whose Evans Hall office is a few door from Zadeh's. "I cannot think of any problem that could not be solved better by ordinary logic." What Zadeh is saying is the same sort of things "Technology got us into this mess and now it can't get us out." Well, technology did not get us into this mess. Greed and weakness and ambivalence got us into this mess. What we need is more logical thinking, not less. The danger of fuzzy theory is that it will encourage the sort of impecise thinking that has brought us so much trouble."

What Lord Kelvin, Rudolph Kalman, William Kahan, and many other brilliant minds did not appreciate is the fundamental importance of the remarkable human capability to perform a wide variety of physical and mental tasks without any measurements and any computations. Familiar examples of such tasks are parking a car; driving in heavy traffic; playing golf; understandable speech, and summarizing a story.

Underlying this remarkable ability is the brain's crucial ability to manipulate perceptions - perceptions of size, distance, weight, speed, time, direction, smell, color, shape, force, likelihood, truth and intent, among others. A fundamental difference between measurements and perceptions is that, in general, measurements are crisp numbers whereas perceptions are fuzzy numbers or, more generally, fuzzy granules, that is, clumps of objects in which the transition from membership to nonmembership is gradual rather than abrupt

The fuzziness of perceptions reflects infinite ability of sensory organs and the brain to resolve detail and store information. A concominant of fuzziness of perceptions is the preponderant partiality of human concepts in the sense that the validity of most human concepts is a matter of degree. For example, we have partial knowledge, partial understanding, partial certainty, partial belief and accept partial solutions, partial truth and partial causality. Furthermore, most human concepts have a granular structure and are context-dependent.

In essence, a granule is a clump of physical or mental objects (points) drawn together by indistinguishability, similarity, proximity or functionality (Fig. 1). A granule may be crisp or fuzzy, depending on whether its boundaries are or are not sharply defined. For example, age may be granulated crisply into years and granulated fuzzily into fuzzy intervals labeled very young, young, middle-aged, old, and very old (Fig. 2). A partial taxonomy of granulation is shown in Fig. 3(a) and (b).

informal: a granule is a clump of objects (points) drawn together by indistinguishability, similarity, proximity or functionality

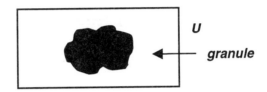

formal: a granule is a clump of objects (points) defined by a generalized constraint

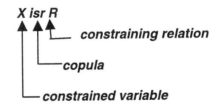

Fig. 1. Informal and formal definitions of a granule

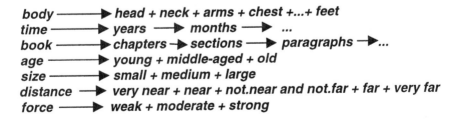

Fig. 2. Examples of crisp and fuzzy granulation

8

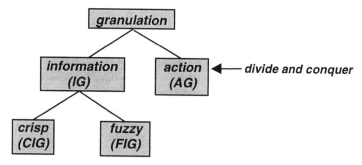

CIG: time ⟶ years ⟶ months ⟶ weeks ⟶ days...
FIG: age ⟶ very.young + young + middle-aged + old + very.old

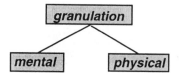

mental granulation: body ⟶ head + neck + left.arm + chest
+right.arm+...
physical granulation: speech
                      walking
                      eating

(a)

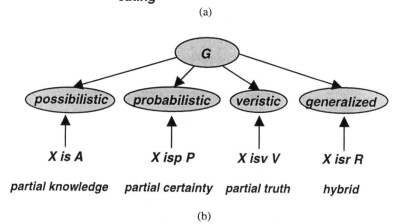

(b)

Fig. 3. (a) Partial taxonomy of granulation. (b) Principal types of granules

In a very broad sense, granulation involves a partitioning of whole into parts. Modes of information granulation (IG) in which granules are crisp play important roles in a wide variety of methods, approaches and techniques. Among them are: interval analysis, quantization, chunking, rough set theory, diakoptics, divide and conquer, Dempster-Shafer theory, machine learning from examples, qualitative process theory, decision trees, semantic networks, analog-to-digital conversion, constraint programming, image segmentation, cluster analysis and many others.

Important though it is, crisp IG has a major blind spot. More specifically, it fails to reflect the fact that most human perceptions are fuzzy rather than crisp. For example, when we mentally granulate the human body into fuzzy granules labeled head, neck, chest, arms, legs, etc., the length of neck is a fuzzy attribute whose value is a fuzzy number. Fuzziness of granules, their attributes and their values is characteristic of ways in which human concepts are formed, organized and manipulated. In effect, fuzzy information granulation (fuzzy IG) may be viewed as a human way of employing data compression for reasoning and, more particularly, making rational decisions in an environment of imprecision, uncertainty and partial truth.

The traditon of pursuit of crispness and precision in scientific theories can be credited with brilliant successes. We have sent men to the moon; we can build computers that are capable of performing billions of computations per second; we have constructed telescopes that can explore the far reaches of the ununiverse; and we can date the age of rocks that are millions of years old. But alongside the brilliant successes stand conspicuous underachievements and outright failures. We cannot build robots which can move with the agility of animals or humans; we cannot automate driving in heavy traffic; we cannot translate from one language to another at the level of a human interpreter; we cannot create programs which can summarize nontrivial stories; our ability to model the behavior of economic systems leaves much to be desired; and we cannot build machines that can compete with children in the performance of a wide variety of physical and cognitive tasks.

What is the explanation for the disparity between the successes and failures? What can be done to advance the frontiers of science and technology beyond where they are today, especially in the realms of machine intelligence and automation of decision processes? In my view, the failures are conspicuous in those areas in which the objects of manipulation are, in the main, perceptions rather than measurements. Thus, what we need are ways of dealing with perceptions, in addition to the many tools which we have for dealing with measurements. In essence, it it this need that motivated the development of the methodology of computing with words (CW) - a methodology in which words play the role of labels of perceptions.

CW provides a methodology for what may be called a *computational theory of perceptions* (CTP) (Fig. 4). However, the potential impact of the methodology of computing with words is much broader. Basically, there are four principal rationales for the use of CW.

1) *The don't know rationale:* In this case, the values of variables and /or parameters are not known with sufficient precision to justify the use of conventional methods of numerical computing. An example is decision- making with poorly defined probabilities and utilities.

2) *The don't need rationale:* In this case, there is a tolerance for imprecision which can be exploited to achieve tractability, robustness, low solution cost, and better rapport with reality. An example is the problem of parking a car.

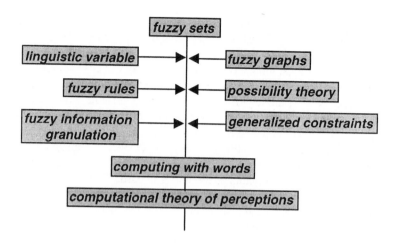

Fig. 4. Conceptual structure of computational theory of perceptions

3) *The can't solve rationale:* In this case, the problem cannot be solved through the use of numerical computing. An example is the problem of automation of driving in city traffic.

4) *The can't define rationale:* In this case, a concept that we wish to define is too complex to admit of definition in terms of a set of numerical criteria. A case in point is the concept of causality. Causality is an instance of what may be called an amorphic concept.

The basic idea underlying the relationship between CW and CTP is conceptually simple. More specifically, in CTP perceptions and queries are expressed as propositions in a natural language. Then the propositions and queries are processed by CW-based methods to yield answers to queries. Simple examples of linguistic characterization of perceptions drawn from everyday experiences are:

> Robert is highly intelligent
> Carol is very attractive
> Hans loves wine
> Overeating causes obesity
> Most Swedes are tall Berjely is more lively than Paolo Alto
> It is likely to rain tomorrow
> It is very unlikely that there will be a significant increase in the
> price of oil in the near future

Examples of correct conclusions drawn from perceptions through the use of CW-based methods are shown in Fig. 5(a). Examples of incorrect conclusions are shown in Fig. 5(b).

Perceptions have long been an object of study in psychology. However, the idea of linking perceptions to computing with words is in a different spirit. An interesting system-theoretic approach to perceptions is described in a recent work of R. Vallee [31]. A logic of perceptions has been described by H. Rasiowa [26]. These approaches are not related to the approach described in our paper.

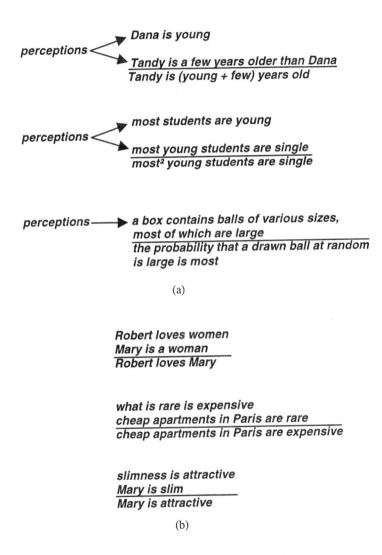

(a)

Robert loves women
Mary is a woman

Robert loves Mary

what is rare is expensive
cheap apartments in Paris are rare
---
cheap apartments in Paris are expensive

slimness is attractive
Mary is slim

Mary is attractive

(b)

Fig. 5. (a) Examples of reasoning with perceptions. (b) Examples of incorrect reasoning

An important point that should be noted is that classical logical systems such as propositional logic, predical logic and modal logic, as well as AI-based techniques for natural language processing and knowledge representation, are concerned in a fundamental way with propositions expressed in a natural language. The main difference between such approaches and CW is that the methodology of CW - which is based on fuzzy logic - provides much more expressive language for knowledge representation and much more versatile machinery for reasoning and computation.

In the final analysis, the role model for computing with words is the human mind and its remarkable ability to manipulate both measurements and perceptions. What should be stressed, however, is that although words are less precise than numbers, the methodology of computing with words rests on a mathematical foundation. An exposition of the basic concepts and techniques of computing with words is presented in the following sections. The linkage of CW and CTP is discussed very briefly because the computational theory of perceptions is still in its early stage of development.

2 What is CW

In this traditional sense, computing involves for the most part manipulation of numbers and symbols. By contrast, humans employ mostly words in computing and reasoning, arriving at conclusions expressed as words from premises expressed in a natural language or having the form of mental perceptions. As used by humans, words have fuzzy denotations. The same applies to the role played by words in CW.

The concept of CW is rooted in several papers starting with my 1973 paper "Outline of a New Approach to the Analysis of Complex Systems and Decision Processes," [37] in which the concepts of a linguistic variable and granulation were introduced. The concepts of a fuzzy constraint and fuzy constraint propagation were introduced in "Calculus of Fuzzy Restrictions," [39], and developed more fully in "A Theory of Approximate Reasoning," [45] and "Outline of a Computational Approach to Meaning and Knowledge Representation Based on a Concept of a Generalized Assignment Statement," [49]. Application of fuzzy logic to meaning representation and its role in test-score semantics are discussed in "PRUF - A Meaning Representation Language for Natural Languages," [43], and " Test-Score Semantics for Natural Languages and Meaning-Representation via PRUF," [46]. The close relationship between CW and fuzzy information granulation is discussed in "Toward a Theory of Fuzzy Information Granulation and its Centrality in Human Reasoning and Fuzzy Logic" [53].

Although the foundations of computing with words were laid some time ago, its evolution into a distinct methodology in its own right reflects many advances in our understanding of fuzzy logic and soft computing - advances which took place within the past few years. (See references and related papers). A key aspect of CW is that it involves a fusion of natural languages and computation with fuzzy variables. It is the fusion that is likely to result in an evolution of CW into a basic methodology in its own right, with wide-ranging ramifications and applications.

We begin our exposition of CW with a few definitions. It should be understood that the definitions are dispositional, that is, admit of exceptions.

As was stated earlier, a concept which plays a pivotal role in CW is that of a granule. Typically, a granule is a fuzzy set of points drawn together by similarity (Fig. 1). A word may be atomic, as in *young*, or composite, as in *not very young* (Fig. 6). Unless stated to the contrary, a word will be assumed to be composite. The denotation of a word may be a higher order predicate, as in Montague grammar [12,23].

- **a word is a label of a fuzzy set**

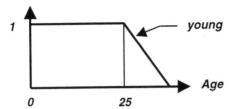

- **a string of words is a label of a function of fuzzy sets**
  - **not very young ➤ (^{2}young)**
- **a word is a description of a constraint on a variable**
  - **Mary is young ➤ Age(Mary) is young**

Fig. 6. Words as labels of fuzzy sets.

In CW, a granule, g, which is the denotation of a word, w, is viewed as a fuzzy constraint on a variable. A pivotal role in CW is played by fuzzy constraint propagation from premises to conclusions. It should be noted that, as a basic technique, constraint propagation plays important roles in many methodologies, especially in mathematical programming and logic programming. (See references and related papers).

As a simple illustration, consider the proposition *Mary is young*, which may be a linguistic characterization of a perception. In this case, *young* is the label of a granule young. (Note that for simplicity the same symbol is used both for a word and its denotation). The fuzzy set *young* plays the role of a fuzzy constraint on the age of Mary (Fig. 6).

As a further example consider the propositions

$$p_1 = \text{Carol lives near Mary}$$

and

$$p_2 = \text{Mary lives near Pat.}$$

In this case, the words *lives near* in p_1 and p_2 play the role of fuzzy constraints on the distances between the residences of Carol and Mary., respectively. If the query is: How far is Carol from Pat?, an answer yielded by fuzzy constraint propagation might be expressed as p_3, where

$$p_3 = \text{Carol lives not far from Pat.}$$

More about fuzzy constraint propagation will be said at a later point.

A basic assumption in CW is that information is conveyed by constraining the values of variables. Furthermore, information is assumed to consist of a collection of

14

propositions expressed in natural or synthetic language. Typically, such propositions play the role of linguistic characterization of perceptions.

A basic generic problem in CW is the following.

We are given a collection of propositions expressed in a natural language which constitute the *initial data set*, or IDS for short.

From the initial data set we wish to infer an answer to a query expressed in a natural language. The answer, also expressed in anatural language, is referred to as the *terminal data set*, or TDS for short. The problem is to derive TDS from IDS (Fig. 7).

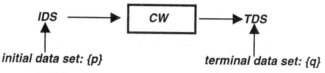

IDS ⟶ **CW** ⟶ **TDS**

initial data set: {p}   *terminal data set: {q}*

p,q: propositional expressed in natural language
{p} = {most students are young, most students are single}
{q} = {most² students are single}

Fig. 7. Computing with words as a transformation of an initial data set (IDS) into a terminal data set (TDS).

A few problems will serve to illustrate these concepts. At this juncture, the problems will be formulated by not solved.

1) Assume that a function, f, $f\colon U \longrightarrow V$, $X \in U$, $Y \in V$, is described in words by the fuzzy if - then rules

$$f\colon \quad \text{if } X \text{ is } small \text{ then } Y \text{ is } small$$
$$\text{if } X \text{ is } medium \text{ then } Y \text{ is } large$$
$$\text{if } X \text{ is } large \text{ then } Y \text{ is } small.$$

What this implies is that f is approximated by a fuzzy graph f^* (Fig. 8), where

$$f^* = small \times small + medium \times large + large \times small.$$

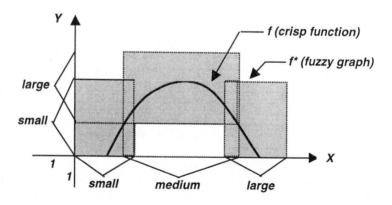

Fig. 8. Fuzzy graph of a function

In f^*, $+$ and $\times$ denote, respectively, the disjunction and Cartesian product. An expression of the form $A \times B$, where A and B are words, will be referred to as a *Cartesian granule*. In this sense, a fuzzy graph may be viewed as a disjunction of cartesian granules. In essence, a fuzzy graph serves as an approximation to a function or a relation [38,51]. Equivalently, it may be viewed as a linguistic characterization of a perception of f (Fig. 9).

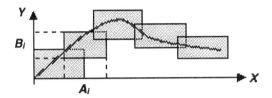

if X is A_i then Y is B_i

Fig. 9. A fuzzy graph of a function represented by a rule-set

In the example under consideration, the IDS consists of the fuzzy rule-set f. The query is: What is the maximum value of f (Fig. 10)?

f: if X is small then Y is small
**   if X is medium then Y is large**
**   if X is large then Y is small**

Problem: <u>maximize f</u>

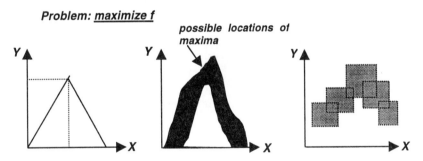

Fig. 10. Fuzzy graph of a function defined by a fuzzy rule-set.

More broadly, the problem is: How can one compute an attribute of a function, f, e.g., its maximum value or its roots if f is described in words as a collection of fuzzy if-then rules? Determination of the maximum value will be discussed in greater detail at a later point.

2) A box contains ten balls of various sizes of which several are large and a few are small. What is the probability that a ball drawn at random is neither larger nor small? In this case, the IDS is a verbal description of the contents of the box; the TDS is the desired probability.

3) A less simple example of computing with words is the following.

Let X and Y be independent random variables taking values in a finite set $V = \{v_1,...v_n\}$ with probabilities $p_1,...p_n$ and $q_1,...q_n$, respectively. For simplification of notation, the same symbols will be used to denote X and Y and their generic values, with p and q denoting the probabilities of X and Y, respectively.

Assume that the probability distributions of X and Y are described in words through the fuzzy if-then rules (Fig. 11):

P: if X is *small* then p is *small*
   if X is *medium* then p is *large*
   if X is *large* then p is *small*

and

Q: if Y is *small* then q is *large*
   if Y is *medium* then q is *small*
   if Y is *large* then q is *large*

where granules *small*, *medium* and *large* are values of linguistic variables X and Y in their respective universes of discourse. In the example under consideration, these rule-sets constitute the IDS. Note that *small* in P need not have the same meaning as *small* in Q, and likewise for *medium* and *large*.

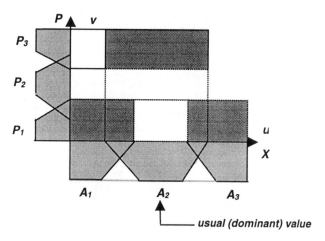

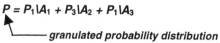

$P = P_1 \backslash A_1 + P_3 \backslash A_2 + P_1 \backslash A_3$

—— *granulated probability distribution*

Fig. 11. A fuzzy graph representation of a granulated probability distribution

The query is: How can we describe in words the joint probability distribution of X and Y? This probability distribution is the TDS.

For convenience, the probability distributions of X and Y may be represented as fuzzy graphs:

$$P: \quad small \times small + medium \times large + large \times small$$

$$Q: \quad small \times large + medium \times small + large \times large$$

with the understanding that the underlying numerical probabilities must add up to unity.

Since X and Y are independent random variables, their joint probability distribution (P,Q) is the product of P and Q. In words, the product may be expressed as [51]

$$(P,Q): \quad small \times small \times (small*large) + small \times medium \times (small*small)$$
$$+ small \times large \times (small*large) + ... + large \times large \times (small*large)$$

where $*$ is the arithmetic product in fuzzy arithmetic [14]. In this example, what we have done, in effect, amounts to a derivation of a linguistic characterization of the joint probability distribution of X and Y starting with linguistic characterizations of the probability distribution of X and the probability distribution of Y.

A few comments are in order. In linguistic characterizations of variables and their dependencies, words serve as values of variables and play the role of fuzzy constraints. In this perspective, the use of words may be viewed as a form of granulation, which in turn may be regarded as a form of fuzzy quantization.

Granulation plays a key role in human cognition. For humans, it serves as a way of achieving data compression. This is one of the pivotal advantages accuring through the use of words in human, machine and man-machine communication.

The point of departure in CW is the premise that the meaning of a proposition, p, in a natural language may be represented as an implicit constraint on an implicit variable. Such a representation is referred to as a *canonical form* of p, denoted as $CF(p)$ (Fig. 12). Thus, a canonical form serves to make explicit the implicit constraint which resides in p. The concept of a canonical form is described in greater detail in the following section.

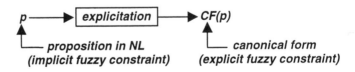

Fig. 12. Canonical form of a proposition.

As a first step in the derivation of TDS from IDS, propositions in IDS are translated into their canonical forms, which collectively represent *antecedent* constraints.

Through the use of rules for constraint propagation, antecedent constraints are transformed into *consequent* constraints.

Finally, consequent constraints are translated into a natural language through the use of *linguistic approximation* [10,18], yielding the terminal data set TDS. This process is schematized in Fig. 13.

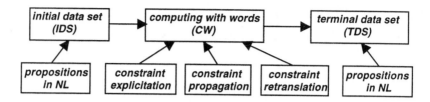

Fig. 13. Conceptual structure of computing with words

In essence, the rationale for computing with words rests on two major imperatives: 1) computing with words is a necessity when the available information is too imprecise to justify the use of numbers and 2) when there is a tolerance for imprecision which can be exploited to achieve tractability, robustness, low solution cost and better rapport with reality.

In computing with words, there are two core issues that arise. First is the issue of representation of fuzzy constraints. More specifically, the question is: How can the fuzzy constraints which are implicit in propositions expressed in a natural language be made explicit. And second is the issue of fuzzy constraint propagation, that is, the question of how can fuzzy constraints in premises, i.e., antecedent constraints, be propagated to conclusions, i.e. consequent constraints.

These are the issues which are addressed in the following.

3 Representation of Fuzzy Constraints, Canonical Forms, and Generalized Constraints

Our approach to the representation of fuzzy constraints is based on test-score semantics [46,47]. In outline, in this semantics, a proposition, p, in a natural language is viewed as a network of fuzzy (elastic) constraints. Upon aggregation, the constraints which are embodied in p result in an overall fuzzy constraint which can be represented as an expression of the form

$$X \text{ is } R$$

where R is a constraining fuzzy relation and X is the constrained variable. The expression in question is the canonical form of p. Basically, the function of a canonical form is to place in evidence the fuzzy constraint which is implicit in p. This is represented schematically as

$$P \rightarrow X \text{ is } R$$

in which the arrow $\rightarrow$ denotes explicitation. The variable X may be vector-valued and/or conditioned.

In this perpective, the meaning of p is defined by two procedures. The first procedure acts on a so-called explanatory database, ED, and returns the constrained variable, X. The second procedure acts on ED and returns the constraining relation, R.

An explanatory database is a collection of relations in terms of which the meaning of p is defined. The relations are empty, that is, they consist of relation names, relations attributes and attributes domains, with no entries in the relations. When there are entries in ED, ED is said to be instantiated and is denoted EDI. EDI may be viewed as a description of a possible world in possible world semantics [6], while ED defines a collection of possible worlds, with each possible world in the collection corresponding to a particular instatiation of ED [47].

As a simple illustration, consider the proposition

$$p = \textit{Mary is not young.}$$

Assume the explanatory database is chosen to be

$$ED = POPULATION[Name;Age] + YOUNG[Age;\mu]$$

in which POPULATION is a relation with arguments Name and Age; YOUNG is a relation with arguments Age and μ; and + is the disjunction. In this case, the constrained variable is the age of Mary, which in terms of ED may be expressed as

$$X = Age(Mary) = {}_{Age} POPULATION[Name = Mary].$$

This expression specifies the procedure which acts on ED and returns X. More specifically, in this procedure, Name is instantiated to Mary and the resulting relation is projected on Age, yielding the age of Mary.

The constraining relation, R, is given by

$$R = ({}^2YOUNG)'$$

which implies that the intensifier *very* is interpreted as a squaring operation, and the negation *not* as the operation of complementation [36].

Equivalently, R may be expressed as

$$R = YOUNG[Age;1 - \mu].$$

As a further example, consider the proposition

$$p = \textit{Carol lives in a small city near San Francisco}$$

and assume that the explanatory database is:

$$ED = POPULATION[name;residence] + SMALL[City;\mu]$$

$$+ NEAR[City1;City2;\mu].$$

In this case,

$$X = residence(Carol) = {}_{Residence}POPULATION[Name = Carol]$$

and

$$R = SMALL[City;\mu] \cap {}_{City1} NEAR[City2 = San\_Francisco].$$

In R, the first constituent is the fuzzy set of small cities; the second constituent is the fuzzy set of cities which are near San Francisco; and $\cap$ denotes the intersection of these sets.

So far we have confined our attention to constraints of the form

$$X \text{ is } R.$$

In fact, constraints can have a variety of forms. In particular, a constraint - expressed as a canonical form - may be conditional, that is, of the form

$$\text{if } X \text{ is } R \text{ then } Y \text{ is } S$$

which may also be written as

$$Y \text{ is } S \text{ if } X \text{ is } R.$$

The constraints in question will be referred to as *basic*.

For purposes of meaning representation, the richness of natural languages necessitates a wide variety of constraints in relation to which the basic constraints form an important though special class. The so-called generalized constraints [49] contain the basic constraints as a special case and are defined as follows. The need for generalized constraints becomes obvious when one attempts to represent the meaning of simple propositions such as

> Robert loves women
> John is very honest
> checkout time is 11 am
> slimness is attractive

in the language of standard logical systems.

A generalized constraint is represented as

$$X \text{ isr } R$$

where isr, pronounced "ezar," is a variable copula which defines the way in which R constrains X. More specifically, the role of R in relation to X is defined by the value of the discrete variable r. The values of r and their interpretations are defined below:

e	equal (abbreviated to =);
d	disjunctive (possibilistic) (abbreviated to blank);
v	veristic;
p	probabilistic;
γ	probability value;
u	usuality;
rs	random set;
rfs	random fuzzy set;
fg	fuzzy graph;
ps	rough set;
.	...

As an illustration, when $r = e$, the constraint is an equality constraint and is abbreviated to =. When r takes the value d, the constraint is *disjunctive* (possiblistic) and isd abbreviated to is, leading to the expression

$$X \text{ is } R$$

in which R is a fuzzy relation which constrains x by playing the role of the possibility distribution of X. More specifically, if X takes values in a universe of discourse, $U = \{u\}$, then $Poss\{X = u\} = \mu_R(u)$, where μ_R is the membership function of R, and Π_X is the possiblity distribution of X, that is, the fuzzy set of its possible values [42]. In schematic form

$$X \text{ is } R \quad \longrightarrow \quad \begin{matrix} \Pi_X = R \\[1em] Poss\{X = u\} = \mu_R(u). \end{matrix}$$

Similarly, when r takes the value v, the constraint is *veristic*.
In the case,

$$X \text{ isv } R$$

means that if grade of membership of u in R is μ, then $X = u$ has truth value μ. For example, a canonical form of the proposition

$$p = \textit{John is proficient in English, French, and German}$$

may be expressed as

$$\text{Proficiency(John) isv (1|English + 0.7|French + 0.6|German)}$$

in which 1.0, 0.7, and 0.6 represent, respectively, the truth values of the propositions *John is proficient in English, John is proficient in French and John is proficient in German*. In a similar vein, the veristic constraint

$$\text{Ethnicity(John) isv (0.5|German + 0.25|French + 0.25|Italian)}$$

represents the meaning of the proposition *John is half German, quarter French and quarter Italian*.
When $r = p$, the constraint is *probabilistic*. In this case,

$$X \text{ isp } R$$

means that R is the probability distribution of X. For example

$$X \text{ isp } N(m,\sigma^2)$$

means that X is normally distributed with mean m and variance σ^2. Similarly,

$$X \text{ isp } (0.2\backslash a + 0.5\backslash b + 0.3\backslash c)$$

means that X is a random variable which takes values a, b, and c with respective probabilities 0.2, 0.5, and 0.3.

The constraint

$$X \text{ isu } R$$

is an abbreviation for

$$usually(X \text{ is } R)$$

which in turn means that

$$Prob\{X \text{ is } R\} \text{ is } usually.$$

In this expression X is R is a fuzzy event and *usually* is its fuzzy probability, that is, the possibility distribution of its crisp probability.

The constraint

$$X \text{ isrs } P$$

is a random set constraint. This constraint is a combination of probabilistic and possibilistic constraints. More specifically, in a schematic form, it is expressed as

$$X \text{ isrs } P$$
$$(X,Y) \text{ is } Q$$
$$\overline{}$$
$$Y \text{ isrs } R$$

where Q is a joint possibilistic constraint on X and Y, and R is a random set. It is of interest to note that the Dempster-Shafer theory of evidence [29] is, in essence, a theory of random set constraints.

In computing with words, the starting point is a collection of propositions which play the role of premises. In many cases, the canonical forms of these propositions are constraints of the basic, possibilistic type. In a more general setting, the constraints are of the genralized type, implying that explicitation of a proposition, p, may be represented as

$$p \rightarrow X \text{ isr } R$$

where X isr R is the canonical form of p (Fig. 14).

As in the case of basic constraints, the canonical form of a proposition may be derived through the use of test-score semantics. In this context, the depth of p is, roughly, a measure of the effort that is needed to explicitate p, that is, to translate p into its canonical form. In this sense, the proposition X isr R is a surface constraint (depth = zero), with the depth of explicitation increasing in the downward direction (Fig. 15). Thus a proposition such as *Mary is young* is a shallow, whereas *it is unlikely that there will be a substantial increase in the price of oil in the near future*, is not.

Once the propositions in the initial data set are expressed in their canonical forms, the groundwork is laid for fuzzy constraint propagation. This is a basic part of CW which is discussed in the following section.

basic premises
- *information is conveyed by constraining - in one way or another - the values which a variable can take*

- *when information is conveyed by propositions in a natural language, a <u>proposition represents an implicit constraint on a variable</u>*

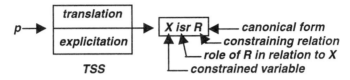

- *the meaning of p is defined by two procedures*
  - *(a) <u>a procedure which identifies X</u>*
  - *(b) <u>a procedure which identifies R and r</u>*

the procedure acts on an explanatory database

Fig. 14. Representation of meaning in test-score semantics

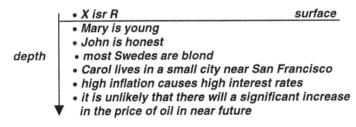

Fig. 15. Depth of explicitation

4 Fuzzy Constraint Propagation and the Rules of Inference in Fuzzy Logic

The rules governing fuzzy constraint propagation are, in effect, the rules of inference in fuzzy logic. In addition to these rules, it is helpful to have rules governing fuzzy constraint modification. The latter rules will be discussed at a later point in this section.

In a summarized form, the rules governing fuzzy constraint propagation are the following [51]. (*A* and *B* are fuzzy relations. Disjunction and conjunction are defined, repsectively, as max and min, with the understanding that, more generally, they could be defined via *t*-norms and *s*-norms [15,24]. The antecedent and consequent constraints are separated by a horinzontal line).

Conjunctive Rule 1:

$$\frac{\begin{array}{l} X \text{ is } A \\ X \text{ is } B \end{array}}{X \text{ is } A \cap B}$$

Conjunctive Rule 2: ($X \in U$, $Y \in B$, $A \subset U$, $B \subset V$)

$$\frac{\begin{array}{l} X \text{ is } A \\ X \text{ is } B \end{array}}{(X,Y) \text{ is } A \times B}$$

Disjunctive Rule 1:

$$X \text{ is } A$$
or
$$\frac{X \text{ is } B}{X \text{ is } A \cup B}$$

Disjunctive Rule 2: ($A \subset U$, $B \subset V$)

$$\frac{\begin{array}{l} X \text{ is } A \\ Y \text{ is } B \end{array}}{(X,Y) \text{ is } A \times V \cup U \times B}$$

where $A \times V$ and $U \times B$ are cylindrical extensions of A and B, respectively.

Conjunctive Rule for isv:

$$\frac{\begin{array}{l} X \text{ isv } A \\ X \text{ isv } B \end{array}}{X \text{ isv } A \cup B}$$

Projective Rule:

$$\frac{(X,Y) \text{ is } A}{Y \text{ is } \text{proj}_V A}$$

Derived Rules:
Compositional Rule:

$$\frac{\begin{array}{l} X \text{ is } A \\ (X,Y) \text{ is } B \end{array}}{Y \text{ is } A \circ B}$$

where $A \circ B$ denotes the composition of A and B.

Extension Principle (mapping rule) [34,40]:

$$\frac{X \text{ is } A}{f(X) \text{ is } f(A)}$$

where $f : U \rightarrow V$, and $f(A)$ is defined by

$$\mu_{f(A)}(v) = \sup_{u | v = f(u)} \mu_A(u)$$

Inverse Mapping Rule:

$$\frac{f(X) \text{ is } A}{X \text{ is } f^{-1}(A)}$$

where $\mu_{f^{-1}(A)}(v) = \mu_A(f(u))$.

Generalized modus ponens:

$$\frac{\begin{array}{c} X \text{ is } A \\ \text{if } X \text{ is } B \text{ then } Y \text{ is } C \end{array}}{Y \text{ is } A \circ \left((\neg B) \oplus C \right)}$$

where the bounded sum $\neg B \oplus C$ represents Lukasiewicz's definition of implication.

Generalized Extension Principle:

$$\frac{f(X) \text{ is } A}{q(X) \text{ is } q(f^{-1}(A))}$$

where

$$\mu_q(v) = \sup_{u | v = f(u)} \mu_A(q(u)).$$

The genralized extension principle plays a pivotal role in fuzzy constraint propagation. However, what is used most frequently in practical applications of fuzzy logic is the *basic interpolation rule*, which is a special case of the compositional rule of inference applied to a function which is defined by a fuzzy graph [38,51]. More specifically, if f is defined by a fuzzy rule set

$$f : \text{if } X \text{ is } A_i \text{ then } Y \text{ is } B_i , \quad i = 1,...,n$$

or equivalently, by a fuzzy graph

$$f \text{ is } \sum_i A_i \times B_i$$

and its argument, X, is defined by the antecedent constraint

$$X \text{ is } A$$

then the consequent constraint on Y may be expressed as

$$Y \text{ is } \sum_i m_i \wedge B_i$$

where m_i is a matching coefficient,

$$m_i = \sup (A_i \cap A)$$

which serves as a measure of the degree to which A matches A_i.

Syllogistic Rule [48]:

$$\frac{\begin{array}{l} Q_1 \text{ } A\text{'s are } B\text{'s} \\ Q_2 \text{ } (A \text{ and } B)\text{'s are } C\text{'s} \end{array}}{(Q_1 \otimes Q_2) \text{ } A\text{'s are } (B \text{ and } C)\text{'s}}$$

where Q_1 and Q_2 are fuzzy quantifiers; A,B and C are fuzzy relations; and $Q_1 \otimes Q_2$ is the product of Q_1 and Q_2 in fuzzy arithmetic.

Constraint Modification Rules [36,43]:

$$X \text{ is } mA \rightarrow X \text{ is } f(A)$$

where m is a modifier such as *not, very, more, or less,* and $f(A)$ defines the way in which m modifies A. Specifically,

$$\text{if } m = not \text{ then } f(A) = A' \text{ (complement)}$$
$$\text{if } m = very \text{ then } f(A) = {}^2A \text{ (left square)}$$

where $\mu_{2A}(u) = (\mu_A(u))^2$. This rule is a convention and should not be constructed as a realistic approximation to the way in which the modifier *very* functions in a natural language.

Probability Qualification Rule [45]:

$$(X \text{ is } A) \text{ is } \Lambda \rightarrow P \text{ is } \Lambda$$

where x is a random variable taking values in U with probability density $p(u)$; Λ is a linguistic probability expressed in words like *likely, not very likely,* etc.; and P is the probability of the fuzzy event X is A, expressed as

$$P = \int_U \mu_A(u) p(u) \, du .$$

The primary purpose of this summary is to underscore the coincidence of the principal rules governing fuzzy constraint propagation with the principal values of inference in fuzzy logic. Of necessity, the summary is not complete and there are many specialized rules which are not included. Furthermore, most of the rules in the summary apply to constraints which are of the basic, possiblistic type. Further development of the rules governing fuzzy constraint propagation will require an extension of the rules of inference to generalized constraints.

As was alluded to in the summary, the principal rule governing constraint propagation is the generalized extension principle which in a schematic form may be represented as

$$\frac{f(X_1,...,X_n) \text{ is } A}{q(X_1,...,X_n) \text{ is } q(f^1(A))} .$$

In this expression, $X_1,...,X_n$ are database variables; the term above the line represents the constraint induced by the IDS; and the term below the line is the TDS expressed as aconstraint on the query $q(X_1,...,X_n)$. In the latter constraint, $f^1(A)$ denotes the preimage of the fuzzy relation A under the mapping $f: U \to V$, where A is a fuzzy subset of V and U is the domain of $f(X_1,...,X_n)$.

Expressed in terms of the meberership functions of A and $q(f^1(A))$, the generalized extension principle reduces the derivation of the TDS to the solution of the constrained maximization problem

$$\mu_{q(X_1,...,X_n)}(v) = \sup_{(u_1,...,u_n)} \mu_A\big(f(u_1,...,u_n)\big)$$

in which $u_1,...,u_n$ are constrained by

$$v = q(u_1,...,u_n) .$$

The generalized extension principle is simpler than it appears. An illustration of its use is provided by the following example.

The IDS is:

most Swedes are tall

The query is: *What is the average height of Swedes?*

The explanatory database consists of a population of N Swedes, $Name_1,...,Name_N$. The database variables are $h_1,...,h_N$, where h_i is the height of $Name_i$, and the grade of membership of $Name_i$ in *tall* is $\mu_{tall}(h_i)$, $i = 1,...,N$.

The proportion of Swedes who are tall is given by the sigma-count [43]

$$\sum Count(tall \cdot Swedes/Swedes) = \frac{1}{N}\sum_i \mu_{tall}(h_i)$$

from which it follows that the constraint on the database variables induced by the IDS is

$$\frac{1}{N}\sum_i \mu_{tall}(h_i) \text{ is } most.$$

In terms of the database variables $h_1,...,h_N$, the average height of Swedes is given by

$$h_{ave} = \frac{1}{N}\sum_i h_i .$$

Since the IDS is a fuzzy proposition, h_{ave} is a fuzzy set whose determination reduces to the constrained maximization problem

$$\mu_{h_{ave}}(v) = \sup_{h_1,\ldots,h_N} \mu_{most}\left(\frac{1}{N}\sum_i \mu_{tall}(h_i)\right)$$

subject to the constraint

$$v = \frac{1}{N}\sum_i h_i .$$

It is possible that approximate solutions to problems of this type might be obtainable through the use of neurocomputing or evolutionary-computing-based methods.

As a further example, we will return to a problem stated in an earlier section, namely, maximization of a function, f, which is described in words by its fuzzy graph, f^* (Fig. 10). More specifically, consider the standard problem of maximization of an objective function in decision analysis. Let us assume as is frequently the case in real-world problems - that the objective function, f, is not well-defined and that what we know about f can be expressed as a fuzzy rule set

$$f: \quad if\ X\ is\ A_1\ then\ Y\ is\ B_1$$
$$if\ X\ is\ A_2\ then\ Y\ is\ B_2$$
$$\ldots\ldots\ldots\ldots\ldots\ldots\ldots\ldots\ldots\ldots\ldots$$
$$if\ X\ is\ A_n\ then\ Y\ is\ B_n$$

or equivalently, as a fuzzy graph

$$f\ is\ \sum_i A_i \times B_i .$$

The question is: What is the point or, more generally, the maximizing set [54] at which f is maximized, and what is the maximum value of f?

The problem can be solved by employing the technique of α-cuts [34,40].

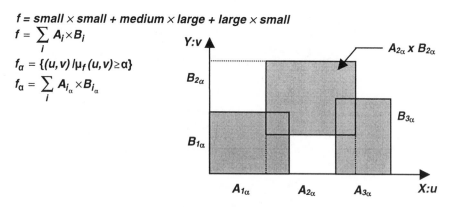

$f = small \times small + medium \times large + large \times small$

$f = \sum_i A_i \times B_i$

$f_\alpha = \{(u,v)\ |\mu_f(u,v) \geq \alpha\}$

$f_\alpha = \sum_i A_{i_\alpha} \times B_{i_\alpha}$

Fig. 16. α-Cuts of a function described by a fuzzy graph

With reference to Fig. 16, if A_{i_α} and B_{i_α} are α-cuts of A_i and B_i, respectively, then the corresponding α-cut of f^* is given by

$$f_\alpha^* = \sum_i A_{i_\alpha} \times B_{i_\alpha}.$$

From this expression, the maximizing fuzzy set, the maximum fuzzy set and maximum value fuzzy set can readily be derived, as shown in Figs. 16 and 17.

A key point which is brought out by these examples and preceding discussions is that explicitation and constraint propagation play pivotal roles in CW. This role can be concretized by viewing explicitation and constraint propagation as translation of propositions expressed in a natural language into what might be called the *generalized constraint language* (GCL) and applying rules of constraint propagation to expressions in this language - expressions which are typically canonical forms of propositions expressed in a natural language. This process is schematized in Fig. 18.

The conceptual framework of GCL is substantively differently from that of conventional logical systems, e.g., predicate logic. But what matters most is that the expressive power of GCL - which is based on fuzzy logic - is much greater than that of standard logical calculi. As an illustration of this point, consider the following problem.

A box contains ten balls of various sizes of which several are large and a few are small. What is the probability that a ball drawn at random is neither large nor small?

To be able to answer this question it is necessary to be able to define the meanings of *large, small, several large balls, few small balls*, and *neither large nor small*. This is a problem in semantics which falls outside of probability theory, neurocomputing and other methodologies.

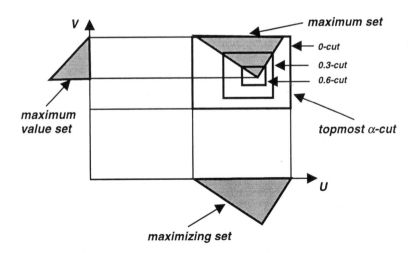

Fig. 17. Computation of maximizing set, maximum set, and maximum value set.

CW

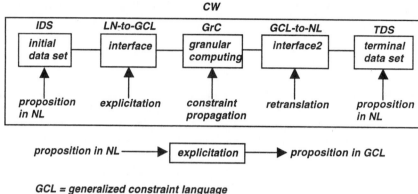

proposition in NL ———▶ explicitation ———▶ proposition in GCL

GCL = generalized constraint language

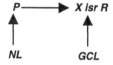

Fig. 18. Conceptual structure of computing with words.

An important application area for computing with words and manipulation of perceptions is decision analysis since in most realistic settings the underlying probabilities and utilities are not known with sufficient precision to jusitfy the use of numerical valuations. There exists an extensive literature on the use of fuzzy probabilities and fuzzy utilities in decision analysis. In what follows, we shall restrict our discussion to two very simple examples which illustrate the use of perceptions.

First, consider a box which contains black balls and white balls (Fig. 19).

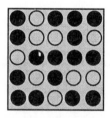

perception: most balls are black
question: what is the probability, P, that a ball
drawn at random is black?

Fig. 19. A box with black and white balls.

If we could count the number of black balls and white balls, the probability of picking a black ball at random would be equal to the proportion, r, of black balls in the box.

Now suppose that we cannot count the number of black balls in the box but our perception is that most of the balls are black. What then, is the probability, p, that a ball drawn at random is black?

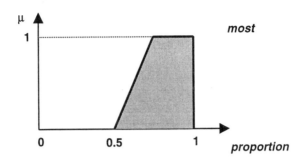

Fig. 20. Membership function of *most*.

Assume that *most* is characterized by its possibility distribution (Fig. 20). In this case, p is a fuzzy number whose possibility distribution is *most*, that is,

$$p \text{ is } most.$$

Next, assume that there is a reward of a dollars if the ball drawn at random is black and a penalty of b dollars if the ball is white. In this case, if p were known as a number, the expected value of the gain would be

$$e = ap - b(1 - p).$$

Since we know not p but its possibility distribution, the problem is to compute the value of e when p is *most*. For this purpose, we can employ the extension principle [34,40], which implies that the possibility distribution, E, of e is a fuzzy number which may be expressed as

$$E = a \, most - b(1 - most).$$

For simplicity, assume that *most* has a trapezoidal possibility distribution (Fig. 20). In this case, the trapezoidal possibility distribution of E can be computed as shown in Fig. 21.

It is of interest to observe that if the support of E is an interval $[\alpha, \beta]$ which straddles 0 (Fig. 22), then there is no noncontroversial decision principle which can be employed to answer the question: Would it be advantageous to play a game in which a ball is picked at random from a box in which most balls are black, and a and b are such that the support of E contains 0. Next, consider a box in which the balls $b_1, \ldots, b_n$ have the same color but vary in size, with b_i, $i = 1, \ldots, n$ having the grade of membership μ_i in the fuzzy set of large balls (Fig. 23). The question is: What is the probability that a ball drawn at random is large, given the perception that most balls are large?

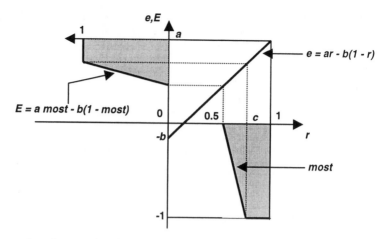

Fig. 21. Computation of expectation through the use of the extension principle.

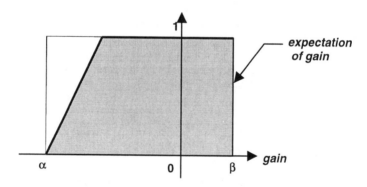

Fig. 22. Expectation of gain.

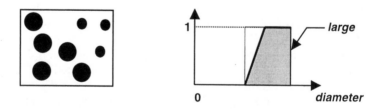

Fig. 23. A box with balls of various sizes and a definition of large ball.

The difference between this example and the preceding one is that the event *the ball drawn at random is large* is a fuzzy event, in contrast to the crisp event *the ball drawn at random is black*.

The probability of drawing b_i is $1/n$. Since the grade of membership of b_i in the fuzzy set of large balls is μ_i, the probability of the fuzzy event *the ball drawn at random is large* is given by [35]

$$P = \frac{1}{n}\sum \mu_i .$$

On the other hand, the proportion of large balls in the box is given by the relative sigma-count [40,43]

$$\sum Count(large \cdot balls / balls \cdot in \cdot box) = \frac{1}{n}\sum \mu_i .$$

Consequently, the canonical form of the perception *most balls are large* may be expressed as

$$\frac{1}{n}\sum \mu_i \text{ is } most$$

which leads to the conclusion that

$$P \text{ is } most.$$

It is of interest to observe that the possibility distribution of P is the same as in the preceding example.

If the question were: What is the probability that a ball drawn at random is *small*, the answer would be

$$P \text{ is } \frac{1}{n}\sum v_i$$

where v_i, $i = 1,...,n$ is the grade of mebership of b_i in the fuzzy set of small balls, given that

$$\frac{1}{n}\sum \mu_i \text{ is } most.$$

What is involved in this case is constraint propagation from the antecedent constraint on the μ_i to a consequent constraint on the v_i. This problem reduces to the solution of a nonlinear program.

What this example points to is that in using fuzzy constraint propagation rules, application of the extension principle reduces, in general, to the solution of a nonlinear program. What we need - and do not have at present - are approximate methods of solving such programs which are capable of exploiting the tolerance for imprecision. Without such methods, the cost of solutions may be excessive in relation to the imprecision which is intrinsic in the use of words. In this connection, an intriguing possibility is to use neurocomputing and evolutionary computing techniques to arrive at approximate solutions to constrained maximization problems. The use of such techniques may provide a closer approximation to the ways in which human manipulate perceptions.

5 Concluding Remarks

In our quest for machines which have a high degree of machine intelligence (high MIQ), we are developing a better understanding of the fundamental importance of the remarkable human capacitiy to perform a wide variety of physical and mental tasks without any measurement and any computations. Underlying this remarkable capability is the brain's crucial ability to manipulate perceptions - perceptions of distance, size, weight, force, color, numbers, likelihood, truth and other characteristics of physical and mental objects. A basic difference between perception and measurement is that, in general, measurements are crisp whereas perceptions are fuzzy. In a fundamental way, this is the reason why to deal with perceptions it is necessary to employ a logical system that fuzzy rather than crisp.

Humans employ words to describe perceptions. It is this obvious observation that is the point of departure for the theory outlined in preceding sections.

When perceptions are described in words, manipulation of perceptions is reduced to computing with words (CW). In CW, the objects of computation are words or, more generally, propositions drawn from a natural language. A basic premise in CW is that the meaning of a proposition, p, may be expressed as a generalized constraint in which the constrained variable and the constraining relation are, in general, implicit in p.

In coming years, computing with words and perception is likely to emerge as an important direction in science and technology. In a reversal of long-standing attitudes, manipulation of perceptions and words which describe them is destined to gain respectability. This is certain to happen because it is becoming increasingly clear that in dealing with real world problems there is much to be gained by exploiting the tolerance for imprecision, uncertainty and partial truth. This is the primary motivation for the methodology of computing with words (CW) and the computational theory of perceptions (CTP) which is outlined in this paper.

Ackowledgement

The author ackowledges Prof. Michio Sugeno, who has contributed so much and in so many ways to the development of fuzzy logic and its applications.

References

1. Berenji, H.R.: Fuzzy Reinforcement Learning and Dynamic Programming. In: Ralescu, A.L.(ed.): Fuzzy Logic in Artificial Intelligence. Proc. IJCAI '93 Workshop. Springer-Verlag, Berlin (1994) 1-9
2. Black, M.: Reasoning with Loose Concepts. Dialog 2 (1963) 1-12
3. Bosch, P.: Vagueness, Ambiguity and All the Rest. In: van de Velde M., Vandeweghe, W. (eds.): Sprachstruktur, Individuum und Gesellschaft. Niemeyer, Tubingen (1978)

4. Bowen, J., Lai, R., Bahler, D.: Fuzzy Semantics and Fuzzy Constraint networks. In: Proc. 1st IEEE Conf. on Fuzzy Systems, San Francisco (1992) 1009-1016

5. Bowen, J., Lai, R., Bahler, D.: Lexical Imprecision in Fuzzy Constraint Networks. In: Proc. Nat. Conf. on Artificial Intelligence (1992) 616-620

6. Cresswell, M.J.: Logic and Languages. Methuen, London (1973)

7. Dubois, D., Fargier, H., Prade, H.: Propagation and Satisfaction of Flexible Constraints. In: Yager, R.R., Zadeh, L.A. (eds.): Fuzzy Sets, Neural Networks, and Soft Computing. Von Nostrand Reinhold, New York (1994) 166-187.

8. Dubois, D., Fargier, H., Prade, H.: Possibility Theory in Constraint Satisfaction Problems: Handling Priority, Preference, and Uncertainty. J. Appl. Intell., to be published

9. Dubois, D., Fargier, H., Prade, H.: The Calculus of Fuzzy Restrictions as a Basis for Flexible Constraint Satisfaction. In: Proc. 2nd IEEE Conf. on Fuzzy Systems, San Francisco (1993) 1131-1136

10. Freuder, E.C., Snow, P.: Improved Relaxation and Search Methods for Approximate Constraint Satisfaction with a Maximum Criterion. In: Proc. 8th Biennial Conf. on the Canadian Society for Computational Studies of Intelligfence, Ontario (1990) 227-230

11. Goguen, J.A.: The Logic of Inexact Concepts. Synthese 19 (1969) 325-373

12. Hobbs, J.R.: Making Compuational Sense of Montague's Intensional Logic. Artif. Intell. 9 (1978) 287-306

13. Katai, O. et al.: Synergetic Computation for Constraint Satisfaction Problems Involving Continuous and Fuzzy Variables by Using Occam. In: Noguchi, S., Umedo, H. (eds.): Transputer/Occam, Proc. 4th Transputer/occam Int. Conf. IOS Press, Amsterdam (1992) 146-160

14. Kaufmann, A., Gupta, M.M.: Introduction to Fuzzy Arithmetic: Theory and Applications. Von Nostrand, New York (1985)

15. Klir, G., Yuan, B.: Fuzzy Sets and Fuzzy Logic. Prentice Hall, Englewood Cliffs (1995)

16. Lano, K.: A Constraint-Based Fuzzy Inference System. In: Barahona, P., Pereira, L.M., Porto, A. (eds.): EPIA 91, 5th Portuguese Conf. on Artificial Intelligence. Springer-Verlag, Berlin (1991) 45-59

17. Lodwick, W.A.: Analysis of Structure in Fuzzy Linear programs. Fuzzy Sets and Systems 38 (1990) 15-26

18. Mamdani, E.H., Gaines, B.R. (eds.): Fuzzy reasoning and Its Applications, London (1981)

19. Mares, M.: Computation Over Fuzzy Quantities. CRC, Boca Raton (1994)

20. Novak, V.: Fuzzy Logic, Fuzzy Sets, and Natural Languages. Int. J. Gen. Syst. 20 (1991) 83-97

21. Novak, V., Ramik, M., Cerny, M., Nekola, J. (eds.): Fuzzy Approach to Reasoning and Decision-Making. Kluwer, Boston (1992)

22. Oshan, M.S., Saad, O.M., Hassan, A.G.: On the Solution of Fuzzy Multiobjective Integer Linear Programming Problems with a Parametric Study. Adv. Modeling Anal. A 24 (1995) 49-64

23. Partee, B.: Montague Grammar. Academic, New York (1976)

24. Pedrycz, W., Gomide, F.: Introduction to Fuzzy Sets. MIT Press, Cambridge, MA (1998)

25. Qi, G., Friedrich, G.: Extending Constraint Satisfaction Problem Solving in Structural Design. In: Belli, F., Rademacher, F.J. (eds.): Industrial and Engineering Applications of Artificial Intelligence and Expert Systems, 5th Int. Conf., IEA/AIE-92. Springer-Verlag, Berlin (1992) 341-350

26. Rasiowa, H., Marek, M.: On Reaching Consensus by Groups of Intelligent Agents. In: Ras, Z.W. (eds.): Methodologies for Intelligent Systems. North-Holland, Amsterdam (1989) 234-243

27. Rosenfeld, A., Hummel, A., Zucker, S.W.: Scene Labeling by Relaxation Operations. IEEE Trans. Syst. Man, Cybern. 6 (1976) 420-433

28. Sakawa, M., Sakawa, K., Inuiguchi, M.: A Fuzzy Satisficing Method for Large-Scale Linear Programming Problems with Block Angular Structure. European J. Oper. res. 81 (1995) 399-409
29. Shafer, G.: A Mathematical Theory of Evidence. Princeton Univ. Press, Princeton (1976)
30. Tong, S.C.: Interval Number and Fuzzy Number Linear programming. Adv. in Modeling Anal. A 20 (1994) 51-56
31. Vallee, R.: Cognition et Systeme. L'Intersisciplinaire Systeme(s), Paris (1995)
32. Yager, R.R.: Some Extensions of Constraint Propagation of Label Sets. Int. J. Approximate Reasoning 3 (1989) 417-435
33. Zadeh, L.A.: From Circuit Theory to System Theory. Proc. IRE 50 (1961) 856-865
34. Zadeh, L.A.: Fuzzy Sets. Inform. Contr. 8 (1965) 338-353
35. Zadeh, L.A.: Probability Measures of Fuzzy Events. J. Math. Anal. Appl. 23 (1968) 421-427
36. Zadeh, L.A.: A Fuzzy-Set Theoretic Interpretation of Linguistic Hedges. J. Cybern. 2 (1972) 4-34
37. Zadeh, L.A.: Outline of a New Approach to the Analysis of Complex System and Decision processes. IEEE trans. Syst., Man, Cybern. SMC-3 (1973) 28-44
38. Zadeh, L.A.: On the Analysis of Large-Scale Systems. In: Gottinger, H. (ed.): Systems Approaches and Environment Problems. Vandenhoeck and Ruprecht, Goettingen (1974) 23-37
39. Zadeh, L.A.: Calculus of Fuzzy Restrictions. In: Zadeh, L.A., Fu, K.S., Shimura, M. (eds.): Fuzzy Sets and Their Application to Cognitive and Decision Processes. Academic, New York (1975) 1-39
40. Zadeh, L.A.: The Concept of a Linguistic Variable and Its Application to Approximate Reasoning. Parts I-III. Inf. Sci. 8 (1975) 199-249, 8 (1975) 301-357, 9 (1975) 43-80
41. Zadeh, L.A.: A Fuzzy-Algorithmic Approach to the Definition of Complex or Imprecise Concepts. Int. J. Man-Machine Studies 8 (1976) 249-291
42. Zadeh, L.A.: Fuzzy Sets as a Basis for a Theory of Possibility. Fuzzy sets and Systems 1 (1978) 3-28
43. Zadeh, L.A.: PRUF - A Meaning Representation Language for Natural Languages. Int. J. Man-Machine Studies 10 (1978) 395-460
44. Zadeh, L.A.: Fuzzy Sets and Information Granularity. In: Gupta, M., Ragade, R., Yager, R.R. (eds.): Advances in Fuzzy Set Theory and Applications. North-Holland, Amstaerdam (1979) 3-18
45. Zadeh, L.A.: A Theory of Approximate Reasoning. In: Hayes, J., Michie, D., Mikulich, L.I. (eds.): Machine Intelligence, vol..9. Halstead, New York (1979) 149-194
46. Zadeh, L.A.: Test-Score Semantics for Natural Languages and Meaning Representation via PRUF. In: Rieger, B. (ed.): Empirical Semantics. Brockmeyer, Germany (1981) 281-349
47. Zadeh, L.A.: Test-Score Semantics for Natural Languages. In: Proc. 9th Int. Conf. on Computational Linguistics. Prague (1982) 425-430
48. Zadeh, L.A.: Syllogistic Reasoning in Fuzzy logic and Its Application to Reasoning with Dispositions. In: Proc. 1984 Int. Symp. on Multiple-Valued Logic, Winnipeg (1984) 148-153
49. Zadeh, L.A.: Outline of a Computational Approach to Meaning and Knowledge Representation Based on the Concept of a generalized Assignment Statement. In: Thoma, M., Wyner, A. (eds.): Proc. Int. Seminar on Artificial Intelligence and Man-Machine Systems. Springer-Verlag, Heidelberg (1986) 198-211
50. Zadeh, L.A.: Fuzzy Logic, Neural Networks, and Soft Computing. Commun. ACM 37 (1994) 77-84
51. Zadeh, L.A.: Fuzzy Logic and the Calculi of Fuzzy Rules and Fuzzy Graphs: A Precis. Multiple Valued Logic 1 (1996) 1-38

52. Zadeh, L.A.: Fuzzy Logic = Computing with Words. IEEE Trans. Fuzzy Syst. 4 (1996) 103-111
53. Zadeh, L.A.: Toward a Theory of Fuzzy Information Granulation and Its Centrality in Human Reasoning and Fuzzy Logic. Fuzzy Sets and Systems 90 (1997) 111-127
54. Zadeh, L.A.: Maximizing Sets and Fuzzy Markoff Algorithms. IEEE Trans. Syst., Man, Cybern. C. 28 (1998) 9-15

Fractal Logics Versus Fuzzy Logics

Manfred Peschel

D - 02779 Großschönau, Hofeweg 3

Abstract. The fractal logics is as well as the fuzzy logics orientated on the aim to reflect phenomena of the real world in a better way as it can be done by binary logics. Within the paper fuzzy logics and fractal logics will be compared based on logical properties. It will be shown that there is no essential difference between the fractal logics and the (non, min, max)-fuzzy logics. The notion of fractal logics uses the notion of logograms which can be decomposed into binary logics. Using logical networks built up of binary logic and fractal logic relations as a construction tool we can generalise the elementary notion of fractal logics to logics on a given halforder relation obeying some simple restrictive conditions. The interpretation of fractal logics leads to very interesting consequences for system sciences especially for model-building and simulation, some of them will be demonstrated in this paper.

1 Logogram and Basic Properties of the (Non, Min, Max)-Fuzzy Logics

A fuzzy logic variable y has an unique negated fuzzy logic variable $non(y)$. In contrary to binary logics in the fuzzy logics does not exist the either-or relation between y and $non(y)$ (Statement of the excluded Third). There is an unique universe element S and an unique impossibility element U. But the following relations, being characteristic for the binary logics, in the fuzzy logics no longer hold

$$y \vee non(y) = S \qquad y \, \& \, non(y) = U \qquad (1)$$

if we realize disjunction operation $\vee$ by *max* in the language of values, if we realize conjunction & by *min* in the language of values. But both relations hold, if we realize the disjunction by next successor and if we realize conjunction by last predecessor of both fuzzy variables y and $non(y)$. Therefore from the point of view of isolated considered fuzzy logical events we meet the following logogram

$$S$$

$$y \qquad\qquad non(y)$$

$$U$$

Fig. 1. Logogram of isolated fuzzy logical events

The decisive difference to binary logics occurs by evaluation that means if we assign truth-values to the considered fuzzy variables y and *non(y)*.

In the binary logics only 0 and 1 are admitted truth-values, but in the fuzzy case all numbers from the closed interval [0, 1] can be truth-values, usually we have grey tone values between 0 and 1.

The truth-values of *non(y)* are generally defined as follows

$$\text{val } (\text{non}(y)) = 1 - \text{val } (y). \tag{2}$$

In the language of values in the fuzzy logics disjunction and conjunction relations are generally defined in the following way

$$\text{val } (y_1) \vee \text{val } (y_2) = \max (\text{ val } (y_1), \text{val } (y_2))$$
$$\text{val } (y_1) \, \& \, \text{val } (y_2) = \min (\text{ val } (y_1), \text{val } (y_2)). \tag{3}$$

Therefore in the value-language the following diagram is no longer a holding logogram

1

val (y)   **val (non(y))**

0

Fig. 2. Diagram in the value-language of fuzzy logics

Of great importance for the evaluation of the fuzzy variables is the method of so-called membership-functions, where according to the concrete conditions of the analyzed problem every fuzzy variable y will be assigned a concrete function *f(y)* defined on the closed interval [0, 1] with values also in this interval. The function *f(y)* can always be considered as a concrete evaluation model for the corresponding fuzzy variable y. Corresponding to the considerations above the following diagram again will not be a valid logogram

1

f (y)   **f (non(y)) = 1 - f (y)**

0

Fig. 3. Diagram in the value-language of fuzzy logics using membership-functions

Usually   max $(f(y), 1 - f(y))$ will be smaller than 1 and   min $(f(y), 1 - f(y))$ will be larger than 0.

It is essential to remark that all properties of the binary logics, which are independent from the „Statement of the excluded Third" preserve their validity in the fuzzy logics expressed in the language of values as there are for example:

1. Laws of commutativity and associativity of $\vee$ and $\&$

2. Idempotent law of negation non (non(y)) = y

3. The formulas of de Morgan

$$A \vee B = non(non(A) \& non(B))$$
$$A \& B = non(non(A) \vee non(B))$$

4. Reproduction laws for $\vee$ and $\&$

$$A = A \& S \qquad A = A \vee U$$

5. Idempotent laws for $\vee$ and $\&$

$$A = A \vee A \qquad A = A \& A$$

6. Distribution laws for $\vee$ and $\&$

$$(A \& B) \vee C = (A \vee C) \& (B \vee C)$$
$$(A \vee B) \& C = (A \& C) \vee (B \& C)$$

Furthermore it is essential to remark, that these properties hold for fuzzy logic logogram ($\vee$ - common next successor, $\&$ - common last predecessor) as well as after evaluation (max, min, fuzzy logic).

2   Logogram and Basic Properties of the Fractal Logics

The fractal logics is as well as the fuzzy logics orientated on the aim to reflect phenomena of the real world in a better way as it can be done by the binary logics firstly all dependencies in the sense of logical interrelationships [1-4].

The fractal logics with its defining logogram originates from religion (taoism), depth-psychology (paradox logics of S. Freud, C. C. Jung, E. Fromm and others).

The defining logogram of the fractal logics is

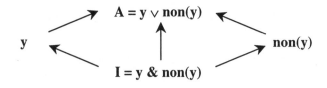

Fig. 4. Logogram of fractal logics

Here $\vee$ in general is the operation of the next successor in the structure and $\&$ in general is the operation of the last common predecessor of the properties or events y and $non(y)$.

The basic notion is here the dialectic polarity between y and $non(y)$. Here A in general has not the significance of an universe S and I in general does not assign the impossibility U.

The bright-dark polarity of the famous YIN & YANG - symbol is a good example of the logogram of the fractal logics.

Fig. 5. YIN & YANG - symbol

In the defining logogram of the fractal logics the horizontal variables and the vertical variables have a quite different character. The horizontal variables belong to the both sides of a dialectic polarity which can be observed in the real world (as an area of cognition, of states etc.). The vertical variables are defined by integration of the variables of the polarity and have therefore the significance of a wholeness (Holons in the sense of A. Koestler), they belong to the spiritual world, to the world of aim, to the will (as area of volition).

This kind of interpretation of the logogram is quite important for all applications of the fractal logics. A little more general the logogram of the fractal logics can be represented as follows

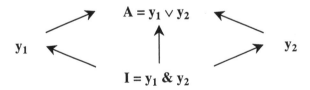

Fig. 6. General representation of the fractal logics logogram

Very interesting is the representation of the fractal logics with the help of binary vectors (y_1, y_2) of length 2 and defining $\vee$ and $\&$ as bit-wise operations of the usual binary logics disjunction $\vee$ and conjunction $\&$. Then we deal with the following logogram exploring us the essential logical properties of the logogram of the fractal logics in general.

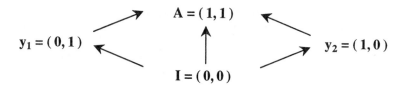

$$A = (1,1)$$

$$y_1 = (0,1)$$

$$y_2 = (1,0)$$

$$I = (0,0)$$

Fig. 7. Fractal logics logogram with binary logic relations

The arrows in this logogram show that there are 5 usual binary logic relations within this structure, namely (I, y_1), (I, y_2), (y_1, A), (y_2, A), and (I, A), whereas (y_1, y_2) no longer is a binary logics.

It is very important to remark that all logical properties 1 - 6 as mentioned above for the case of the fuzzy logics preserve their validity also for the introduced fractal logics, that means, there is no essential difference between the fractal logics and the (non, min, max)-fuzzy logics.

3 Generalisation of the Fractal Logics to Halforder Logics

I refer to the proceedings of the international conference in Zittau „Fuzzy 96" especially to the plenary paper „Fractal Logics of Nature" [4]. We extend now the fundamental logogram of the fractal logics to similar but much more general logical structures with the same basic properties as the fractal logics (properties 1 - 6 without using negations).

There are some disadvantages of the fractal logics reducing the generality of possible applications. Below will be characterized some of these disadvantages.

D1: On the neutral level between A and I, the position of the polarity in fractal logics, there might exist more than two different states symmetrically but not in a polarity interrelationship.

D2: Maybe there exists a whole hierarchy of neutral levels between A and I, and the states on one of the neutral levels can be tree-like splitted up in some consecutive states on a higher neutral level or some different states also on different neutral levels can be fusioned (integrated) to a wholeness between A and I.

With the help of the fractal logics we had up to now not the possibility to describe such generalized logical structures spanned between integrated variables (Holons) and usual states.

But it is quite easy to extend fractal logics on much more general structures arising from some kind of halforders.

We follow according to the rules:

1. On a finite number of elements, called states, a halforder is introduced. Given two different states y_1 and y_2 in the halforder under consideration the following statements are possible:
   * y_1 smaller (lower) than y_2
   * y_2 smaller (lower) than y_1
   * y_1 and y_2 are not in the immediate halforder-relation (are incomparable)

   Such kind of halforder relations are very well known in connection with Pareto-optimality in multi-objective optimization.

2. We assume that the halforder under consideration has an unique minimum element, designated with I (lower compromise) and an unique maximum element A (upper compromise).

According to such a halforder all states between I and A can be uniquely linearly ordered onto levels. All states which are immediate successors to a given state get a reward 1. Proceeding the halforder structure from below a state can accumulate some rewards from immediate predecessors. Thus we get a linear structure of level sets of states.

Usually there is no negation operation in the halforder logics. In the halforder logics can be introduced, in an unique way, the important pair-wise relations disjunction ∨ and conjunction & in the following way:

For a pair of two different states y_1 and y_2 in the halforder structure disjunction ∨ and conjunction & are defined as follows:

$$y_1 \vee y_2 = \text{the smallest common successor in the halforder structure}$$
$$y_1 \,\&\, y_2 = \text{the largest common predecessor in the halforder structure.}$$

(4)

There is an important exclusion property: Both elements (successor and predecessor) must be uniquely defined, as it is for example the case for two trees, one growing downwards and one growing upwards, with the end-branches linked to each other.

In a halforder structure considered we meet exact two states of maximum integration, namely the global maximum A and the global minimum I.

General we admit in the following mixed structures with distributed Holons and distributed states, but the ordering principle for all these structures will be the just introduced halforder logics.

4 The Structural Semantics of Halforder Logics

The structural semantics of fractal logics depends on the interpretation of the integrated states A and I as expressions for a will (volition) aiming to differentiate respectively to integrate different states (state variable submitted to cognition). Thus in the integrated states above (A) and below (I) originate such opposing operations as for example:

Enfolding	&	Contracting
Differentiation	&	Integration
Distribution	&	Accumulation
Breathe out	&	Inhale
Evaporate	&	Condense
Mutation	&	Selection

These interpretations are of importance for diverse fields of application of the fractal logics. The halforder logics with the two Holons A and I admit a corresponding interpretation with the only difference that we now observe a whole tree of diversification or on the contrary of unification. This leads us to a more general structure concept, the logic structure for volition and cognition using occasionally halforder logics as substructure elements. Let us define the structural concept for volition and cognition in a way which promises interesting applications for model-building and simulation of systems behaviour.

5 Networking with Holons (Volition) and States (Cognition)

We consider networks with two kinds of nodes: Holons and states. As a general modelling concept we propose corresponding (fractal respective fuzzy) Petri-Nets. Here the Holons are the transitions and the states are the counter-places.

At first we preserve the structure of usual or generalized Petri-Nets, where different counter-places are always separated from each other by transitions and two different transitions are separated by counter-places. In our extended Petri-Net concept we allow for the possibility to insert between two transitions (A and I) a whole half-order structure according to the logogram of halforder logics.

5.1 A Holon-State Structure Analogous to Petri-Nets

In this structure the transitions are realized by Holons and the counter-places by states. Spread we call a Holon which distributes its actions to a finite set of immediately following state-nodes as successors. Concentrator we call a Holon which accumulates inputs from a finite set of immediate predecessors being all state-nodes. This network is made up from Holons which are simultaneously spreads and concentrators. The whole structure built up from Holons (spread-concentrator) as transitions and states as counter-places is completely identical to Petri-Net structure.

5.2 A Generalized Holon-State Structure

In this structure the Holons again play the role of transitions of the usual Petri-Nets. But now the Holons do not possess single predecessor states. In the structure according to Sect. 5.1 we meet everywhere linear structure-elements of the form

HOLON-STATE-HOLON

In the generalized Holon-State-Holon structure instead of one state now between two different consecutive Holons may be inserted a whole halforder logogram structure.

6 Fuzzification of All Proposed Structural Concepts

In Sect. 1 we saw that from the logical point of view fuzzy logics is identical with fractal logics in the most simple case, but the most important special properties of the fuzzy logics result from the corresponding evaluation concept with the help of membership-functions. Then the basic logical structure is

S

y_1 y_2

U

Fig. 8. Fuzzy logics interpreted as fractal logics

This will be transformed in the following way by singletons

1

f(y₁)   f(y₂)

0

Fig. 9. Result of the evaluation of the fuzzy logic logogram (Fig. 8)

where the states y are assigned specially chosen singletons (membership-functions) and the operations $\vee$ and & are defined by

$$y_1 \vee y_2 = \max (f(y_1), f(y_2))$$
$$y_1 \,\&\, y_2 = \min (f(y_1), f(y_2))$$

(5)

The result is not a state-variable but a corresponding truth-value of the states variable integration by $\vee$ respective &.

The aim of fuzzification of states by using membership-evaluation is to have the possibility of model-building for studying systems behaviour. There are no limits of the fantasy for the fuzzification of the logical structures. We will study here only some simple possibilities quite usual in the common system theory.

6.1 Fuzzification with Singletons and Value Transformation

A singleton is a map

$$x = f(y) \tag{6}$$

defined on $y \in [0, 1]$ with values $x \in [0, 1]$.

We consider a halforder logogram. $I = y$ is the overall input and $x = A$ the overall output of the structure.

Every variable y_i (inclusive y and x) of this structure is assigned a singleton

$$x_i = f_i(y_i). \tag{7}$$

For immediate successors y_i and y_{i+1} we put

$$y_{i+1} = x_i. \tag{8}$$

The immediate predecessors of A maybe

$$z_1, z_2, \ldots, z_k. \tag{9}$$

In general we use for the computation of the value of x a map $F(z_1, \ldots, z_k)$ from

$$[0, 1] \times [0, 1] \times \ldots \times [0, 1] \text{ into } [0, 1]. \tag{10}$$

A special possibility in this context is the fuzzy conjunction or fuzzy disjunction

$$x = \min (f(z_1), f(z_2), \ldots, f(z_k)) \quad \text{respective}$$
$$x = \max (f(z_1), f(z_2), \ldots, f(z_k)). \tag{11}$$

What we get by this method is a static evaluation model of the halforder logic structure.

6.2 Fuzzification Similar to Usual Transitions in Petri-Nets

Petri-Nets are mainly used to model dynamic behaviour of parallel processing systems. We propose now a fuzzification concept which preserves in essential features the behaviour of usual Petri-Nets. We have to differentiate between the predecessor behaviour and the successor behaviour.

Predecessor-behaviour

z_1, z_2, ..., z_k are the immediate predecessors of a spread-concentrator T (transition). With the corresponding singletons $f_i(z_i)$ we compute decrements

$$\Delta z_i = f_i(z_i). \tag{12}$$

The option is to reduce the corresponding state z_i

$$z_i := z_i - \Delta z_i \tag{13}$$

under the condition that the transition T is allowed to fire.

If it occurs, that one of the new states z_i will be smaller than 0, then the transition T is forbidden to fire. If it is not the case, then the *pre-condition* for firing of T is fulfilled.

Successor-behaviour

Maybe x_1, x_2, ..., x_l are the immediate successors of the spread-concentrator T. With the corresponding singletons $f_i(x_i)$ we compute increments

$$\Delta x_i = f_i(x_i). \tag{14}$$

The option is to increase the corresponding state x_i

$$x_i := x_i + \Delta x_i \tag{15}$$

under the condition that the transition T is allowed to fire.

If it occurs, that one of the new states x_i will be larger than 1, then the transition T is forbidden to fire. If it is not the case, then the *post-condition* for firing of T is fulfilled.

The transition T is allowed for firing (action) exactly in the case that both pre-condition and post-condition are fulfilled.

The concept is, that the whole Petri-Net is parallely acting according to the following procedure:

1. For all transitions T_j the corresponding decrements and increments are computed.

2. It is checked which of the transitions T_j are allowed for firing.

3. In parallel for all firing transitions T_j the predecessor states z_i and the successor states x_i are dynamically adjusted according to the computed decrements and increments

$$z_i := z_i - \Delta z_i$$
$$x_i := x_i + \Delta x_i .$$

(16)

7 Modelling of a Lotka-Volterra-Like Behaviour of Ecological Systems

An universal concept for modelling bio-ecological systems was proposed by V. Volterra with the so-called Lotka-Volterra equations

$$\frac{d \ln x_i}{dt} = \sum G_{ij} \cdot x_j .$$

(17)

It was shown by Mende/Peschel [9,10] that all systems described by ordinary differential equations can be transformed in such a way that we get the Lotka-Volterra equations (in the non-autonomous form)

$$\frac{d \ln x_i}{dt} = \sum G_{ij} \cdot x_j + \sum H_{ir} \cdot y_r .$$

(18)

It was shown by Peschel [7,8] that these equations can be iteratively simulated by the following iteration equations

$$x_i' = x_i \cdot \frac{\prod_{G_{ij}>0}(1+G_{ij} \cdot x_i) \cdot \prod_{H_{ir}>0}(1+H_{ir} \cdot y_r)}{\prod_{G_{ij}<0}(1+|G_{ij}| \cdot x_i) \cdot \prod_{H_{ir}<0}(1+|H_{ir}| \cdot y_r)} .$$

(19)

We will study in the following how this kind of general systems behaviour can be modelled by the fuzzification concept proposed in this paper. For this purpose we will use in principle the language of the fuzzified Petri-Nets.

Every species x_i is assigned a spread-concentrator transition T_i. We restrict our consideration on autonomous ecological systems, that means $H_{ir} = 0$. Immediate predecessors of T_i are counter-places of the species x_j with $G_{ij} \neq 0$. Immediate successors of T_i are counter-places of the species x_j with $G_{ji} \neq 0$. That means to every species x_i belongs on one hand a transition T_i and on the other hand a counter-place in the network, therefore all species are in the network double-represented.

We shall see that the fuzzified Petri-Net proposed in this paper offers a more realistic concept for modelling ecological systems than the original Lotka-Volterra equations do.

We shall now study the possibilities how to take into account pre-conditions and post-conditions of the transitions T_i of our Petri-Net. This can easily be done by help of singleton-like mappings. Here the pre-condition will be realized by an existence-preserving mapping, while the post-conditions will be realized by a capacity-preserving mapping. Both functions can be implemented in one type of mappings.

A mapping-type for continuous firing of all transitions should be defined on $[-\infty, \infty]$. On the negative values of y in f(y) the mapping should assume small positive values as a bonus for permanent existence (persistence), for positive values up to the prescribed capacity K of y the map f(y) could have a linear slope and for values of y larger than the capacity K the map f(y) should give values moderate smaller than K.

This concept is more realistic than the usual Lotka-Volterra approach because of the action in discrete time, of the continuous parallely functioning and of the preserving of existence (persistence) and taking into account the capacity restriction.

The behaviour of the Lotka-Volterra equations will be according to our concept implemented in the following way:

$$\Delta x_i := x_i \cdot \sum_{G_{ij} \neq 0} f_{ij}(x_j) \cdot \text{sgn}(G_{ij}) \qquad \textit{accumulation.} \qquad (20)$$

The consequences for the successors are given by

$$x_j' := x_j + f_{ji}(x_i) \cdot \text{sgn}(G_{ji}) \qquad \text{for all } G_{ji} \neq 0 \qquad \textit{distribution} \qquad (21)$$

Remark 1. For modelling the original Lotka-Volterra behaviour the accumulation statements would be sufficient.

8 Demonstration of the Fuzzy Petri-Net Modelling Method for the Most Simple Predator-Prey System (Verhulst 1841)

The simplest predator-prey model describes the interaction of one predator (for example foxes, f) with one prey (for example hares, h) with the following Lotka-Volterra system

$$\frac{d\ln h}{dt} = a - b \cdot f \qquad \frac{d\ln f}{dt} = c \cdot h - d \qquad (22)$$

The hares h are fed by a constant recourse a and are consumed by the foxes f with the rate b. In consequence of eating hares the fox-population increases with the growth rate c, the foxes f then die with the rate d in a constant rate usually.

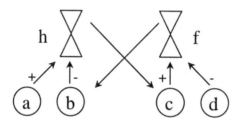

Fig. 10. Predator-prey interaction

According to our approach we get immediately the following model

$$h' = h \cdot [1 + f(a) - f(f)]$$
$$f' = f \cdot [1 + f(h) - f(d)] \qquad (23)$$

Here $f(a) = a$ is the usual assumption, of interest is also $f(a) = \chi \cdot a$, where χ smaller than 1 means, that the food-base for the hares h will be worse from year to year (tact to tact), larger than 1 shows that the food situation improves from year to year.

If we define $f(d) = \mu \cdot d$ we get the usual assumption for $\mu = 1$, $\mu < 1$ means that the danger of dying is less where for $\mu > 1$ the danger of dying is increased.

The mapping f(f) and f(h) guarantee the persistence and following the capacity condition, and take into account possible proportionalities b respective c.

With this improved model for the most simple predator-prey model we quite easily meet chaotic behaviour of the bio-ecological model similar to Feigenbaum behaviour according May's iteration [5,6]

$$x := r \cdot x \cdot (1 - x) \, . \tag{24}$$

9   Conclusions

1. The fractal logics respective halforder logics generalise the well-known possibilities of the fuzzy logics.

2. Petri-Net concepts based on fractal logics or halforder logics represent a general tool for realistic model-building.

3. Models should be made up of two system concepts:
    −   one network for the Holons representing control aims and optimisation objective functions (transitions),
    −   one network for the updating of all state-variables (counter-places).

References

1. Peschel, M.: Der Taoismus in Religion und Wissenschaft. Seminar Systemwissenschaften, Report 9. TU Chemnitz 1995/1999, Chair of System Theory Prof. Dr. St. Bocklisch
2. Gotzmann, J.; Peschel, M.: Nichtlineare Dynamik in der Systemwissenschaft. Seminar Systemwissenschaften, Report 10. TU Chemnitz 1996, Chair of System Theory Prof. Dr. St. Bocklisch
3. Ferstl, F. (ed.): Jacob Böhme - seine Zeit und unsere Zeit. Manuskriptsammlung des Jakob-Böhme-Workshop, HTWS Zittau/Görlitz. November 1995
4. Peschel, M.: Fractal Logics of Nature. Proceedings of Fuzzy 96. Fuzzy Logic in Engineering and Natural Sciences. HTWS Zittau/Görlitz. September 1996
5. Mandelbrot, B.B.: The Fractal Geometry of Nature. W. H. Freeman and Company. New York 1977
6. Peitgen, H.-O., Richter P.H.: The Beauty of Fractals. Springer-Verlag Berlin-Heidelberg-New York-Tokyo 1986
7. Peschel, M.: Rechnergestützte Analyse regelungstechnischer Systeme. Akademie-Verlag Berlin 1992
8. Peschel, M.: Regelungstechnik auf dem Personalcomputer. VEB Verlag Technik Berlin 1988
9. Peschel, M., Mende, W: The Predator-Prey Model, Do we live in a Volterra World? Akademie-Verlag Berlin und Springer Verlag Wien 1986
10. Peschel, M., Breitenecker, F.: Kreisdynamik. Akademie-Verlag Berlin 1990

Knowledge Representation Using Fuzzy Logic Based Characteristics

Nasredin Chaker, Rainer Hampel

Institute of Process Technique, Process
Automation and Measuring Technique (IPM)
University of Applied Sciences Zittau/Görlitz
Theodor-Körner-Allee 16
D-02763 Zittau, Germany
{r.hampel,n.chaker}@htw-zittau.de
http://www.hs-zigr.de/ipm/

Abstract. Rule based systems for Fuzzy Control, Fuzzy Supervising and Fuzzy Diagnosis are non-linear multi-input, multi-output systems. The result of the signal processing within the system is a high dimensional characteristic field. For optimizing the behaviour of the system, we have a large number of degrees of freedom. The reconstruction of the knowledge base from the characteristic field as starting point is impossible. With this background, the paper describes the quality of a two dimensional Fuzzy Controller characteristic field in connection with the necessary deformation for the compensation of non-linear effects. We will demonstrate that one should define two restrictions for the characteristic field: continuity without local extrema and differentiability. Under such conditions we need only two free degrees for optimizing the Controller behaviour using the characteristic field deformation. With help of this experience, the high dimensional Controller will be cascaded. A subjective decision for it is to distinguish the type of the input variables in dominant, non-dominant, and optimization variables. The quality of the cascade is depending on the final characteristic field and the completeness of the rule base. An example will demonstrate the effects of cascading. The cascading concept is realized in the so called High Speed Matrix Controller (HSMC), whose structure is described in this paper, as well.

1 Introduction

The application of Fuzzy Logic in process automation and process diagnosis has firmly established itself in the dating back years. While small applications in the consumer goods branch (e. g. camera, washing machines, etc.) have been accepted without reservations, fewer applications in energy and process engineering field are known. Causes for it could be:

- The requirements on monitoring, automation and diagnosis are very complex regarding the equipment and process specifications, so that a slight repetition degree for a technical realization is given.
- There often do exist reservations against the application of new methods of process automation by the process' experts and users.

- Real-time requirements.
- Software reliability.

The above schedule is not complete and the items are of equal relevance. However, during the introduction of new techniques, there are both a phase of euphoria as well objective and subjective hindrance reasons.

Concepts and formulations like

- linguistic values of input variables
- incomplete knowledge base
- unknown physical and/or analytical models

lead to the situation that engineers interpret the Fuzzy Logic as

Fuzzy information processing

and make doubtful the reproducible relationship between input and output variables.
Therefore, it is necessary to emphasize that Fuzzy Logic is a

method for processing Fuzzy signals and relations

with reproducible algorithms, so that for the same (similar) input data set, the same (similar) output data set is determined. In this relation as well as in the computing expenditure, Fuzzy Logic does not differ from other modern methods of signal processing in process automation.

The condition for including further application areas is to provide efficient methods and procedures for

- choice of structure
- parameterization
- optimization

for the application of Fuzzy Logic in process automation and diagnosis.

With this aim, the following tasks were carried out at the IPM at the University of Zittau/Görlitz:

- Description of Fuzzy Controllers by characteristic fields.
- Minimization of the degrees of freedom for the parameterization and optimization.
- Cascading high dimensional Fuzzy Controllers.

The following report summarizes our results. Experience with the application of these results has been gained and can be exchanged.

Fuzzy Controllers for
- Steam turbine control
- Position control for magnetic borne rotating shafts
- Waste water neutralization
- Optimal operation of flue gas desulfurizing plants

Model based measuring methods for the level measurement
- Fuzzy modelling of multidimensional non-linear processes - design and analysis of structures [20]
- Fuzzy modelling of dynamic non-linear processes applied for water level measurement [19]

Fuzzy supported diagnosis systems for
 - Level measurement according to the hydrostatic measuring principle
 - State diagnosis for magnetic borne rotating machines.

2   Description of Fuzzy Controllers by Characteristic Fields

Classical PID Controllers with different complements (adaptation, disturbance and setpoint feedforward) have a high acceptance in process automation. Hence, we aim at a comparison between the quality of Fuzzy Controllers and classical PID Controllers [1,4,7,11-14,17].

Figure 1 shows that the Fuzzy Controller in the process shows the same interfaces like the classical PID Controller. Independently from the chosen structure, the dynamic is determined by differentiation and integration of the system deviation outside the Fuzzy Controller module. With it, the Fuzzy-PID-Controller is provided with the input variables. The dimension still rises if additional disturbance and setpoint variables are feedforwarded [18]. As a result, the system becomes rather complicated.

The strategy developed at the IPM aims at the decription of multi input - single output (MISO) systems by structures with two-dimensional Fuzzy Controllers that can be optimized and parameterized individually.

Figure 2 exemplarily shows a Fuzzy-PI-Controller for which the basis rule has the form.

$$\text{IF } X1 \text{ AND } X2 \text{ THEN } Y. \tag{1}$$

X1 - integral of system deviation
X2 - system deviation
 Y - manipulated variable.

The characteristic field for the PI-Controller presents two areas

 - non-deformed characteristic field according to the classical non-adjustable PI-Controller,
 - deformed characteristic field according to the non-linear Fuzzy-PI-Controller.

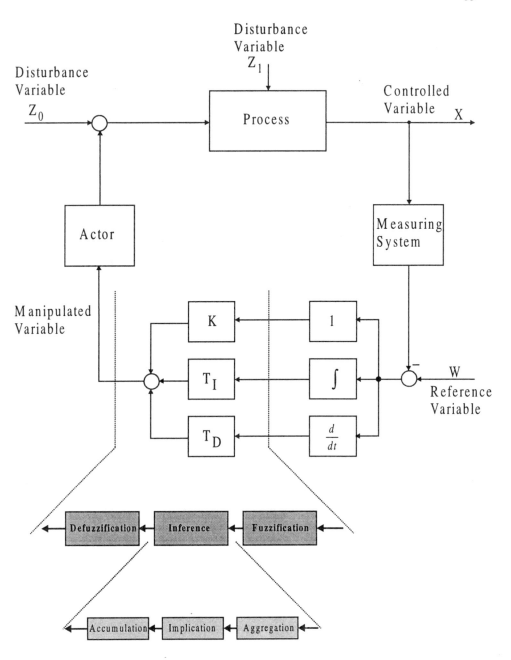

Fig. 1. Fuzzy closed-loop control system

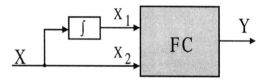

Fuzzy Controller Structure

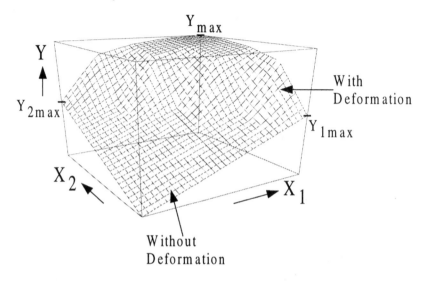

Fig. 2. Fuzzy Controller characteristic field of a non-linear PI-Controller

For the parameterization, the values

$$Y_{1max}, Y_{2max} \text{ and } Y_{max}$$

have to be determined. The optimization results from deforming the characteristic field. By a simple example for a mass flow rate control with an adaptive PI-Controller (compressible medium), it is shown that the necessary deformation of the characteristic field lies in narrow boundaries and can be approximated by experience very well.

Under the assumption that through adaptation of the setting value K_I and K_P an optimal behaviour can be achieved according to the classical control theory, the characteristic fields

$$K_I = f_1(p, \dot{m}) \tag{2}$$

$$K_P = f_2(p, \dot{m}) \tag{3}$$

in Figure 4 must have a deformation which corresponds to this adaptation. For the calculation of the values the method of gain optimum was used. All properties of the Controller are reproducibly stored in the characteristic field.

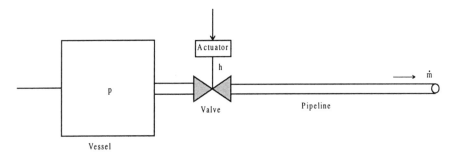

a) Functional block diagram

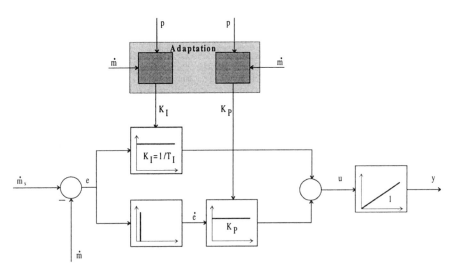

b) Structure of the classical adaptive PI-Controller

Fig. 3. Principle and functional block diagram of the controlled system

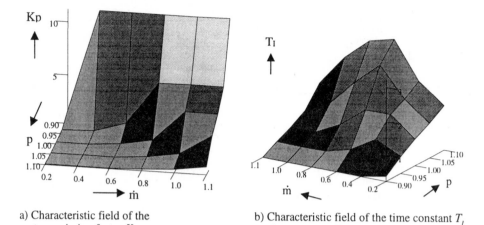

a) Characteristic field of the
transmission factor K_p

b) Characteristic field of the time constant T_I
time constant T_I

Fig. 4. Optimal Controller setting values (according to gain optimum)

3 Degrees of Freedom for the Parameterization and Optimization of Two-Dimensional Fuzzy Controllers

For the parameterization and optimization of the Fuzzy Controller characteristic field there exist many degrees of freedom (Table 1). This diversity, however, makes the user feel more insecure than necessary. Results of an extensive sensibility analysis which was carried out by Chaker [16], show that by fixing some boundary conditions, for many applications the number of free parameters can be reduced to the one.

Table 1. Free variable parameters for the optimization of Fuzzy Controllers

Fuzzification (MSF)	Inference (Operator)	Defuzzification
• Number of sets	• T-Norm	• Center-Of-Gravity
• Membership function	Min	• Singleton
• Distribution	Prod	• Maximum
symmetrical	Bounded Difference	• Mean of Maximum
non symmetrical	• S-Norm	
equidistant	Max	
non equidistant	Sum	
• Overlapping	Bounded Sum	
• Spread		
left spread		
right spread		

MSF - membership function

The following quality criteria for the shape and deformation of the characteristic field are used:

- Differentiability (smooth characteristic by changeable transmission coefficients)
- Slight waviness
- Ability to deformation (maximal necessary deviation from the linear characteristic field).

Differentiability and waviness are essentially determined by the combination operator - membership function. Well suited are:

- Operator: Sum-Prod for λ-sets
- Operator: Max-Min for Gaussian-sets.

For these cases it is sufficient to vary the number and distribution of the Fuzzy Sets for the deformation of the characteristic field.

The optimal behaviour of the Fuzzy Controller is copied in the deformation of the characteristic field. Figures 5 to 7 demonstrate these results.

Figure 5 shows as reference sample a linear characteristic field which was generated by λ-sets, the operator Sum-Prod and the Singleton method for defuzzification. This characteristic field fulfills also all quality criteria cited above.

By changing the distribution of the sets (Figure 6) with fixed overlap degree, a deformation of the characteristic field will be obtained so that in the center the transmission coefficient is slighter than at the edge. So, tolerance ranges can be generated . In reverse way, an increase of the transmission coefficient can be realized by spreading the sets of Y.

Figure 7 shows a variant which is unsuitable for optimal control due to high waviness. In the worst case, many local maxima and minima appear and a sign change for the transmission factor can occur. With it, local band-limited instabilities appear. The quality criterion of the waviness is not fulfilled.

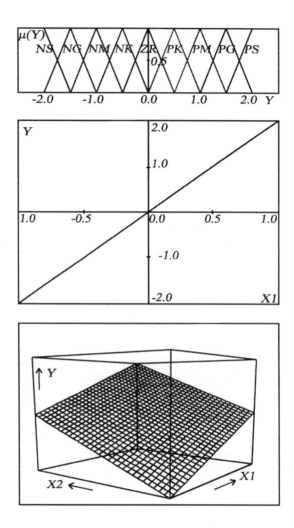

Fig. 5. Characteristic of a linear Fuzzy Controller

Inference: Sum-Prod
Defuzzification: Singleton

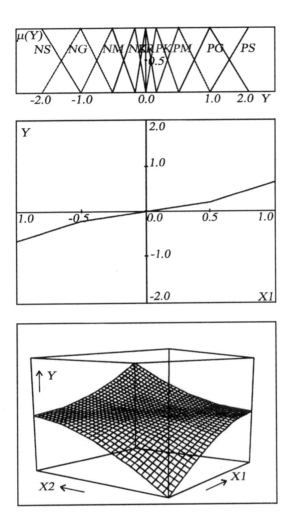

Fig. 6. Characteristic of a Fuzzy Controller

Inference: Sum-Prod
Defuzzification: Singleton
Variation of Distribution of Fuzzy Sets

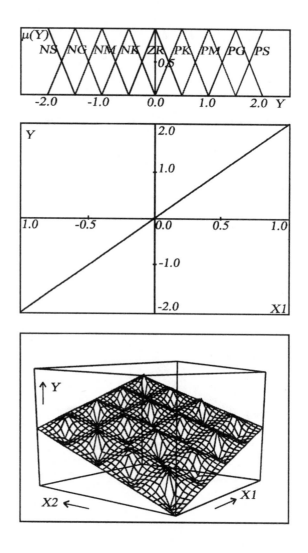

Fig. 7. Characteristic of a Fuzzy Controller

Inference: Sum-Bounded Difference
Defuzzification: Singleton

4 Cascading High-Dimensional Fuzzy Controllers

The transformation of structure for high-dimensional Fuzzy Controllers can result in two directions:

- cascaded Fuzzy Controllers,
- hierarchical Fuzzy Controllers.

Figure 8 shows both structures consisting of two-dimensional Fuzzy Controllers. The hierarchical Controller is a special case of the cascaded one. Therefore, in the following only cascaded Controller will be considered.

For the 4-dimensional Controller presented in Figure 9 the following holds

Basis - Rule

$$\text{IF [OP Xi]} \qquad \text{THEN} \quad \text{Y} \qquad i \in [1, 4] \tag{4}$$

Individual - Rule

$$\text{IF [OP Xij]} \qquad \text{THEN} \quad \text{Yk} \qquad i \in [1, 4] \tag{5}$$
$$j \in [1, 5]$$
$$k \in [1, 5]$$

The number of individual rules Rm for the complete description of the Controller is

$$R_m = \prod_{i=1}^{n} m_i \tag{6}$$

n - number of input variables
m_i - number of sets of the i-th input variable

For the Controller in Figure 9, the number of individual rules is 225. This number is rather large and the rule matrix becomes unhandy. Hence, requirements will be derived for

- Reduction of the number of the individual rules;
- Increase of lucidity.

These requirements will be fulfilled by cascading the Fuzzy Controller.

For the Controllers described in Figure 9, one obtains the structure and rule base presented in Figure 10. In this case, the number of rules is only 55.

a) High Dimensional Fuzzy Controller

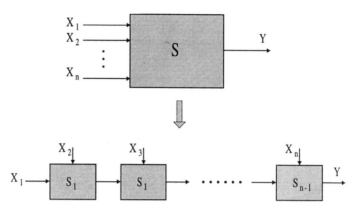

b) Cascaded Fuzzy Controller

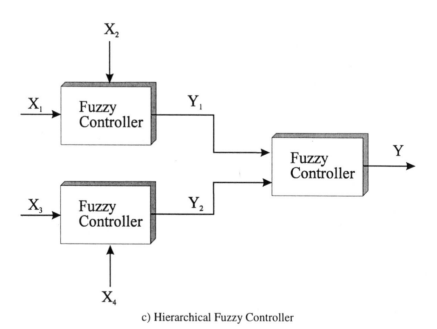

c) Hierarchical Fuzzy Controller

Fig. 8. Fuzzy Controller structures

Y				X_2														
				K					M					G				
				X_4														
				NG	NM	ZR	PM	PG	NG	NM	ZR	PM	PG	NG	NM	ZR	PM	PG
	G		NG	NG	NG	NM	NK	ZR	NG	NG	NG	NM	NK	NG	NG	NM	NK	ZR
			NK	NG	NM	NK	ZR	PK	NG	NG	NM	NK	ZR	NG	NM	NK	ZR	PK
			ZR	NM	NK	ZR	PK	PM	NG	NM	NK	ZR	PK	NM	NK	ZR	PK	PM
			PK	NK	ZR	PK	PM	PG	NM	NK	ZR	PK	PM	NK	ZR	PK	PM	PG
			PG	ZR	PK	PM	PG	PG	NK	ZR	PK	PM	PG	ZR	PK	PM	PG	PG
X_1	M	X_3	NG	NG	NM	NK	ZR	PK	NG	NG	NM	NK	ZR	NG	NG	NG	NM	NK
			NK	NM	NK	ZR	PK	PM	NG	NM	NK	ZR	PK	NG	NG	NM	NK	ZR
			ZR	NK	ZR	PK	PM	PG	NM	NK	ZR	PK	PM	NG	NM	NK	ZR	PK
			PK	ZR	PK	PM	PG	PG	NK	ZR	PK	PM	PG	NM	NK	ZR	PK	PM
			PG	PK	PM	PG	PG	PG	ZR	PK	PM	PG	PG	NK	ZR	PK	PM	PG
	K		NG	NM	NK	ZR	PK	PM	NG	NM	NK	ZR	PK	NG	NG	NM	NK	ZR
			NK	NK	ZR	PK	PM	PG	NM	NK	ZR	PK	PM	NG	NM	NK	ZR	PK
			ZR	ZR	PK	PM	PG	PG	NK	ZR	PK	PM	PG	NM	NK	ZR	PK	PM
			PK	PK	PM	PG	PG	PG	ZR	PK	PM	PG	PG	NK	ZR	PK	PM	PG
			PG	PM	PG	PG	PG	PG	PK	PM	PG	PG	PG	ZR	PK	PM	PG	PG

Fig. 9. Structure and rule basis of a 4-dimensional Fuzzy Controller

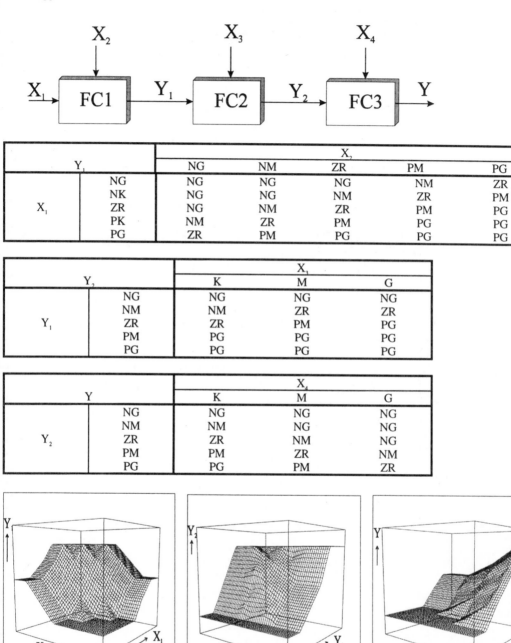

	Y_1		X_2			
		NG	NM	ZR	PM	PG
X_1	NG	NG	NG	NG	NM	ZR
	NK	NG	NG	NM	ZR	PM
	ZR	NG	NM	ZR	PM	PG
	PK	NM	ZR	PM	PG	PG
	PG	ZR	PM	PG	PG	PG

	Y_2		X_3	
		K	M	G
Y_1	NG	NG	NG	NG
	NM	NM	ZR	ZR
	ZR	ZR	PM	PG
	PM	PG	PG	PG
	PG	PG	PG	PG

	Y		X_4	
		K	M	G
Y_2	NG	NG	NG	NG
	NM	NM	NG	NG
	ZR	ZR	NM	NG
	PM	PM	ZR	NM
	PG	PG	PM	ZR

Fig. 10. Structure and rule basis of a 4-dimensional cascaded Fuzzy Controller

The condition for the reliability of the transformation is the fulfillment of the associative law

$$X_1 \cap X_2 \cap X_3 \cap X_4 = ((X_1 \cap X_2) \cap X_3) \cap X_4. \qquad (7)$$

This means also that the individual rules of the multidimensional Controller must agree with that of the cascaded Controller.

As basis of the physical-technical founded cascading, the following agreements are necessary:

Type of the basic rules:

Type 1   basic rules for describing the response characteristic of the Controller

Type 2   basic rules for the adaptation of the response characteristic of the Controller by changing the rule base and/or free optimizing parameters

Type of the input variables:

Dominant input variable
     Input variable in strong influence on the form (deformation) of the characteristic field

Non-dominant input variable
     Input variables with slight influence on the deformation of the characteristic field

Figure 11 serves as demonstration of the results. X_1 and X_2 are dominating input variables of the type 1. X_3 is a non-dominating input variable. Its influence on the Controller will be clear through the adaptation rules

IF $X_3 = L$ displacment of Y_1 in L direction
IF $X_3 = H$ displacment of Y_1 in H direction

The comparison of the rule matrix of the complete three-dimensional Controller (Figure 11a) with the cascaded Controller matrix (Figure 11b) shows a mismatch at two positions (highlighted in Figure 11a and 11b). By adding two sets to the virtual auxiliary variables Yv_1 (Figure 11c), complete consistency can be obtained.

Under these conditions, the possible reduction of the number of the individual rules and the number of the calculating operations AND and OR shows Figure 12.

Y_M				X_1		
				L	N	H
			L	L	L	L
	L		N	L	L	N
			H	L	N	H
			L	L	L	N
X_3	N	X_2	N	L	N	H
			H	N	H	H
			L	L	N	H
	H		N	N	H	H
			H	H	H	H

a) Complete rule matrix for the 3-dimensional Fuzzy Controller FC

Y_{VI}		X_1		
		L	N	H
	L	L	L	N
X_2	N	L	N	H
	H	N	H	H

FC1

Y_C		X_3		
		L	N	H
	L	L	L	N
Y_{VI}	N	L	N	H
	H	N	H	H

FC2

b) Complete rule matrix for the cascaded Fuzzy Controller

Y_{VI}		X_1		
		L	N	H
	L	VL	L	N
X_2	N	L	N	H
	H	N	H	VH

Y_C		X_3		
		L	N	H
	VL	L	L	L
	L	L	L	N
Y_{VI}	N	L	N	H
	H	N	H	H
	VH	H	H	H

c) Rule matrix for the improved 3-dimensional cascaded Fuzzy Controller
(with 5 Sets for Y_{VI})

Fig. 11. Rule basis of a cascaded Fuzzy Controller

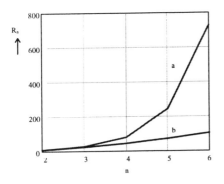

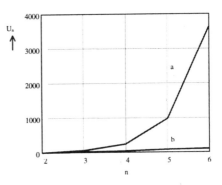

a) Number of rules
  a : high dimensional FC
  b : cascaded FC

b) Number of AND-Operations
  a : high dimensional FC
  b : cascaded FC

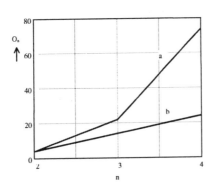

c) Number of OR-Operations
  a : high dimensional FC
  b : cascaded FC

Fig. 12. Comparison of rule number and operations between high dimensional and cascaded Fuzzy Controllers

n : number of input variables
$m = 3$: number of Fuzzy Sets per input variable

5   High Speed Matrix Controller (HSMC)

With the cascaded structure it is possible to reduce the amount of the rules and computing time considerably, and to increase clearness, as well [6].

In [16], approaches related to the structure transformation, parameterization, and optimization of the characteristic fields were presented.

As shown in Figure 13, time critical processes are limited to the signal processing in the Fuzzy Controller which is represented by the characteristic field, i.e. to the sampling period.

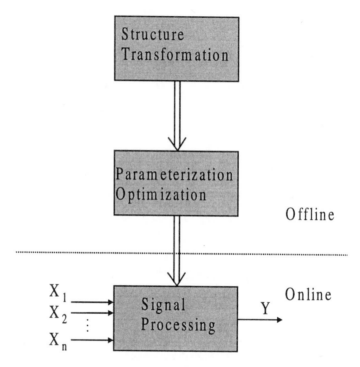

Fig. 13. Design steps of the Fuzzy Controller

Signal processing for the Fuzzy Control algorithm is relatively expensive (fuzzification, inference with aggregation, implication and accumulation, defuzzification). The properties of the Fuzzy Controller are reproducible stored in the characteristic field. Thereby, it is possible to subdivide the signal processing in off-line and real-time processes [21] (Figure 14).

After successfully structuring the Fuzzy Controller, the Fuzzy Characteristic field is calculated on the basis of relative input variables $\overline{X}_1$, $\overline{X}_2$, ..., $\overline{X}_n$ in off-line process. Because of that, it is possible to take advantage of all degrees of freedom for the optimization of the structure and form of the characteristic field. The calculated characteristic field is downloaded in a memory (RAM, EPROM).

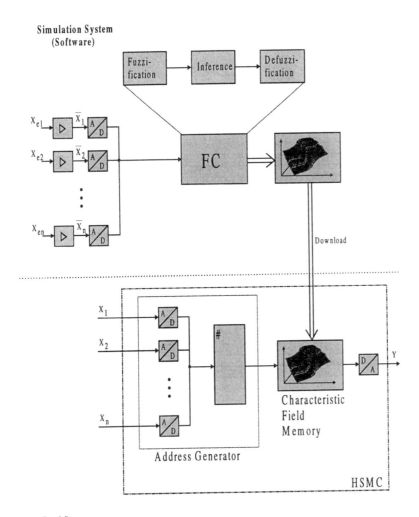

Fig. 14. Structure of the High Speed Matrix Controller (HSMC)

Xe - input signal
$\overline{X}$ - relative input signal
X - process signal

The principal item of the real-time signal processing in the HSMC is the characteristic field memory. The characteristic field is point-to-point deposited with a given resolution for the input variables. For example, 10^4 memory locations with two input variables or 10^6 memory locations with three input variables are required for a resolution of 1%. The resolution of the output signal is determined depending on the memory word length (i.e. 10 bits), the resolution of the D/A-converter and on the process requirements.

The address generator forms the input addresses for the memory unit from the output variables of n A/D-converters with e.g. 8 bits resolution each.

Outgoing from this structure, the HSMC presents the following features:

1. The sampling period is determined by the components

   - Analog-digital conversion;
   - Address generator;
   - Memory access time;
   - Digital-analog conversion.

2. The resolution ability is determined by the memory capacity.

3. The sampling period is independent from the Fuzzy Controller algorithm and number of input variables.

Regarding the limited memory capacity and the clearness, it seems to be senseful to use cascaded Controllers with only two input variables instead of multi-dimensional ones.

6 Conclusion

The paper presented the main results of the investigations in the field of Fuzzy Logic at the IPM. The Fuzzy Controller substitutes the classical one with the same interface to the process. The response characteristic of a Fuzzy Controller can be described by characteristic fields with only two input variables each. Consequently, high dimensional Fuzzy Controllers can be transformed into cadscaded two-dimensional ones. Thereby, the knowledge representation of Fuzzy Controllers is made clearer and the optimization simpler. The optimal setting of the Fuzzy Controller is reflected in the deformation of the non-linear characteristic field. This can be achieved by changing only two degrees of freedom among a multitude of them.

The cascading concept is realized in the so called High Speed Matrix Controller (HSMC), a real-time able and validable hardware component. It is suitable for the open loop, closed loop, and signal processing in safety relevant and time critical applications. It is composed of n A/D-converters for the digitization of the input signals, an electronic memory for mapping the characteristic fields, and a D/A-converter for the manipulated variable. Because of the special codification of the input signals and generation of the memory addresses, the time for the processing of the algorithm is minimized. The HSMC Controller has been patented.

References

1. Kuccera, T.: Hierarchical Fuzzy Controllers (Conventional PID Controller and Fuzzy Logic Controllers FLC). In: Proc. Conference on Fuzzy Logic in Engineering and Natural Sciences, Zittau, Germany (1996) 80-82

2. Hampel, R., Chaker, N.: Structure Analysis for Fuzzy-Controller. In: Proc. Conference on Fuzzy Logic in Engineering and Natural Sciences, Zittau, Germany (1996) 83-90

3. Steinkogler, A., Koch, J.: Genetic Programming Designs Hierarchie Fuzzy Logic Controller. In: Proc. Conference on Fuzzy Logic in Engineering and Natural Sciences, Zittau, Germany (1996) 150-159

4. Pivonka, P., Breijl, M.: Use of PID Controllers in Fuzzy Control of coal power plants. In: Proc. Conference on Fuzzy Logic in Engineering and Natural Sciences, Zittau, Germany (1996) 441-448

5. Czogala, E., Leski, J.: On Destructive Fuzzy Logic Controllers. In: Proc. 5th Zittau Fuzzy-Colloquium, Zittau, Germany (1997) 8-12

6. Hampel, R., Chaker, N.: Cascading of Multi-Dimensional Fuzzy Controllers. In: Proc. 5th Zittau Fuzzy-Colloquium, Zittau, Germany (1997) 17-31

7. Vogrin, P., Halang, W. A.: Approximation of Conventional Controllers by Fuzzy Controllers with Equal Discribing Functions. In: Proc. 5th Zittau Fuzzy-Colloquium, Zittau, Germany (1997) 51-65

8. Czogala, E., Henzel, N., Leski, J.: The Equality of Inference Results Using Fuzzy Implication and Conjunctive Interpretations of the IF-THEN Rules under Defuzzification. In: Proc. 6th Zittau Fuzzy-Colloquium, Zittau, Germany (1998) 1-6

9. Halang, W. A., Colnaric, M., Vogrin, P.: Safety Licensable Inference Controller. In: Proc. 6th Zittau Fuzzy-Colloquium, Zittau, Germany (1998) 18-23

10. Wagenknecht, M., Chaker, N.: Towards an Algorithmic Cascading of Fuzzy Rules. In: Proc. 6th Zittau Fuzzy-Colloquium, Zittau, Germany (1998) 56-61

11. Pivonka, P., Sidlo, M.: Fuzzy PI + PD Controller with a Normalised Universe-The Exact Solution for Setting of Parameters. In: Proc. 6th Zittau Fuzzy-Colloquium, Zittau, Germany (1998) 62-67

12. Arakeljan, E., Panko, M., Usenko, V.: Comperative Analysis of Classical and Fuzzy PID Algorithms. In: Proc. 6th Zittau Fuzzy-Colloquium, Zittau, Germany (1998) 68-73

13. Pacyna, K., Pieczynski, A.: Influence of Changes of Membership Function on PID Fuzzy Logic Control. In: Proc. 6th Zittau Fuzzy-Colloquium, Zittau, Germany (1998) 80-85

14. Rotach, V.: On Connection Between Traditional and Fuzzy PID Regulators. In: Proc. 6th Zittau Fuzzy-Colloquium, Zittau, Germany (1998) 86-90

15. Hampel, R., Keil, A., Gierth, L.: Fuzzy Drehzahlregelung. atp 3 (1999) 37-42

16. Hampel, R., Chaker, N.: Minimizing the number of variable parameters for optimizing the Fuzzy Controller. Fuzzy Sets and Systems 100 (1998) 131-142

17. Hampel, R., Chaker, N., Gierth, L.: Adaptive Dampfturbinenregelung mit Fuzzy Logik zur Beherrschung von Lastabwürfen in Heizkraftwerken. atp 2 (1998) 42-49

18. Hampel, R., Chaker, N.: Application of Fuzzy Logic in Control and Limitation Systems Using Industrial Hardware. In: Proc. Mendel 97, Brno (1997) 291-298

19. Traichel, A., Kästner, W., Hampel, R.: Fuzzy Modeling of Dynamic Non-Linear Processes-Applied for Water Level Measurement. In: Proc. 7th Zittau Fuzzy Colloquium, Zittau, Germany (1999) 119-134

20. Pieczynski, A., Kästner, W., Hampel, R.: Fuzzy Modeling of Multidimensional Non-Linear Processes-Design and Analysis of Structures. In: Proc. 7th Zittau Fuzzy Colloquium, Zittau, Germany (1999) 85-101

21. Hampel, R., Chaker, N., Stegemann, H.: High Speed Matrix Controller for Safety Related Applications. In: Proc. Mendel 99, Brno (1999) 243-248

Fuzzy Sets Theory

On t-Norms as Basic Connectives for Fuzzy Logic

Siegfried Gottwald

Institut für Logik und Wissenschaftstheorie
Leipzig University, D-04109 Leipzig, Germany
gottwald@rz.uni-leipzig.de

Abstract. In fuzzy logic in wider sense, i.e. in the field of fuzzy sets applications like fuzzy control and approximate reasoning, t-norms have reached a core position in recent times. And from a theoretical point of view fuzzy logic in the narrow sense, i.e. many-valued logic with a graded notion of entailment, is the main background theory e.g. for fuzzy reasoning.

In pure many-valued logic, the LUKASIEWICZ systems, the GÖDEL sytems, and also the quite recent product logic all are t-norm based systems in the sense that the basic connectives of these systems can be defined starting with a suitable t-norm.

The present paper discusses the problem of the adequate axiomatizability for such t-norm based logical systems in general, surveying results of the last years. The main emphasis in the present paper is on propositional logic.

Keywords: fuzzy logic, many-valued logic, t-norms, t-norm based connectives.

1 t-norm based logical systems

Many-valued propositional logic differs from classical propositional logic in so far as each of its systems S has some set $\mathcal{W}^S$ of truth degrees which is (or at least: may be) different from the set $\{\top, \bot\}$ of (standard) truth values. Like classical logic, however, it is based on the basic *principle of compositionality*. And this means for every system of many-valued logic that each of its connectives is completely determined by its associated truth degree function.

By a *t-norm* one means some binary operation t in the real unit interval $[0, 1]$ which is associative, commutative, and non-decreasing (in each argument), and which has 1 as a neutral element, i.e. satisfies $t(x, 1) = x$ for each $x \in [0, 1]$. As an immediate corollary one has $t(x, 0) = 0$ for all $x \in [0, 1]$ because of $t(x, 0) \le t(1, 0) = 0$.

From the point of view of many-valued logic, such a t-norm is a suitable candidate for a truth degree function of some conjunction connective. Accepting this, one essentially is concerned with systems S of many-valued logic with infinite truth degree set $\mathcal{W}^S = [0, 1]$. And additionally one prefers to consider such systems S which have the truth degree 1 as the only designated truth degree.[1]

Such a system S_t of many-valued logic is called *t-norm based* (upon some particular t-norm t) iff all the other connectives of S_t have associated truth degree functions which are defined from t (using possibly some truth degree constants). Usually one considers together with the conjunction connective $\&_t$ with truth degree function t an implication connective $\to_t$ with truth degree function seq_t characterized by

$$\text{seq}_t(u,v) = \sup\{z \mid t(u,z) \le v\}, \tag{1}$$

and a negation connective $-_t$ with truth degree function non_t, given as

$$\text{non}_t(u) = \text{seq}_t(u,0).$$

The definition (1) determines a reasonable implication function just in the case that the t-norm t is left continuous.[2] Here "reasonable" essentially means that $\to_t$ satisfies a suitable version of the rule of detachment.

In more technical terms it means, that for left continuous t-norms condition (1) defines a so-called φ-operator, cf. [6]. And it means also, under this assumption of left continuity of t, that condition (1) is equivalent to the *adjointness condition*

$$w \le \text{seq}_t(u,v) \quad \Leftrightarrow \quad t(u,w) \le v, \tag{2}$$

i.e. that the operations t and seq_t form an *adjoint pair*.

Forced by these results one usually restricts, in this logical context, the considerations to left continuous – or even to continuous – t-norms.

But together with this restriction of the t-norms, a generalization of the possible truth degree sets sometimes is useful: one may accept each subset of the unit interval $[0,1]$ as a truth degree set which is closed under the particular t-norm t and all their derived operations.

The restriction to *continuous*[3] t-norms enables even the definition of the operations max and min, which make $[0,1]$ a lattice. On the one hand one has from straigthforward calculations that always

$$\min\{u,v\} = t(u, \text{seq}_t(u,v)).$$

And on the other hand one gets always

$$\max\{u,v\} = \min\{\text{seq}_t(\text{seq}_t(u,v),v), \text{seq}_t(\text{seq}_t(v,u),u)\},$$

---

[1] This means e.g. that a formula of the language of S counts as logically valid just in case it always assumes this designated truth degree 1. This notion, as well as the other notions from many-valued logic are explained in detail in [7].

[2] We call a function of two variables left continuous iff it is left continuous in both of its arguments (always with the other one fixed as a parameter).

[3] For t-norms one gets their continuity as functions of two variables already from their continuity (as unary functions) in each argument, cf. [7, 13].

cf. e.g. [8, 9]. If one denotes by $\wedge, \vee$ the connectives with truth degree functions min, max, respectively, then these definitions can equivalently be given as

$$\varphi \wedge \psi =_{\mathrm{df}} \varphi \,\&_t (\varphi \to_t \psi) \,,$$
$$\varphi \vee \psi =_{\mathrm{df}} ((\varphi \to_t \psi) \to_t \psi) \wedge ((\psi \to_t \varphi) \to_t \varphi) \,,$$

where φ, ψ are formulas of the language of S_t.

2 Some facts on t-norms

The class of t-norms is a very large class of binary functions, which contains rather difficult examples, cf. e.g. [13]. Also with the restriction to left continuous, and even with the restriction to continuous ones, these classes remain large.

For continuous Archimedean t-norms one has some important structural results which say that these t-norms fall into two classes. Here a continuous t-norm t is called *Archimedean* iff one has

$$t(u, u) < u \qquad \text{for all } 0 < u < 1.$$

Theorem 1 *Each continuous and Archimedean t-norm is either strict, i.e. satisfies for all $u, v \in [0, 1]$ the strong monotonicity condition*

$$u < v \quad \Rightarrow \quad t(u, w) < t(v, w) \,,$$

or it has zero divisors, i.e. there exist $0 < \hat{u}, \hat{v} < 1$ with $t(\hat{u}, \hat{v}) = 0$.

Theorem 2 *(i) A continuous Archimedean t-norm t is strict iff there exists an automorphism h of the unit interval with*

$$t(u, v) = h^{-1}(h(u) \cdot h(v)) = h^{-1}(\mathrm{et}_3(h(u), h(v))) \,. \tag{3}$$

(ii) A continuous Archimedean t-norm t has zero divisors iff there exists an automorphism h of the unit interval with

$$t(u, v) = h^{-1}(\max\{0, h(u) + h(v) - 1\}) = h^{-1}(\mathrm{et}_2(h(u), h(v))) \,. \tag{4}$$

Corollary 3 *If t is a continuous Archimedean t-norm then the (lattice) ordered semigroup $\langle [0, 1], t, \leq \rangle$ is either isomorphic with the LUKASIEWICZ semigroup $\langle [0, 1], \mathrm{et}_2, \leq \rangle$, or with the product semigroup $\langle [0, 1], \mathrm{et}_3, \leq \rangle$.*

Furthermore, one has even for the continuous t-norms in general a kind of overall picture, expressed by the following result which essentially was proved in [15], cf. also [13].

Theorem 4 *Each continuous t-norm t is an ordinal sum of the minimum t-norm and of isomorphic copies of the LUKASIEWICZ t-norm and the product t-norm.*

Here an *ordinal sum* of some countable family $(t_i)_{i \in I}$ of t-norms w.r.t. some countable family of non-overlapping[4] subintervals $([a_i, b_i])_{i \in I}$ of the unit interval $[0, 1]$ is the t-norm

$$T(u, v) = \begin{cases} a_k + (b_k - a_k) \cdot t_k(\frac{u - a_k}{b_k - a_k}, \frac{v - a_k}{b_k - a_k}), & \text{if } (u, v) \in [a_k, b_k] \\ \min\{u, v\} & \text{otherwise.} \end{cases}$$

Because one knows adequate axiomatizations of the t-norms based logical systems which have as their basic t-norm either the minimum, or the the LUKASIEWICZ t-norm, or of the product t-norm, one knows axiomatizations for all the ordinal summands. However, there is no general method to get from this an adequate axiomatization for the logical system which is based upon the ordinal sum as basic t-norm. Even more, for left continuous t-norms such structural results are lacking at all.

So at all it does not seem appropriate to look for each particular t-norm based system for an adequate axiomatization. Instead we will consider the problem to axiomatize the classes of all t-norm based systems, either w.r.t. left continuous t-norms, or w.r.t. continuous t-norms.

2.1 The continuous Archimedean case

As a first remark let us mention that the well known axiomatizations of the infinite-valued LUKASIEWICZ system L_∞ and of the infinite-valued product logic Π, which both are t-norm based systems, cover an even wider range of t-norm based systems. The essential result for such an extension is Corollary 3, which also implies that all the additional truth degree functions one can define in $[0, 1]$ using either the LUKASIEWICZ t-norm et_2 or the product t-norm et_3, can similarly be defined in the lattice ordered semigroups $\langle [0, 1], t, \leq \rangle$ with either a continuous Archimedean t-norm with zero divisors, or with a continuous and strict t-norm.

So one has the following two results.

Proposition 5 *The standard axiomatization of the product logic Π*

$$\varphi \to (\psi \to \varphi),$$
$$(\varphi \to \psi) \to ((\psi \to \chi) \to (\varphi \to \chi)),$$
$$0 \to \varphi,$$
$$\varphi \& \psi \to \psi \& \varphi,$$
$$\varphi \& (\psi \& \chi) \to (\varphi \& \psi) \& \chi,$$
$$(\varphi \& \psi) \& \chi \to \varphi \& (\psi \& \chi),$$
$$((\varphi \& \psi) \to \chi) \to (\varphi \to (\psi \to \chi)),$$
$$(\varphi \to (\psi \to \chi)) \to ((\varphi \& \psi) \to \chi),$$

---

[4] Because we are considering closed intervals here, this means that the intersection of two such intervals is at most a singleton.

$$(\varphi \to \psi) \to (\varphi \& \chi \to \psi \& \chi),$$
$$\sim\sim \chi \to ((\varphi \& \chi \to \psi \& \chi) \to (\varphi \to \psi)),$$
$$(\varphi \to \psi) \to ((\varphi \to \chi) \to (\varphi \to \psi \wedge \chi)),$$
$$(\varphi \to \chi) \to ((\psi \to \chi) \to (\varphi \vee \psi \to \chi)),$$
$$(\varphi \to \psi) \vee (\psi \to \varphi),$$
$$\varphi \wedge \sim \varphi \to 0,$$

gives an adequate axiomatization for each t-norm based system S_t which is based upon a strict continuous t-norm t, and formulated in the language with the primitive connectives $\&_t, \to_t$ and the truth degree constant 0.

Proof. Because of Theorem 2(i) one has that the algebraic structures $\langle [0,1], t, \leq \rangle$ and $\langle [0,1], et_3, \leq \rangle$ are isomorphic. Hence also the definitionally extended algebraic structures $\langle [0,1], t, \mathsf{seq}\,_t, \mathsf{non}_t, \leq \rangle$ and $\langle [0,1], et_3, \mathsf{seq}\,_\Pi, \mathsf{non}_0, \leq \rangle$ are isomorphic, and thus also the structures $\langle [0,1], t, \mathsf{seq}\,_t, 0 \rangle$ and $\langle [0,1], et_3, \mathsf{seq}\,_\Pi, 0 \rangle$. Furthermore one can define in $\langle [0,1], t, \mathsf{seq}\,_t, 0 \rangle$ the operation non_0 and the (lattice) ordering $\leq$.

This means all together that the t-norm based system S_t can be based on the logical matrix $\langle [0,1], t, \mathsf{seq}\,_t, 0 \rangle$, and that this logical matrix is isomorphic to the logical matrix for the product logic. Therefore one immediately gets by routine calculations that the logical systems S_t and Π have the same entailment relation, and hence can be adequately axiomatized by the same logical calculus.

Proposition 6 *The standard axiomatization of the infinite valued* Lukasiewicz *system* L_∞:

$$\varphi \to (\psi \to \varphi),$$
$$(\varphi \to \psi) \to ((\psi \to \chi) \to (\varphi \to \chi)),$$
$$(\neg\psi \to \neg\varphi) \to (\varphi \to \psi),$$
$$((\varphi \to \psi) \to \psi) \to ((\psi \to \varphi) \to \varphi),$$

gives an adequate axiomatization for each t-norm based system S_t which is based upon a continuous Archimedean t-norm t with zero divisors (and formulated in the language with the primitive connectives $\&_t, \to_t, -_t$).

Proof. Because of Theorem 2(ii) one has that the algebraic structures $\langle [0,1], t, \leq \rangle$ and $\langle [0,1], et_2, \leq \rangle$ are isomorphic. Now one argues as in the proof of the preceeding proposition.

3 Approaching the problem via algebraic semantics

For the problem of adequate axiomatization of (classes of) t-norm based systems of many-valued logic there is an important difference to the standard approach toward semantically based systems of many-valued logic: here there is no single, "standard" semantical matrix for the general approach.

The most appropriate way out of this situation is: to find suitable class(es) of algebraic structures which can be used to characterize these logical systems, and which preferably should be algebraic varieties, i.e. equationally definable.

From an algebraic point of view, the following conditions seem to be structurally important for t-norms:
- $\langle [0,1], t, 1 \rangle$ is a commutative semigroup with a neutral element, i.e. a commutative monoid,
- $\leq$ is a (lattice) ordering in $[0,1]$ which has a universal lower bound and a universal upper bound,
- both structures "fit together": t is non-decreasing w.r.t. this lattice ordering.

Thus it seems reasonable to consider (complete) abelian lattice-ordered monoids as the truth degree structures for the t-norm based systems.

In general, however, abelian lattice-ordered monoids may have different elements as the universal upper bound of the lattice and as the neutral element of the monoid. This is not the case for the t-norm based systems, they make $[0, 1]$ into an *integral* abelian lattice-ordered monoid as truth degree structure, i.e. one in which the universal upper bound of the lattice ordering and the neutral element of the monoidal structure coincide.

Furthermore one also likes to have the t-norm t combined with another operations, its R-implication operator, which forms together with t an adjoint pair: i.e. the abelian lattice-ordered monoid formed by the truth degree structure has also to be a *residuated* one.

At all, hence, we are going to consider *residuated lattices*, i.e. algebraic structures $\langle L, \cap, \cup, *, \rightarrowtail, 0, 1 \rangle$ such that L is a lattice under $\cap, \cup$ with zero 0 and unit 1, and an abelian lattice-ordered monoid under $*$ with neutral element 1, and such that the operations $*$ and $\rightarrowtail$ form an adjoint pair, i.e. satisfy

$$ z \leqslant (x \rightarrowtail y) \quad \Leftrightarrow \quad x * z \leqslant y . $$

It is nice to recognize that the adjointness condition is in the present setting the suitable algebraic equivalent of the analytical notion of left continuity.

Proposition 7 *For any t-norm t one has that t and its R-implication* seq $_t$ *form an adjoint pair iff t is left continuous (in both arguments).*

And also the continuity of the basic t-norm has an algebraic equivalent: the property of divisibility for the commutative lattice-ordered monoids.

Definition 1 *A lattice ordered monoid $\langle L, *, 1, \leqslant \rangle$ is divisible iff for all $a, b \in L$ with $a \leqslant b$ there exists some $c \in L$ with $a = b * c$.*

For residuated lattices one has another nice and useful characterization of divisibility: a residuated lattice $\langle L, \cap, \cup, *, \rightarrowtail, 0, 1 \rangle$ is divisible iff one has $a \cap b = a * (a \rightarrowtail b)$ for all $a, b \in L$.

Proposition 8 *A t-norm based residuated lattice* $\langle [0,1], \min, \max, t, \mathrm{seq}_t, 0, 1 \rangle$ *is divisible iff t is continuous.*

A further, independent restriction seems to be suitable because each t-norm based residuated lattice is linearly ordered, and thus makes the wff $(\varphi \to_t \psi) \vee (\psi \to_t \varphi)$ valid. As an algebraic correspondence we consider the *pre-linearity* condition: $(x \rightarrowtail y) \cup (y \rightarrowtail x) = 1$.

It is however suitable, to refer in the later axiomatizations to an equivalent version:

Proposition 9 *In residuated lattices the following two conditions are equivalent:*

$$(i) \ (x \rightarrowtail y) \cup (y \rightarrowtail x) = 1,$$

$$(ii) \ ((x \rightarrowtail y) \rightarrowtail z) * ((y \rightarrowtail x) \rightarrowtail z) \leqslant z.$$

So finally, we have three different approaches:

- following HÖHLE [11, 12] to consider the residuated lattices as the fundamental algebraic structures;
- following ESTEVA/GODO [5] to consider the pre-linear residuated lattices as the fundamental algebraic structures;
- following HÁJEK [8, 9], to consider the pre-linear and divisible residuated lattices as the fundamental algebraic structures, also called *BL-algebras*.

4 Syntactic approaches

Now we consider the problems of syntactic characterizations of the classes of all wffs which are valid in all residuated lattices, in all pre-linear residuated lattices, or in all BL-algebras. For more details the reader may e.g. consult [7].

4.1 Monoidal logic

The axiom system $\mathsf{Ax_{ML}}$ of HÖHLE [11, 12] for the class of wffs which are valid in all residuated lattices has the axiom schemata:

$$(\varphi \to \psi) \to ((\psi \to \chi) \to (\varphi \to \chi)),$$
$$\varphi \,\&\, \psi \to \varphi,$$
$$\varphi \,\&\, \psi \to \psi \,\&\, \varphi,$$
$$(\varphi \to (\psi \to \chi)) \to (\varphi \,\&\, \psi \to \chi),$$
$$(\varphi \,\&\, \psi \to \chi) \to (\varphi \to (\psi \to \chi)),$$
$$\varphi \wedge \psi \to \varphi,$$
$$\varphi \wedge \psi \to \psi \wedge \varphi,$$
$$\varphi \,\&\, (\varphi \to \psi) \to \varphi \wedge \psi,$$

$$0 \to \varphi,$$
$$\varphi \to \varphi \vee \psi,$$
$$\psi \to \varphi \vee \psi,$$
$$(\varphi \to \chi) \to ((\psi \to \chi) \to (\varphi \vee \psi \to \chi)),$$

and has as its (only) inference rule the rule of detachment (w.r.t. the implication connective $\to$).

The logical calculus which is constituted by this axiom system and its inference rule, and which has the standard notion of derivation, shall be denoted by $\mathbb{K}_{\mathsf{ML}}$.

Proposition 10 *The calculus* $\mathbb{K}_{\mathsf{ML}}$ *is sound, i.e. proves only such formulas which are valid in all residuated lattices.*

Corollary 11 *The* LINDENBAUM *algebra of the calculus* $\mathbb{K}_{\mathsf{ML}}$ *is a residuated lattice.*

Theorem 12 (Completeness) *For a wff* φ *of* $\mathcal{L}_t$ *there are equivalent:*
(i) φ *is derivable within the logical calculus* $\mathbb{K}_{\mathsf{ML}}$*;*
(ii) φ *is valid in all residuated lattices.*

Proof. The part $(i) \Rightarrow (ii)$ here is just the soundness of the calculus $\mathbb{K}_{\mathsf{ML}}$. The part $(ii) \Rightarrow (i)$ on the other hand results from the fact that the LINDENBAUM algebra of $\mathbb{K}_{\mathsf{ML}}$ is a residuated lattice and has the class of all $\mathbb{K}_{\mathsf{ML}}$-derivable wffs as its unit element, i.e. as the universal upper bound of its lattice part and at the same time as the neutral element of its monoidal part. Therefore each non-$\mathbb{K}_{\mathsf{ML}}$-derivable wff is not valid in at least one residuated lattice.

4.2 Monoidal t-norm logic

The axiomatization of ESTEVA/GODO [5] for the monoidal t-norm logic, i.e. for the class of wffs which are valid in all residuated lattices which satisfy the pre-linearity condition, is given in a language $\mathcal{L}_{\mathsf{MTL}}$ which has as its basic vocabulary the sets

$$\mathcal{J}^{\mathsf{MTL}} = \{\to, \&, \wedge\}, \qquad \mathcal{K}^{\mathsf{MTL}} = \{0\}$$

of connectives and of truth degree constants, interpreted in each such residuated lattice $\langle L, \cap, \cup, *, \mapsto, 0, 1 \rangle$ as the operations $\mapsto, *, \cap$ and the truth degree 0, respectively.

This monoidal t-norm logic has a possible set of axioms $\mathrm{Ax}_{\mathsf{MTL}}$ determined by the following list of axiom schemata:

$$(\varphi \to \psi) \to ((\psi \to \chi) \to (\varphi \to \chi)),$$
$$\varphi \& \psi \to \varphi,$$

$$\varphi \& \psi \to \psi \& \varphi,$$
$$(\varphi \to (\psi \to \chi)) \to (\varphi \& \psi \to \chi),$$
$$(\varphi \& \psi \to \chi) \to (\varphi \to (\psi \to \chi)),$$
$$\varphi \wedge \psi \to \varphi,$$
$$\varphi \wedge \psi \to \psi \wedge \varphi,$$
$$\varphi \& (\varphi \to \psi) \to \varphi \wedge \psi,$$
$$0 \to \varphi,$$
$$((\varphi \to \psi) \to \chi) \to (((\psi \to \varphi) \to \chi) \to \chi),$$

and has as its (only) inference rule the rule of detachment (w.r.t. the implication connective $\to$).

The logical calculus which is constituted by this axiom system and its inference rule, and which has the standard notion of derivation, shall be denoted by $\mathbb{K}_{\mathsf{MTL}}$.

Proposition 13 *The (extended) logical calculus $\mathbb{K}_{\mathsf{MTL}}$ is sound, i.e. derives only such formulas which are valid in all residuated lattices which satisfy the condition of pre-linearity.*

Theorem 14 (Completeness Theorem) *For each wff φ of $\mathcal{L}_{\mathsf{MTL}}$ the following assertions are equivalent:*

(i) φ is derivable within the logical calculus $\mathbb{K}_{\mathsf{MTL}}$;
(ii) φ is valid in all residuated lattices which satisfy the pre-linearity condition.
(iii) φ is valid in all linearly ordered residuated lattices which satisfy the pre-linearity condition.

4.3 Basic t-norm logic

The axiomatization of HÁJEK [9] for the basic logic, i.e. for the class of all wffs which are valid in all BL-algebras, is given in a language which has as basic vocabulary the connectives $\to$, $\&$ and the truth degree constant 0, taken in each BL-algebra $\langle L, \cap, \cup, *, \rightarrowtail, 0, 1 \rangle$ as the operations $\rightarrowtail$, $*$ and the element 0. Then basic logic has as axiom system $\mathsf{Ax}_{\mathsf{BTL}}$ the following schemata:

$$(\varphi \to \psi) \to ((\psi \to \chi) \to (\varphi \to \chi)),$$
$$\varphi \& \psi \to \varphi,$$
$$\varphi \& \psi \to \psi \& \varphi,$$
$$\varphi \& (\varphi \to \psi) \to \psi \& (\psi \to \varphi),$$
$$(\varphi \to (\psi \to \chi)) \to (\varphi \& \psi \to \chi),$$
$$(\varphi \& \psi \to \chi) \to (\varphi \to (\psi \to \chi)),$$
$$((\varphi \to \psi) \to \chi) \to (((\psi \to \varphi) \to \chi) \to \chi),$$
$$0 \to \varphi,$$

and has as its (only) inference rule the rule of detachment (w.r.t. the implication connective $\rightarrow$).

The logical calculus which is constituted by this axiom system and its inference rule, and which has the standard notion of derivation, shall be denoted by $\mathbb{K}_{\mathsf{BTL}}$.

Usually the language is extended by definitions of additional connectives and a further truth degree constant:

$$\varphi \wedge \psi =_{\mathrm{df}} \varphi \,\&\, (\varphi \rightarrow \psi)\,,$$
$$\varphi \vee \psi =_{\mathrm{df}} ((\varphi \rightarrow \psi) \rightarrow \psi) \wedge ((\psi \rightarrow \varphi) \rightarrow \varphi)\,.$$

Calculations (in BL-algebras) show that the additional connectives $\wedge$, $\vee$ just have the BL-algebraic operations $\cap$, $\cup$ as their truth degree functions.

It is a routine matter, but a bit tedious, to check that this logical calculus $\mathbb{K}_{\mathsf{BTL}}$ is sound, i.e. derives only such formulas which are valid in all BL-algebras. A proof is given in [9].

Corollary 15 *The axioms of the monoidal logic as well as those of the monoidal t-norm logic are $\mathbb{K}_{\mathsf{BTL}}$-derivable.*

Corollary 16 *The* LINDENBAUM *algebra of the logical calculus $\mathbb{K}_{\mathsf{BTL}}$ is a BL-algebra.*

Theorem 17 (Completeness) *For a wff φ of $\mathcal{L}_t$ there are equivalent:*
(i) φ is derivable within the logical calculus $\mathbb{K}_{\mathsf{BTL}}$;
(ii) φ is valid in all BL-algebras;
(iii) φ is valid in all linearly ordered BL-algebras, i.e. in all BL-chains.

5 The logic of continuous t-norms

From the preceding considerations it is clear that all the logically valid wffs of monoidal t-norm (resp. basic t-norm) logic are valid in all t-norm based residuated lattices with a left continuous (resp. a continuous) t-norm.

The problem thus arises whether there exist wffs

(i) which are valid in all t-norm based residuated lattices with a left continuous t-norm but are not theorems of monoidal logic, and

(ii) which are valid in all t-norm based residuated lattices with a continuous t-norm but are not theorems of basic logic.

The first problem is actually an open one. But the second problem has a negative answer, given by the following theorem.

Theorem 18 (Completeness) *The class of all wff which are provable in basic logic coincides with the class of all wffs which are logically valid in all t-norm based residuated lattices with a continuous t-norm.*

The main steps in the proof are to show (i) that each BL-algebra is a subdirect product of BL-chains, i.e. of linearly ordered BL-algebras, and (ii) that each BL-chain is the ordinal sum of BL-chains which are either trivial one-element BL-chains, or linearly ordered MV-algebras, or linearly ordered product algebras, such that (iii) each such ordinal summand is locally embeddable into a t-norm based residuated lattice with a continuous t-norm, cf. [3, 8] and again [7].

References

1. Chang, C.C.: Algebraic analysis of many valued logics. Transactions American Mathematical Society 88 (1958) 476–490
2. Cignoli, R., d'Ottaviano, I, Mundici, D.: Algebraic Foundations of Many-Valued Reasoning. Kluwer, Dordrecht (1999)
3. Cignoli, R., Esteva, F., Godo, L., Torrens, A.: Basic Fuzzy Logic is the logic of continuous t-norms and their residua. Soft Computing (1999) (to appear)
4. Dummett, M.: A propositional calculus with denumerable matrix, Journal Symbolic Logic 24 (1959) 97–106
5. Esteva, F., Godo, L.: QBL: towards a logic for left-continuous t-norms, in: European Soc. Fuzzy Logic and Technol.: Eusflat-Estylf Joint Conference, Palma (Mallorca) 1999 Proceedings. Univ. de les Illes Balears (1999) 35–37
6. Gottwald, S.: Fuzzy Sets and Fuzzy Logic. The Foundations of Application – from a Mathematical Point of View. Vieweg, Braunschweig/Wiesbaden and Teknea, Toulouse (1993)
7. Gottwald, S.: A Treatise on Many-Valued Logic. (2000) (book in preparation)
8. Hájek, P.: Basic fuzzy logic and BL-algebras. Soft Computing 2 (1998) 124–128
9. Hájek, P.: Metamathematics of Fuzzy Logic. Kluwer Acad. Publ., Dordrecht (1998)
10. Hájek, P., Godo, L., Esteva, F.: A complete many-valued logic with product-conjunction. Arch. Math. Logic 35 (1996) 191–208
11. Höhle, U.: Monoidal logic, in Kruse/Gebhardt/Palm (eds.), Fuzzy Systems in Computer Science. Vieweg, Wiesbaden (1994) 233–243
12. Höhle, U.: On the fundamentals of fuzzy set theory. J. Math. Anal. Appl. 201 (1996), 786–826
13. Klement, E.P., Mesiar, R., Pap, E.: Triangular Norms. Kluwer, Dordrecht (2000) (to appear)
14. Mangani, P.: Su certe algebre connesse con logiche a piú valori. Boll. Unione Math. Italiana, ser. 8, 4 (1973) 68–78
15. Mostert, P.S., Shields, A.L.: On the structure of semigroups on a compact manifold with boundary. Annals of Mathematics 65 (1957) 117–143
16. Rasiowa, H.: An Algebraic Approach to Non-Classical Logics. PWN, Warsaw and North-Holland Publ. Comp., Amsterdam (1974)

Generalized Parametric Conjunction Operations in Fuzzy Modeling

Ildar Batyrshin

Kazan State Technological University,
K.Marx st., 68, Kazan, 420015, Republic of Tatarstan, Russia
batyr@emntu.kcn.ru

Abstract. An approach to construct optimal fuzzy models based on an optimization of parameters of generalized conjunction operations is discussed. Several approaches to construct parametric classes of conjunction operations simpler than known parametric classes of T-norms are considered. These approaches are based on elimination of associativity and commutativity properties from definition of conjunction operation. A new simplest generalized parametric conjunction operation is introduced. Different approaches to approximation of real valued function by fuzzy models based on optimization of parameters of membership functions and parameters of operations are compared by numerical examples. It is shown that fuzzy models based on new conjunction operation have better performance than previous ones.

1 Introduction

Fuzzy models are usually based on fuzzy rules and define some real valued functions $y = f(x_1, x_2, \ldots x_n)$, where $x_1, x_2, \ldots x_n$ are real valued input variables of model [7,8,10]. Different approaches to fuzzy modeling differ on structures of rules and methods of processing of these rules. But usually, the left parts of rules used in fuzzy models have the similar structure. For simplicity we will consider here first order Sugeno fuzzy model with two input and one output variables. This model has the rules R_i $(i=1,\ldots,n)$ with the following structure:

$$R_i: If\ X_1\ is\ A_{i1}\ and\ X_2\ is\ A_{i2}\ then\ y_i = s_{i1}x_1 + s_{i2}x_2 + r_i, \tag{1}$$

where X_1 and X_2 are names of input variables (for example *PRESSURE* and *VOLUME*), A_{i1} and A_{i2} are fuzzy sets like *HIGH* and *SMALL* defined on the domains of input variables X_1 and X_2 respectively, x_1 and x_2 are numerical values of variables X_1 and X_2, s_{i1}, s_{i2} and r_i are real valued constants. For each real values x_1^* and x_2^* of input variables X_1 and X_2 the real value y^* of function $y=f(x_1,x_2)$ defined by this fuzzy model may be calculated as a result of the following sequential procedures applied to all rules:

P1. A calculation of firing values of rules $w_i = AND(A_{i1}(x_1^*), A_{i2}(x_2^*))$, where *AND* some fuzzy conjunction operation and $A_{i1}(x_1^*)$, $A_{i2}(x_2^*)$ are membership values of x_1^* and x_2^* in fuzzy sets A_{i1} and A_{i2} respectively;

P2. A calculation of conclusions of rules $y_i^* = s_{i1}x_1^* + s_{i2}x_2^* + r_i$;

P3. A calculation of the output of fuzzy model $y^* = f(x_1^*, x_2^*)$ as follows:

$$y^* = \frac{\sum\limits_{i=1}^{n} w_i \cdot y_i^*}{\sum\limits_{i=1}^{n} w_i} . \tag{2}$$

If a fuzzy model must approximate given set of experimental data then optimization of fuzzy model is used for minimization of the error of approximation. Such optimization is usually based on a tuning of membership functions and parameters of right sides of rules [7,8]. For example, if membership functions are defined as bell functions:

$$A(x) = \frac{1}{1 + \left(\dfrac{x-c}{a}\right)^{2b}} , \tag{3}$$

where a,b,c are parameters, then a search of optimal values of parameters a,b,c for each fuzzy set and parameters s_{i1}, s_{i2}, r_i for each fuzzy rule may be done.

A system based on a fuzzy model is usually considered as "intelligent" because fuzzy rules may represent some knowledge about subject area. For example, these rules may formalize expert's knowledge about problem. But after optimization of fuzzy model we can loss this expert's knowledge because fuzzy sets obtained after such optimization may essentially differ from initial fuzzy sets defined by experts. Moreover, "a natural" interpretation of new fuzzy sets in optimal fuzzy model sometimes may be difficult.

Another approach to optimization of fuzzy models may be based on a tuning of operations used in a fuzzy model. For example, conjunction operation may be given parametrically and optimization of these parameters may be used [2-6]. Such approach applied without optimization of membership functions gives possibility to keep expert's information presented in membership functions. From another point of view, optimization of operations together with membership functions will give possibility to build more optimal models than optimization of only one of these classes of functions [3].

As parametric conjunction operations it is possible to consider parametric T-norms but optimization of fuzzy model with such functions is not so easy and quick procedure due to the complexity of known classes of these operations [1,4,6]. Different approaches to definition and generation of parametric conjunction operations simpler than T-norms were considered in [4,5]. These approaches are based on generalizations of the definition of conjunction operation. As result, the more general class of conjunction operations we consider the more simple parametric conjunction operations we can introduce. The second consequence of generalization useful from practical point of view is the following: the less restrictions on the form of parametric conjunction functions are used the more variable these functions may be and, hence, the better fitting of fuzzy model with experimental data may be achieved.

Here we consider a new generalization of the concept of conjunction operation, which gives possibility to build more optimal fuzzy models as result of deleting of some limitations on the class of functions, considered as fuzzy conjunctions. Application of new conjunction operations in fuzzy models may be also considered as application of modifiers of fuzzy sets together with usual conjunctions.

2 Generalized conjunction operations

Usually T-norms are considered as fuzzy conjunction operations [1]. These operations are defined as commutative and associative functions $T:[0,1]\times[0,1]\to[0,1]$ satisfying boundary condition: $T(x,1) = T(1,x) = x$, and monotonicity property: $T(x,y) \le T(u,v)$ if $x \le u,\ y \le v$. For these conjunctions the following restrictions are fulfilled:

$$T_d(x,y) \le T(x,y) \le T_c(x,y),\qquad(4)$$

where $T_d(x,y) = \begin{cases} x, & \text{if } y=1 \\ y, & \text{if } x=1 \\ 0, & \text{otherwise} \end{cases}$ and $T_c(x,y) = min(x,\ y)$.

T-norms may be generated as follows: $T(x,y) = \bullet^{-1}(T_1(\bullet(x,y),\ \bullet(x,y)))$, where $\bullet$ is some increasing bijection $\bullet:[0,1]\to[0,1]$ with $\bullet(0)= 0,\ \bullet(1) = 1$ and T_1 is another T-norm, for example $T_1(x,y) = xy$. Different types of parametric classes of T-norms have been introduced [1, 7] but all these functions are sufficiently complicated for use in optimization procedures. For example, one of the simplest parametric classes of T-norms introduced by Yager has the form:

$$T(x,y) = 1 - min\left(1, \sqrt[p]{(1-x)^p + (1-y)^p}\right).\qquad(5)$$

Generalized conjunction operations as functions satisfying only monotonicity property and boundary condition were considered in [4]. These functions also satisfy to restrictions (4). Generalized conjunction may be generated by $T(x,y) = T_2(T_1(x,y),s(g_1(x),g_2(y)))$, where T_2 and T_1 are some generalized conjunctions, g_1, $g_2:[0,1]\to[0,1]$ are non-decreasing functions such that $g_1(1)= g_2(1) = 1$, and function $s:[0,1]\times[0,1]\to[0,1]$ satisfies the property $s(1,y)=s(x,\ 1)= 1$ and the monotonicity property. One of the simplest generalized parametric conjunctions is following:

$$T(x,y) = (xy)min(1,x^p+y^q).\qquad(6)$$

More general class of G-conjunctions satisfying monotonicity property and properties $T(0,0) = T(0,1) = T(1,0) = 0$, $T(1,1) = 1$ was considered in [5]. These conjunctions may be generated as follows: $T(x,y) = f(T_1(g(x),h(y)))$, where T_1 is some G-conjunction and f, g, h:$[0,1] \rightarrow [0,1]$ are non-decreasing functions such that $f(0) = g(0) = h(0) = 0$, $f(1) = g(1) = h(1) = 1$. This class of conjunctions contains the simplest parametric conjunction operation

$$T(x,y) = x^p y^q, \qquad (p,q > 0) . \tag{7}$$

The examples of optimization of fuzzy models with this parametric G-conjunction operation were discussed in [2,3]. The optimization of parameters of this operation additionally to optimization of parameters of membership functions and right sides of rules gave possibility to minimize the error of approximation of given data by fuzzy model on more than 40% [3].

3 A new approach to processing of fuzzy rules

The parametric G-conjunction operation (7) is the simplest one, but we consider here the further generalization of this function by deleting limitations on the values of parameters p and q. We will suppose here that parameters p and q may have any real values. Since parameters p and q may be now negative we will suppose that membership functions used in rules of fuzzy model have only positive values. This property is fulfilled for membership functions represented by bell functions and gaussian functions [7]. It is also fulfilled for triangular and trapezoidal membership functions whith positive values on all domains of input variables.

Only one property $T(1,1) = 1$, of considered above conjunction operations is fulfilled for this function. Moreover, the function

$$T(x,y) = x^p y^q , \tag{8}$$

where p,q are any real values and $x,y \in (0,1]$, can have any positive value but not only values from $[0,1]$. For simplicity, we will call this function with unconstrained parameters as UG-conjunction. Of course, this function can not be considered as "natural" generalization of conjunction operation, nevertheless, we can use it for processing of fuzzy rules. We should note that several "abnormal" functions have been used successfully in fuzzy modeling. For example, the $sinc$ function $A(x) = sin(x)/x$ with negative values was considered as membership function in [8,9] and a summation of membership values with result greater than 1 is used in standard additive models [8]. Below we will consider an example when fuzzy model obtained after optimization of parameters of UG-conjunction shows better performance than fuzzy model obtained after optimization of parameters of membership functions together with parameters of G-conjunction.

4 Example

Let us consider the nonlinear function with three variables $f(x,y,z)=(1+x^{0.5}+y^{-1}+z^{-1.5})^2$, defined on $[1,6]\times[1,6]\times[1,6]$. This function was used in [7, 3] for testing different modeling approaches. For modeling this function we use here Sugeno fuzzy model with 8 rules and with two bell membership functions assigned to each input variable x,y and z:

R_1: If X is A_1 and Y is B_1 and Z is C_1 then $u_1 = s_{11}x+ s_{12}y+ s_{13}z+r_1$,
R_2: If X is A_1 and Y is B_1 and Z is C_2 then $u_2 = s_{21}x+ s_{22}y+ s_{23}z+r_2$,
R_3: If X is A_1 and Y is B_2 and Z is C_1 then $u_3 = s_{31}x+ s_{32}y+ s_{33}z+r_3$,   (9)
...
R_8: If X is A_2 and Y is B_2 and Z is C_2 then $u_2 = s_{81}x+ s_{82}y+ s_{83}z+r_8$.

As in [7], we used 216 training data and 125 checking data distributed uniformly on $[1,6]\times[1,6]\times[1,6]$ and $[1.5,5.5]\times[1.5,5.5]\times[1.5,5.5]$, respectively. The following performance index

$$APE = \text{average percentage error} = \frac{1}{m}\sum_{j=1}^{m}\frac{|D(j)-U(j)|}{|D(j)|}\cdot 100\% \qquad (10)$$

was considered, where m is a number of data, and $D(j)$ and $U(j)$ are the jth desired output and predicted output, respectively. We will compare performance of fuzzy models obtained after optimization of the following parameters:
A. Parameters of membership functions and right sides of rules;
B. Parameters of operations $T(x,y) = x^p y^q$, $(p,q > 0)$ and right sides of rules;
C. Case A and then case B;
D. Parameters of operations $T(x,y) = x^p y^q$, $(p,q$ are any real values) and right sides of rules;
E. Case A and then case D.
Initial values of parameters of bell membership functions were equal $a=2.5$, $b=2.5$, $c=1$ for A_1, B_1, C_1 and $a=2.5$, $b=2.5$, $c=6$ for A_2, B_2, C_2. The total number of these parameters is equal to $3*2*3=18$. Initial values of parameters of operations and right sides of rules were equal 1. Conjunction $T(x,y)=xy$ was used in case A. We calculated firing values of rules as follows:

$$w_i = A_i(x^*)^{p(i)}\cdot B_i (y^*)^{q(i)}\cdot C_i (z^*)^{r(i)}, \qquad (i=1,...,8), \qquad (11)$$

where each A_i, B_i, C_i takes one of two possible values A_1, A_2, B_1, B_2, C_1,C_2 respectively.
In all cases we have used optimization of parameters during 100 steps of $10*m$ iterations, where m is the total number of optimized parameters. In case A, we optimized 18 parameters of membership functions and $8*4=32$ parameters of right sides of rules. In case B, we optimized $8*3=24$ parameters of operations and 32 parameters

of right-hand sides of rules. Cases C and E uses 30 steps of optimization of 50 parameters as in the case A and then 70 steps of optimization of 56 parameters as in cases B or E. In cases B and C we used constrained optimization of parameters p and q with $p,q > 0$, but in cases D and E an unconstrained optimization was applied.

Results of optimization of fuzzy models are presented in Table 1 and on Fig. 1. These results are compared with other approaches discussed in [10, 7, 3]. The 1st and 2nd approaches considered in [3] are the same as cases A and B considered here, but instead of bell functions the triangular membership functions were used in rules of fuzzy model.

Table 1. Errors of approximation of function $f(x,y,z)=(1 + x^{0.5} + y^{-1} + z^{-1.5})^2$

Model	Training error	Checking error	Parameters number	Training data size	Checking data size
Fuzzy model 1 [10]	1.5	2.1	22	20	20
Fuzzy model 2 [10]	0.59	3.4	32	20	20
1st approach [3]	1.832	3.498	50	216	125
2nd approach [3]	1.053	1.932	50, 56	216	125
Case A	0.361	10.751	50	216	125
Case B	2.640	4.186	56	216	125
Case C	0.296	8.880	50, 56	216	125
Case D	0.115	1.848	56	216	125
Case E	0.035	3.945	50,56	216	125
ANFIS [7]	0.043	1.066	50	216	125

As we can see, the performance of fuzzy model in Case D when only *UG*-conjunction operations are optimized is even better than the performance of fuzzy model in Case C when the optimization of membership functions together with *G*-conjunction operations is used. An optimization of *UG*-conjunctions together with membership functions (Case E) shows better results on training data than results obtained by adaptive neuro-fuzzy inference system ANFIS [7]. Results obtained by ANFIS are based only on optimization of bell functions as in Case A. We see, that results of optimization (Case A and ANFIS) depend on method applied. We can suppose that optimization of *UG*-conjunctions after applying ANFIS will give better results than in Case E. The comparison of different methods of optimization of fuzzy models will be done also in our future work.

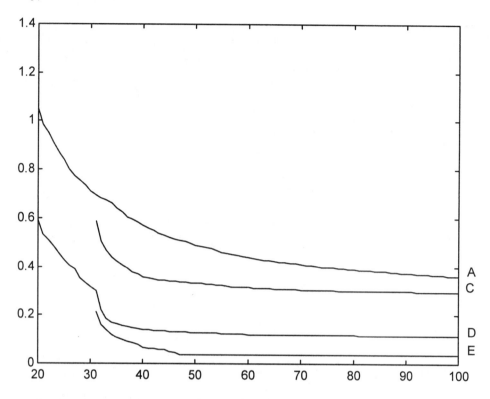

Fig. 1. Errors of approximation of function $f(x,y,z)=(1 + x^{0.5} + y^{-1} + z^{-1.5})^2$ by Sugeno models based on optimizations of the following parameters: A) Parameters of bell membership functions and right sides of rules (100 steps of iteration procedure); C) Case A (30 steps) and then parameters of operations $T(x,y) = x^p y^q$, $(p,q > 0)$ and right sides of rules (70 steps); D) Parameters of operations $T(x,y) = x^p y^q$, $(p,q$ are any real values) and right sides of rules (100 steps); E) Case A (30 steps) and then case D (70 steps).

5 Generalized conjunctions as modifiers

We can also consider *UG*-conjunction from another point of view as the generalization of procedure used for processing of fuzzy rules. Let us replace the procedure P1 for calculation of the value of premises of fuzzy models by the following one:

P1^0. $w_i=AND(g_{i1}(A_{i1}(x_1^*)),g_{i2}(A_{i2}(x_2^*)))$, where g_{i1} and g_{i2} are modifiers of membership functions A_{i1} and A_{i2} respectively and operation *AND* is some conjunction-like function defined on the set of positive real values.

From this point of view, we can consider the *UG*-conjunction $T(x,y)= x^p y^q$ as a composition of conjunction operation $AND(x,y)= xy$ defined on the set of positive real values and modifiers $g_{t1}(x) = x^p$, $g_{t2}(x) = x^q$. These operations were considered by Zadeh in [11].

It should be noted that starting from the idea of tuning operations in fuzzy models instead of (or additionally to) tuning of membership functions, we have introduced different generalizations of conjunction operations and methods for construction of simple parametric conjunction operations [4,5]. But finally, we are coming to such generalization of the conjunction operation, which may be considered as some procedure of modification of membership functions used in fuzzy models. Nevertheless, we can say that new approach gives possibility to keep expert knowledge given in fuzzy sets if only parameters of *UG*-conjunctions (parameters of modifiers) are optimized in fuzzy model. New point of view on *UG*-operations gives us possibility to compare Cases A and D as two approaches to construction of optimal fuzzy models based on a tuning of membership functions.

Initial and optimal membership functions

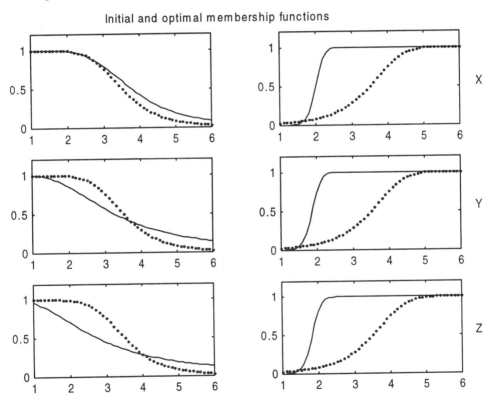

Fig. 2. Optimal and initial (dotted curves) membership functions (Case A).

Fig. 2 shows the optimal membership functions of fuzzy model obtained in the example considered in Case A. Fig. 3 shows the minimal and the maximal modifications of fuzzy sets produced by optimal modifiers. As we can see, the maximal value of modified membership function is greater than 6 and it is achieved for fuzzy set A_2 defined on X. Such modification is obtained by optimal value of parameter $p(8) = -0.5284$ in (9) which modifies the values of A_2 in the rule R_8.

Initial and modified membership functions

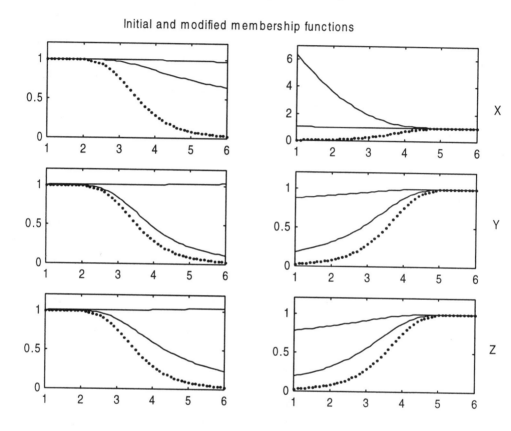

Fig. 3. Modified and initial (dotted curves) membership functions (Case D).

6 Conclusions

Optimization of fuzzy models may be based on a tuning of membership functions and operations given parametrically. The known parametric classes of T-norms are sufficiently complicated for use in optimization procedures and for hardware realization. The construction of simple parametric conjunction operations may be based on generalizations of these operations. The more general class of conjunction operations is considered the more simple parametric conjunction operations can be introduced. The less restrictions on the form of parametric conjunction functions are used the more variable these functions may be and, hence, the better fitting of fuzzy model with experimental data may be achieved. Several difinitions and methods of generation of conjunction operations were considered in this paper and the new parametric fuzzy conjunction operation was introduced.

Different approaches to construction of optimal fuzzy models were compared on example of approximation of real valued function with three variables. The optimal fuzzy model based on the optimization of parameters of new conjunction operation shows the better performance than the optimal fuzzy model based on the optimization of parameters of fuzzy sets.

The new approach to construction of optimal fuzzy models may be considered also as generalized procedure of processing of fuzzy rules based on an optimization of modifiers of fuzzy sets. Modified fuzzy sets obtained after optimization of modifiers differ from optimal fuzzy sets obtained after traditional optimization of fuzzy model.

Acknowledgment

This work was partly supported by the Academy of Sciences of Republic of Tatarstan.

References

1. Alsina, C., Trillas, E., Valverde, L.: On Some Logical Connectives for Fuzzy Sets Theory. J. Math. Anal. Appl. 93 (1983) 15 -26
2. Batyrshin, I., Bikbulatov, A.: Construction of Optimal Fuzzy Models Based on the Tuning of Fuzzy Operations. In: Proceedings of 6th Zittau Fuzzy Colloquium. Zittau (1998) 33-38
3. Batyrshin, I., Bikbulatov, A., Kaynak, O., Rudas, I.: Functions Approximation Based on the Tuning of Generalized Connectives. In: Proceedings of EUROFUSE - SIC '99. Budapest (1999) 556-561
4. Batyrshin, I., Kaynak, O.: Parametric Classes of Generalized Conjunction and Disjunction Operations for Fuzzy Modeling. IEEE Trans. Fuzzy Syst. 7 (1999) 586-596
5. Batyrshin, I., Kaynak, O., Rudas, I.: Generalized Conjunction and Disjunction Operations for Fuzzy Control. In: Proceeding of 6th European Congress on Intelligent Technignes & Soft Computing, EUFIT'98, Vol. 1. Aachen (1998) 52-57
6. Cervinka, O.: Automatic Tuning of Parametric T-norms and T-conorms in Fuzzy Modeling. In: Proceedings of 7th IFSA World Congress, Vol. 1. ACADEMIA, Prague (1997) 416-421
7. Jang, J.-S. R., Sun, C. T., Mizutani, E.: Neuro-Fuzzy and Soft Computing. A Computational Approach to Learning and Machine Intelligence. Prentice-Hall International, New York (1997)
8. Kosko, B.: Fuzzy Engineering. Prentice-Hall, New Jersey (1997)
9. Mitaim, S., Kosko, B.: What is the Best Shape for a Fuzzy Set in Function Approximation? In: Proceeding of 5th IEEE Int. Conf. Fuzzy Systems, FUZZ-96, Vol. 2. (1996) 1237 - 1243
10. Sugeno, M., Kang, G.T.: Structure Identification of Fuzzy Model. Fuzzy Sets and Systems 28 (1988) 15-33
11. Zadeh, L.A.: A Fuzzy-Set-Theoretic Interpretation of Linguistic Hedges. Journal of Cybernetics 2 (1972) 4-34

The Equality of Inference Results Using Fuzzy Implication and Conjunctive Interpretations of the If-Then Rules Under Defuzzification

E.Czogała, N.Henzel and J.Leski
Institute of Electronics
Technical University of Silesia
Akademicka 16, 44-100 Gliwice, Poland
e-mail: jl@biomed.iele.polsl.gliwice.pl

Abstract. A specific type of equivalence of inference results under defuzzification between logical implication interpretation and conjunctive interpretation of the fuzzy if-then rules has been considered in this paper. The theoretical considerations are illustrated by means of numerical examples in the field of fuzzy modeling.

1 Introductory Remarks

A study of inference processes when premises in if-then rules are fuzzy is still a subject of many papers in specialized literature c.f. [1]-[10]. In such processes, a sound and proper choice of logical operators plays an essential role. The theoretical (mathematical) and the practical (computational) behavior of logical operators in inference processes has to be known before such a choice is made.

Some selected logical operators and fuzzy implications were investigated here with respect to their behavior in the inference processes. A specific type of equivalence (equality) of inference results obtained using on one hand a conjunction interpretation of fuzzy if-then rules and on the other hand the interpretation of such rules in the spirit of classical logical implication may be shown. Such equivalence is important regarding the inference algorithms construction. The inference algorithms based on logical implication interpretation of the if-then rules may be replaced by the simpler, faster and more exact algorithms used for the conjunctive interpretation of such rules.

2 The Equality of Approximate Reasoning Results Under Defuzzification

An output fuzzy set $B'(y)$ obtained from inference system based on fuzzy implication interpretation of if-then rules is different from the resulting fuzzy set obtained from inference system based on conjunctive interpretation of fuzzy if-then rules. However in many applications crisp results are required instead of fuzzy ones. Hence, a question arises: whether it is possible to get the same or approximately the same crisp results from inference system when defuzzification is applied. The answer is positive under the respective circumstances. The point of departure of our considerations is the equality expressed in the form:

$$DEFUZ_{mf_L} \left\{ \sup_{\underline{x} \in \underline{X}} \star_{T'} \left[\underline{A}'(\underline{x}), \left[\frac{\bigwedge_{k=1}^{K} T}{\sum_{k=1}^{K}} I(\underline{A}^{(k)}(\underline{x}), B^{(k)}(y)) \right] \right] \right\} =$$

$$= DEFUZ_{mf_R} \left\{ \sup_{\underline{x} \in \underline{X}} \star_{T'} \left[\underline{A}'(\underline{x}), \left[\frac{\bigvee_{k=1}^{K} S}{\sum_{k=1}^{K}} \underline{A}^{(k)}(\underline{x}) \star_T B^{(k)}(y) \right] \right] \right\}. \tag{1}$$

where $\star_T, \star_{T'}$ denote respective t-norms T and T', $\underline{A}^{(k)}(\underline{x}) = A_1^{(k)}(x_1) \star_T \ldots \star_T A_n^{(k)}(x_n)$ and $DEFUZ_{mf}$ stands for defuzzification operator of membership function.

The left-side part of that equality represents the defuzzified output of fuzzy implication I based inference system whereas the right-side part represents an inference system based on Mamdani's composition (conjunctive interpretation of if-then rules).

The problem is to find such fuzzy implications (for the left-side part of the last equality), conjunctive operators (in the right-side of this equality), aggregation operations and defuzzification methods for both sides of the last equality in order to get the same crisp results. Generally, solving such a problem causes difficulties. One of the most important reasons is the different nature of fuzzy implication and conjunction (c.f. different truth table in classical logic). However, in special cases, under some assumptions, pragmatic solutions exist. To show such solutions let us assume for simplicity that the input fuzzy sets $\underline{A}'$ are singletons in $\underline{x}_0$ (in this particular case FATI is equivalent to FITA). The last equality can be rewritten in the simplified form:

$$DEFUZ_{mf_L} \left\{ \left[\frac{\bigwedge_{k=1}^{K} T}{\sum_{k=1}^{K}} I(\underline{A}^{(k)}(\underline{x}_0), B^{(k)}(y)) \right] \right\} =$$

$$= DEFUZ_{mf_R} \left\{ \left[\frac{\bigvee_{k=1}^{K} S}{\sum_{k=1}^{K}} \underline{A}^{(k)}(\underline{x}_0) \star_{T'} B^{(k)}(y) \right] \right\}. \tag{2}$$

Because of the differences in aggregation operations, we accept the same for both sides of the equality aggregation operation i.e. normalized arithmetic sum. Additionally, we assume different defuzzification methods for both sides of the last equality, e.g. $MISD_\alpha$ (modified indexed standard defuzzifier) for the left-side and SD (standard defuzzifier) for the right-side. For our purposes we will use as the left-side defuzzification method the method presented in the previous section named $MICOG_\alpha$. For some fuzzy implications (e.g. Kleene-Dienes (max{1-a,b}), Lukasiewicz (min{1, 1-a+b}), Reichenbach (1-a+ab), Zadeh (max{1-a, min{a,b}}), Fodor (if a+b>1 then 1 else max{1-a,b}) and others) we have:

$$\alpha = \sum_{k=1}^{K} \alpha_k = \sum_{k=1}^{K} \overline{\underline{A}^{(k)}(x_0)} = \sum_{k=1}^{K} \left(1 - \tau^{(k)}\right), \tag{3}$$

where $\tau^{(k)}$ stands for firing degree of k-th rule.

However, for some fuzzy implications (e.g. Gödel (if $a \leq b$ then 1 else b), Rescher (if $a \leq b$ then 1 else 0), Goguen (if $a \neq 0$ then min{1,b/a} else 1) and others) equals zero.

As the right side defuzzification method the well known COG method can be used. Taking into account the above mentioned assumptions and simplifications the last equality may be written as:

$$MICOG_\alpha \left\{ \sum_{k=1}^{K} I\left(\underline{A}^{(k)}(x_0), B^{(k)}(y)\right) \right\} =$$
$$= COG \left\{ \sum_{k=1}^{K} \underline{A}^{(k)}(x_0) *_{T'} B^{(k)}(y) \right\}. \tag{4}$$

It should be pointed out here that the modification in $MICOG_\alpha$ is responsible for elimination of the non-informative part of output membership function $B'(y)$ (if such a part exists).

The specific equivalence (equality) of inference results mentioned above can be seen straight on, if we take e.g. Reichenbach fuzzy implication under assumption that is computed by means of the formula (3), as aggregation operation a normalized arithmetic sum is applied and the defuzzification method $MICOG_\alpha$ is taken as in [2]. On the right side of the formula (4) algebraic product operation is taken in order to get Larsen's inference system with normalized arithmetic sum as aggregation operation and COG as defuzzification method. The above mentioned equivalence can also be describe by the chain of identities:

$$MICOG_\alpha \left[\sum_{k=1}^{K} \left(\overline{\underline{A}^{(k)}(\underline{x}_0)} + \underline{A}^{(k)}(\underline{x}_0)B^{(k)}(y) \right) \right]_{Reichenbach} =$$

$$MICOG_\alpha \left[\sum_{k=1}^{K} (1 - \tau_k) + \sum_{k=1}^{K} \tau_k B^{(k)}(y) \right]_{Reichenbach} =$$

$$MIHM_\alpha \left[\sum_{k=1}^{K} \alpha_k + \sum_{k=1}^{K} (1 - \alpha_k) B^{(k)}(y) \right]_{Reichenbach} = \qquad (5)$$

$$COG \left[\sum_{k=1}^{K} \tau_k B^{(k)}(y) \right]_{Larsen} = HM \left[\sum_{k=1}^{K} \tau_k B^{(k)}(y) \right]_{Larsen} =$$

$$\left[\frac{\sum_{k=1}^{K} \tau_k \, y^{(k)}}{\sum_{k=1}^{K} \tau_k} \right]_{Larsen} .$$

When we consider the Łukasiewicz fuzzy implication under assumption that is also computed using the formula (3), aggregation operation and $MICOG_\alpha$ are the same as in the previous case and on the right side of the formula (4), $*_{T'}$ represents minimum, we get a well-known Mamdani's inference system with normalized arithmetic sum as aggregation operation and COG as defuzzification method. This equivalence can be described as follows:

$$MICOG_\alpha \left\{ \sum_{k=1}^{K} \left[1 \wedge \left(\overline{\underline{A}^{(k)}(\underline{x}_0)} + B^{(k)}(y) \right) \right] \right\}_{Lukasiewicz} =$$

$$MICOG_\alpha \left\{ \sum_{k=1}^{K} \left[1 \wedge ((1 - \tau_k) + B^{(k)}(y)) \right] \right\}_{Lukasiewicz} =$$

$$MICOG_\alpha \left\{ \sum_{k=1}^{K} \left[1 \wedge (\alpha_k + B^{(k)}(y)) \right] \right\}_{Lukasiewicz} = \qquad (6)$$

$$COG \left\{ \sum_{k=1}^{K} \left[1 \wedge (\alpha_k + B^{(k)}(y)) \right] - \sum_{k=1}^{K} \alpha_k \right\}_{Mamdani} =$$

$$COG \left\{ \sum_{k=1}^{K} \left[(1 - \alpha_k) \wedge B^{(k)}(y) \right] \right\}_{Mamdani} =$$

$$COG \left\{ \sum_{k=1}^{K} \left[\tau_k \wedge B^{(k)}(y) \right] \right\}_{Mamdani} .$$

Such formulated equivalence may be also shown using the generalized bounded difference c.f.[10]. In order to illustrate the considerations presented above some numerical examples will be discussed below.

3   Numerical Examples

A fuzzy knowledge base presented in Fig.1 is taken as a basis of numerical calculation carried out here. Such knowledge base consisting of nine fuzzy if-then rules may have practical meaning in many fields e.g. fuzzy control, fuzzy modeling, decision support systems and others. Each rule consists of two premises

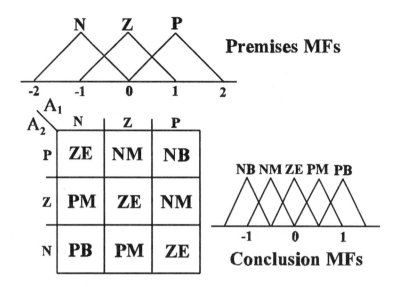

Fig. 1. Fuzzy knowledge base.

(e.g. error and change of error in the antecedent part of the rule and one conclusion (e.g. control) in consequent part. The fuzzy sets representing possible values of the respective linguistic variables are also shown in Fig.1.

Using such a knowledge base we can geometrically show the essence of the above discussed equivalence. For all numerical examples the 'and' connective is considered to be an algebraic product. Considering Reichenbach implication with the normalized arithmetic sum as the aggregation operation and defuzzifier of $MICOG_\alpha$ type (α's are computed from the formula (3)), the equivalence of inference results with those obtained on the basis of Larsen's product is illustrated in Fig. 2. Taking into account the Lukasiewicz fuzzy implication with the same aggregation operation and defuzzifier of $MICOG_\alpha$ type for the same α's as in the previous case, the equivalence of inference results with inference results obtained on the basis of Mamdani's minimum is shown in Fig. 3. If we use the defuzzifier $MIHM_\alpha$ instead $MICOG_\alpha$ in the last case, then we get the same results as we got for Reichenbach implication. It means that the reduced (clipped) conclusion fuzzy sets are transformed into scaled conclusion fuzzy sets. Let us notice that the reduced fuzzy sets obtained using Lukasiewicz fuzzy implication include the scaled fuzzy sets.

In Fig. 4. the inference results obtained on the basis of Kleene-Dienes fuzzy implication are shown. However, the equivalence between the inference results obtained on the basis of this fuzzy implication and inference results obtained using a conjunction is not found but applying the defuzzifier $MIHM_\alpha$ instead of $MICOG_\alpha$ we get the same results as for Reichenbach fuzzy implication (see Fig.4). Because of the inclusion of Kleene-Dienes conclusion fuzzy sets in

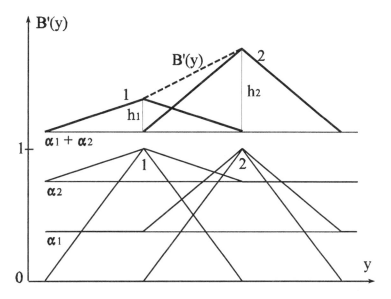

Fig. 2. An illustration of the equivalence of inference results (Reichenbach fuzzy implication versus Larsen's product).

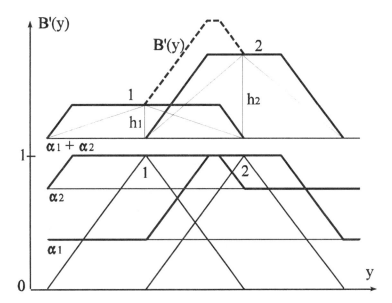

Fig. 3. An illustration of the equivalence of inference results (Łukasiewicz fuzzy implication versus Mamdani's minimum).

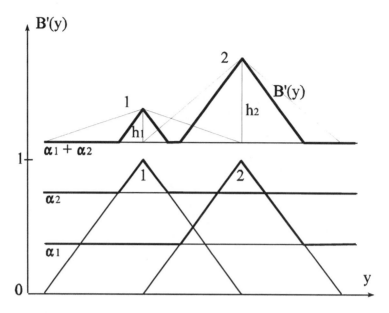

Fig. 4. An illustration of the equivalence of inference results (Kleene-Dienes fuzzy implication versus Reichenbach fuzzy implication).

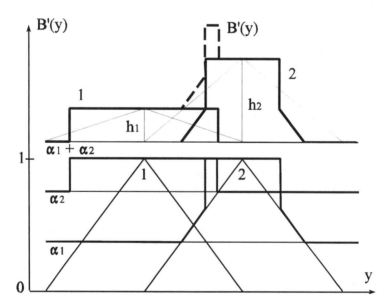

Fig. 5. An illustration of the equivalence of inference results (Fodor fuzzy implication versus Reichenbach fuzzy implication).

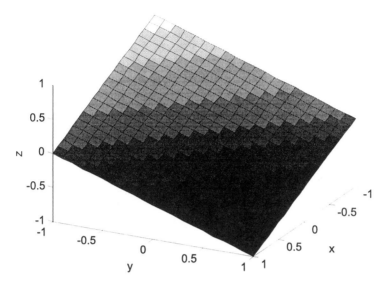

Fig. 6. Output surface for Reichenbach fuzzy implication.

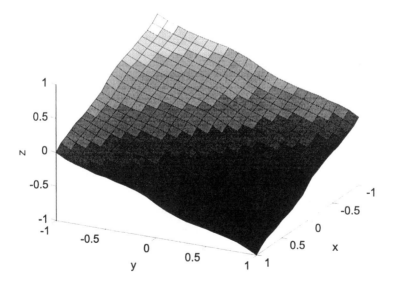

Fig. 7. Output surface for Łukasiewicz fuzzy implication.

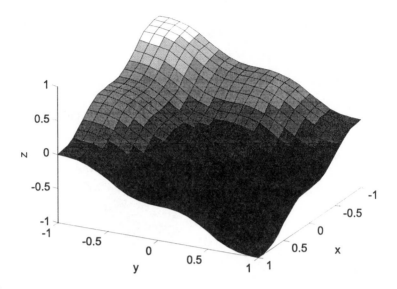

Fig. 8. Output surface for Kleene-Dienes fuzzy implication.

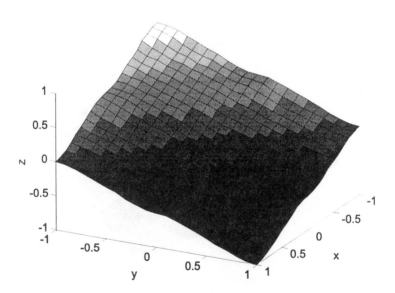

Fig. 9. Output surface for Fodor fuzzy implication.

Implication Name	λ_{MSE}	λ_{MAX}
Kleene-Dienes	0.005428	0.1463
Reichenbach	0	0
Lukasiewicz	0.000524	0.0446
Fodor	0.000287	0.0418

Table 1. Values of quality indexes.

Reichenbach conclusion fuzzy sets the Kleene-Dienes conclusion fuzzy sets have to be respectively rescaled in order to get the last ones.

An analogous situation occurs if Fodor fuzzy implication is considered. The equivalence between the inference results obtained on the basis of this implication and results obtained by means of a conjunction is also not found. However, using $MIHM_\alpha$ defuzzifier instead of $MICOG_\alpha$ we get the same results as for Reichenbach fuzzy implication as well. Additionally it should be also pointed out that conclusion fuzzy sets obtained on the basis of Fodor fuzzy implication are not included in conclusion fuzzy sets obtained on the basis of Reichenbach fuzzy implication.

The rule base depicted in Fig. 1 was tested by means of complete set of singletons for the reason of fuzzy modeling of the function $z = f(x, y) = -(x + y)/2$.

Four fuzzy implications (Reichenbach, Lukasiewicz, Kleene-Dienes and Fodor) were applied in inference system with the normalized arithmetic sum as an aggregation operation and defuzzifier of $MICOG_\alpha$ type, where is computed from (3) as well.

For the above mentioned fuzzy implications the resulting surfaces are presented in Figs 6-9. These surfaces correspond to the surfaces obtained from the conjunctive based inference system. The quality indexes ($\lambda_{MSE}, \lambda_{MAX}$) characterizing the quality of inference system may be defined as follows:

$$
\begin{aligned}
\lambda_{MSE} &= \frac{1}{IJ} \sum_{i=1}^{I} \sum_{j=1}^{J} \left[I\left(x_i, y_j\right) - \frac{-(x_i+y_j)}{2} \right]^2 \\
\lambda_{MAX} &= \max_{i,j} \left| I\left(x_i, y_j\right) - \frac{-(x_i+y_j)}{2} \right|
\end{aligned}
\tag{7}
$$

The values of the above mentioned indexes are gathered in Table 1. It should be pointed out that both indexes for Reichenbach implication are equal to zero and for Lukasiewicz implication the values of indexes are the same as for Mamdani's minimum.

The programming tool applied to these investigations was FDSS Fuzzy-Flou and Fuzzy Logic toolbox for MATLAB® systems. Special procedures (m-files) concerning fuzzy implications and modified COG methods were written into the last mentioned toolbox.

4 Conclusions

In this paper a specific type of equivalence of inference results using fuzzy implication (Reichenbach and Lukasiewicz) interpretation and the respective conjunctive (Larsen and Mamdani) interpretation of the if-then rules has been discussed. Such equivalence is important regarding the inference algorithm construction. The inference algorithms based on conjunctive operators in some cases seem to be faster, simpler and more exact than the algorithms of fuzzy implication based inference system. However, the interpretation of the fuzzy if-then rules based on fuzzy implications is sounder from logical point of view. For future research, the equivalence mentioned above should be considered more deeply from mathematical point of view.

References

1. O.Cordon, F. Herrera, A. Peregrin: Applicability of the fuzzy operators in the design of fuzzy logic controllers, Fuzzy Sets and Systems 86 (1997)15-41
2. E.Czogała, J.Łeski: An equivalence of approximate reasoning under defuzzification, BUSEFAL 74 (1998) 83-92
3. E. Czogała, R. Kowalczyk: Investigation of selected fuzzy operations and implications in engineering, Fifth IEEE Int. Conf. on Fuzzy Systems (1996) 879-885
4. D. Dubois, H. Prade: Fuzzy sets in approximate reasoning, Part 1: Inference with possibility distributions, Fuzzy Sets and Systems 40 (1991) 143-202
5. D. Dubois, H. Prade: What are fuzzy rules and how to use them, Fuzzy Sets and Systems 84 (1996) 169-185
6. J.C. Fodor: On fuzzy implication operators, Fuzzy Sets and Systems 42 (1991) 293-300
7. E. Kerre: A comparative study of the behavior of some popular fuzzy implication operators on the generalized modus ponens, in: L. Zadeh and J. Kacprzyk, Eds., Fuzzy Logic for the Management of Uncertainty, Wiley, New York (1992) 281-295
8. H.Maeda: An investigation on the spread of fuzziness in multi-fold multi-stage approximate reasoning by pictorial representation - under sup-min composition and triangular type membership function, Fuzzy Sets and Systems 80 (1996) 133 - 148
9. S. Weber: A general concept of fuzzy connectives, negations and implications based on t-norms and t-conorms, Fuzzy Sets and Systems 11 (1983) 115-134
10. R. R. Yager: On the interpretation of fuzzy if-then rules, Applied Intelligence 6 (1996) 141-151

Adaptive Information System for Data Approximation Problems

Witold Kosiński[1,2,3] and Martyna Weigl[2,4]

[1] Polish–Japanese Institute of Information Technology
ul. Koszykowa 86 , PL-02-008 Warszawa, Poland
wkos@pjwstk.waw.pl
[2] Center of Mechanics and Information Technology
Inst. Fund. Technol. Research, Polish Academy of Sciences
IPPT PAN, PL-00-049 Warszawa, Poland
wkos@ippt.gov.pl
[3] Department of Mathematics, Technology and Natural Sciences
Pedagogical University of Bydgoszcz, PL-85-064 Bydgoszcz
[4] Polska Telefonia Cyfrowa, ERA GSM, PL-02-222 Warszawa
mweigl@eragsm.com.pl

Abstract. An adaptive information system is constructed for the approximation of a multidimensional data base. Such a construction problem can be regarded as an approximation of a multivariate real-valued function that is known in a discrete number of points. That set of points forming the multidimensional data base is called a training set TRE for the adaptive information system. The constructed system contains a module of one-conditional fuzzy rules consequent parts of which are artificial feed-forward neural networks. The numerical data contained in TRE are used to construct premise parts of each rule and membership functions of fuzzy sets involved in them. They are specified with the help of a clustering analysis of TRE. For each fuzzy rule the neural network appearing as its consequent part is trained on the corresponding cluster as on its subdomain. In that way the cluster analysis can supply a knowledge in designing the approximator.

1 Introduction

Several methods have been proposed in recent years to construct universal approximators. Most of those methods are based on some heuristic approaches. Some of them are grounded with theoretical results concerning the proof of the existence problem, others are equipped with adaptation schemes. General existence results have been obtained within fuzzy systems (cf. [2, 8]) or neural networks (cf.[3, 4, 7, 16, 18]).

The present authors have approached such problems using two different tools: one based on an adaptive fuzzy inference system, the other – on feed-forward neural networks [12, 20–23]. It was also shown there that a non-homogeneous M-Delta neural networks as well as an adaptive fuzzy inference system equipped with a generalized Takagi-Sugeno fuzzy rules are universal approximators ([12, 22]).

Universal tools constructed in the above publications do not possess, unfortunately, any universality as far as their real approximation capabilities are concerned. While constructing and working with neural networks and fuzzy inference systems our experience shows that their real approximation capabilities can be improved by an appropriate pre-processing of sample data introduced for such systems. For example our observations made for an adaptive fuzzy inference system show that they are very sensitive to the definition of the membership functions of the fuzzy sets and their support used in the covering of input domain. It is difficult to adjust the overall output of the system to the sample data by adapting the generalized weight vector and parameters of the membership functions, if the data are of a particular type. For example, if on a part of the domain the function to be approximated is constant, while on another part the function has very high gradient.

In the present paper we are changing the order of construction. Knowing a particular multidimensional data base TRE that entries can be regarded as points from a graph of a multivariate real-valued (continuous) function (that is known in a discrete number of points), we start our construction procedure with a process of data mining. In that initial process the knowledge about the type and the behaviour of the function to be approximated is extracted. This knowledge will be used in the actual designing process. The process of knowledge extraction from numerical multidimensional data can be regarded as a kind of clustering analysis during which partitioning of the data base into a union of subsets is performed. Since each entry to the data base can be treated as an order pair in which the first component represents a value of a multidimensional input vector, while the second is a real–valued output to that input value, the set TRE, called at other occasions the training set, can be regarded as a subset of the cartesian product of multidimensional input space, say $\mathcal{X}$ and 1D output one, say $\mathcal{Y}$. Hence together with that partitioning of the data base one can perform the projection on the space $\mathcal{X}$ to get a splitting of the input domain into subdomains, or clusters.

After this first stage of construction procedure the next one appears in which a module of one-conditional rules for a fuzzy inference system can be constructed, in such a way that each cluster leads to a single rule. Thanks to the clustering of data the premise parts of the rules can be constructed together with the form of the membership functions of the fuzzy sets involved. We are assuming only that all membership functions are generalized Gaussian functions, and then the clustering analysis supplies us with the knowledge about the form of their exponent.

Now the next stage arises in which forms of consequent parts of those rules are fixed. In this paper we make the next step beyond our previous approach with generalized Takagi-Sugeno fuzzy rules [19]. Now each consequent part of the rule is a single mapping neural network SIMNN constructed for data from the corresponding cluster. Each network is trained on the cluster as on its input domain.

After this stage the overall output of constructed information system has to specified. Notice that till now constructed modules have grounded rather well the term a fuzzy–neural system for our information system. In the paper, as a natural solution, we assume that the overall output of the system is a convex combination of the outputs of all individual networks, where the coefficients of the combination are based on the level of activity of the individual rules calculated from weighted values of their membership functions.

Now the terminal stage of the construction procedure follows which is a final adaptation process. In this final stage a learning process is performed on the whole set TRE during which some free parameters of the membership functions involved in the fuzzy sets of the fuzzy rules can be tuned.

In the author's opinion proposed type of procedure is essential for the efficiency and accuracy of the constructed system: an adaptive fuzzy–neural inference system. Moreover, it can reduce greatly the noise contained in the numerical data.

For knowledge extraction from data one can use different methods. We can suggest two of them proposed recently in our previous publications. One is based on a seed-growing methods [9, 11], another uses an evolutionary programming, that makes use of a problem-oriented genetic algorithms [13].

2   Knowledge extraction from data

In the standard approach when dealing with an approximation problem the knowledge about the function to be approximated is contained in a set of the so-called training pairs:

$$\mathsf{TRE} = \{p^q = (x^q, y^q) \in \mathbb{R}^{n+1} : q = 1, 2, ...P\} \tag{1}$$

that represents a discrete number of points from the graph of the function. Here in each pair $p^q = (x^q, y^q)$, the value y^q is the desired value corresponding to the input vector x^q.

In the Appendix we describe shortly two methods of extracting knowledge from the set TRE which forms our multidimensional data base, for which the information system for approximation is constructed. Both methods deal with an appropriate cluster analysis of data in TRE. At the end of either method a partitioning of TRE into clusters is for our disposal. In other words we have ready the so-called fuzzy covering $\mathcal{C}$ of TRE, i.e. a family of clusters $\{\mathcal{K}_1, \mathcal{K}_2, ..., \mathcal{K}_M\}$ such that

$$\mathsf{TRE} = \bigcup_{h=1}^{M} \mathcal{K}_h \tag{2}$$

where each $\mathcal{K}_h \subset \mathbb{R}^{n+1}$.

In the recent paper [11] the authors have presented a seed growing approach for clustering problem of a large numerical multidimensional data set. In our most recent publication [13] an evolutionary approach has been sketched to the clustering problem for an adaptive inference system.

For each cluster $\mathcal{K}_h$ its centroid $p^h = (a^h, d^h) \in \mathcal{X}$ is by defined in similar way to (3).

$$p^h = \frac{1}{N_h} \sum_{j=1}^{N} p_j^h \tag{3}$$

is a mean value of all N_h training pairs $p_j^h = (x_j^h, y_j^h)$, $j = 1, 2, ..., N_h$ belonging to the cluster $\mathcal{K}_h$.

To each cluster $\mathcal{K}_h$ we relate its scatter (variance-covariance) matrix W^h calculated according to the formula [1]:

$$W^h = \frac{1}{N_h} \sum_{j=1}^{N_h} (p_j^h - p^h) \otimes (p_j^h - p^h), \tag{4}$$

where $\otimes$ denotes the tensor product of two vectors. The scatter matrix can be used to measure the efficiency of the clustering analysis (cf. the fitness function defined in [13]). Now the projection of each $\mathcal{K}_h \subset \mathbb{R}^{n+1}$ on the input space $\mathcal{X} \subset \mathbb{R}^n$ forms a family $\{X_1, X_2, ... X_M\}$ of subdomains (input clusters) that forms a covering of the input data $x's$.

Fuzzy sets A_h involved in premise parts of fuzzy rules will be related to clusters X_h and their functional (crisp) consequent parts in the form of neural networks that will be trained on the same subdomains X_h

Hence, if for a cluster $\mathcal{K}_h$, a point x_j^h is a projection of $p_j^h \in \mathcal{K}_h$ on the input space $\mathcal{X}$ with $j = 1, 2, ..., N_h$, then in a similar way to (4), we can define the corresponding scatter matrix S_h of each input cluster X_h, where $h = 1, 2, ..., M$,

$$S_h = \frac{1}{N_h} \sum_{j=1}^{N_h} (x_j^h - a^h) \otimes (x_j^h - a^h). \tag{5}$$

Let us notice that for each h the matrix S_h, as well W^h are symmetric and positive semi-definite, i.e. some number of the eigenvalues, say r are positive and the remainder are zero. The eigenvectors corresponding to the nonzero eigenvalues form the so-called axis of inertia of the cluster (in the cluster analysis they are called the principal directions, cf. [1, 5]), and the corresponding eigenvalues determine the length of the axis (and the maximal diameters along this axis). The eigenvectors corresponding to the vanishing eigenvalues determine the directions of the vanishing „thickness" of the cluster. These observations will be used in further publication in reducing the data supplied to the inference system and to define appropriate fuzzy rules.

In what follows we assume that each of matrices $S_h, h = 1, 2, ..., M$ is non-singular. The case of vanishing eigenvalues of the scatter matrix will be consider in the further publications. Now we point out that in practical examples of large numerical data set the singular scatter matrix is a very seldom case.

One can go beyond the present case and admit more than one fuzzy domain covering as results of different tools implemented in the process of the extraction of knowledge from the data base. Such a case has been described by the author of thesis [15]. She has assumed that on the input domain two families of fuzzy rules have been constructed: the first family contains multiconditional (generalized) fuzzy rules of Takagi–Sugeno and the other – one-conditional rules, similar to the described below with SIMNN as their consequent parts.

3   Adaptive fuzzy inference system

Constructing a generalized adaptive fuzzy inference system for approximation problems we consider a family $\{\mathcal{R}_h : h = 1, 2, ..., M\}$ of one-conditional rules of the form

$$\text{if } x \text{ is } A_h \text{ then } y \text{ is } B_h, \qquad (6)$$

in which the consequent part B_h is not a fuzzy set but a function relation of the type $y = f_h(x, \Omega_h)$, where x is the input variable and Ω_h represents a set of parameters, exactly a generalized weight vector of an h-th single mapping neural network SIMNN designed for the training pairs $p_j^h, j = 1, 2, ..., N_h$ from the cluster $\mathcal{K}_h$ in the next section. This type of generalized Takagi-Sugeno's fuzzy rule will lead in the final construction stage to an adaptive fuzzy-neural inference system AFNIS.

For each rule $\mathcal{R}_h$ the value of membership function μ_{A_h} of the fuzzy set A_h involved in its premise part can be calculated, if the form of the membership function will be given. Now the system of one - condition fuzzy rules will be equipped with their premise parts and the fuzzy sets. The general form of their membership functions will be similar to those from our previous publications ([20, 21, 12]), a generalized Gaussian functions. However, we are admitting one - conditional rules and the membership function needs to reflect this fact. Hence we assume it in the general from

$$\mu_{A_h}(x) = d^h \exp(-0.5((x - a^h) \cdot S_h^{-1}(x - a^h))^{b_h}), \qquad (7)$$

where the inverse of the matrix S^h from (5) appears. In the learning process the adaptation will undergo the parameters d^h and b_h. (If necessary we could adapt the center a^h, then in view of (5) – the matrix S_h will be adapted.)

In the case of multi - conditional fuzzy rules, on the other hand, employed in our previous papers (cf.for example [12]), each coordinate $x_i, i = 1, 2, ..., n$ has been covered by fuzzy sets. In our previous publications ([12, 20–23]) it was shown that by introducing two additional adaptable parameters d^h and b^h one makes the system more flexible. The parameter d^h has to be non-negative and such that the maximum value of the membership function does not exceed 1.

A crucial adaptive features is contained in exponent b^h. Depending on its value (i.e. whether it is smaller or bigger than 1, or even non-negative) we can reach for a particular membership function practically a constant value or a singleton. The negative exponent b^h is also possible.

Notice that each matrix S_h represents a part of the knowledge about the type and the behaviour of the function to be approximated and which is contained in samples entering the corresponding cluster $\mathcal{K}_h$ as a subset of the numerical data set TRE. Just over that cluster as its domain the fuzzy set A_h is constructed. For given set of training pairs TRE the process of the knowledge extraction constructs M clusters and scatter matrices $S_h, h = 1, 2, ..., M$, and finally the initial form of the membership functions μ_{A_h} of fuzzy sets A_h in (7); the parameters: d^h and b_h for each $h = 1, 2, ..., M$ will be then adapted to the training data from TRE in the final stage of the training process. Corresponding to those clusters individual functional consequent parts $y = f_h(x, \Omega_h)$ will be proposed in the next section.

The membership function of the set A_h leads to the definition of the so-called *normalized level of activity* of an h-th fuzzy rule $\mathcal{R}_h$ at x and denoted by $v_h(x)$:

$$v_h(x) = \mu_{A_h}(x)\{\sum_{k=1}^{M} \mu_{A_k}(x)\}^{-1}. \tag{8}$$

Having this calculated, the premise parts of all rules are checked as far as their activity.

The overall output $z(x)$ of the constructed system of one-conditional rules of the constructed system at x will be a convex combination of all outputs from the functional consequent parts. This will be given in the next section. However, if more than one fuzzy covering is for our disposal at the end of in the first stage of construction (cf. the remark at the end of Section 2) then consequent parts corresponding to the premise parts founded on the other covering have to be defined and their individual outputs should be incorporated in the overall output [15].

4   Adaptive fuzzy-neural inference system AFNIS

A standard back–propagation perceptron - type neural network cannot well approximate given set of training pairs and leads to rather high values of an error function. In the present paper, after the extracting knowledge from the set TRE we build a family of mapping neural networks. Each h-th element of the family is a single network SIMNN and composed of one hidden layer. The number of neurons in the input layer is n, while the number of neurons in the hidden layer is l_h. We restrict ourself to a perceptron type neural network, which is a universal approximator according to the theorems proved in [16, 17]. Hence in the h-th network each neuron of hidden layer is equipped with an activation function, say σ_h, which is not a polynomial. The activation function can be taken from the family of two–parameter generalized sigmoidal functions, implemented by the present authors in a number of publications (cf. [6, 12, 20–23]):

$$\sigma_h(z) = \frac{m_h}{1 + \exp(-\delta_h z)}. \tag{9}$$

Both parameters m_h and δ_h give more flexibility in the adaptation process. Moreover, their appearance has given a possibility to design a corrected adaptation algoritm for neural network weight vector [6].

More complex neuron networks are also possible. However, in the present paper output layer nodes (neurons) have the identity activation function, and hence neurons can be characterized by constants, only. Hence the output from the h-th network, denoted by z_h can written as

$$z_h = f_h(x, \Omega_h) = \sum_{j=0}^{l} \omega_{hj}^{II} \sigma_h \left(\sum_{i=0}^{n} \omega_{hji}^{I} x_i \right) \tag{10}$$

where ω_{hj}^{II}, $j = 1, ..., l$, and ω_{hji}^{I}, $i = 1, ..., n$ are constant components of the weight vectors $\omega_h{}^{II}$ and $\omega_{hj}{}^{II}$, respectively. Here the zero component x_0 of the input variable x is equal to 1 and was introduced to incorporate the bias ω_{hj0}^{I} under one summation sign.

The overall output of the adaptive fuzzy-neural inference system AFNIS is the convex combination of outputs of all M fuzzy rules where the consequent parts are formed from output functions of single mapping neural networks SIMNN, and is given by

$$z = f(x, \Theta, \Omega) = \sum_{h=1}^{M} v_h(x) f_h(x, \Omega_h), \tag{11}$$

since we assume that at each x the overall activity of all one-conditional rules is normalized to one (cf.(8)). Here Ω is a collection of all individual vectors $\Omega_h, h = 1, 2, .., M$, and $\Theta = \{(d^h, b_h) : h = 1, 2, ..., M\}$ forms a vector of parameters, that will be adapted in the final stage of the learning process. It is worthwhile to mention that in a full case when two different fuzzy domain coverings are constructed, one can assume that additional to the one-conditional fuzzy rules, multi-conditional ones are contained in the module [11, 15].

5   Final stage of learning and conclusions

The individual outputs $z_h = f_h, h1, 2, ..., M$ from the single neural networks trained on the corresponding clusters $\mathcal{K}_h$ give rise to the overall output $z = f(x, \Theta, \Omega)$. Constructed in the last sections AFNIS presented in the form of (11) needs the last stage of adaptation of the coefficients v_h of the convex combination. Since only free parameters now are (d^h, b_h) we have to define a new error function.

The form a new error function will be

$$E(\Theta, x, y) := \frac{1}{M}|f(x, \Theta, \Omega) - y|^2 = \frac{1}{M}|\sum_{h=1}^{M} v_h(x)f_h(x, \Omega_h) - y|^2, \quad (12)$$

where $\Theta = \{(d^h, b_h) : h = 1, 2, ..., M\}$ is a vector parameter to be adapted. Now the terminal stage of the construction follows: the adaptation of the weight vector Θ. To this end the error function (12) will be minimize over all points (x, y) taken from TRE. The gradient descent method or a genetic algorithm can be implemented for this purpose, since the error function is non-quadratic in the variables Θ. As initial values for each d^h one can take the y-component of the centroid (cf.(3)) of the cluster $\mathcal{K}_h$ while initial value for b_h could be taken 1. The implementation of the algorithm constructed on the basis of the present approach is under development and numerical simulations. Their results will be presented in the near future [15].

6 Appendix

In the recent paper [9, 11] the authors have presented a seed growing approach. In our most recent publication [13] an evolutionary approach has been sketched to the clustering problem for an adaptive inference system. Now we present shortly both approaches.

6.1 Seed growth clustering algorithm

In the recent paper [9] the authors have presented a seed growing approach for clustering problem of a large numerical multidimensional data set. The approach used in [9] has been based on an image segmentation known in the digital picture segmentation. The segmentation process helps to reduce amount of data receiving by a man from a computer. The main aim is an extraction of the significant information from the image in order to improve the interpretation process realized by the end-user, e.g. a physician.

In digital processing of large numerical multidimensional data the clustering analysis corresponds some how to the image segmentation. In particular, in the medical image seed growth algorithm the number of possible classes is given by the user (a physician) explicitly together with the seed pixel for each class. In the clustering problem, on the other hand, the seed pixel cannot be given by the end-user; however, it can be chosen in some sense automatically by the algorithm.

The main difference between digital pictures and large multidimensional data, regarded as points from a graph of a scalar-valued multivariate function is the lack of a natural neighbouring topology of the images that is based on the concept of a 4- or 8-connetedness in the image.

The algorithm is composed of two parts:

– Rough clustering according to y in which a fuzzy histogram of the variability of y is used, and is divided into the following steps:
1. Create a histogram of the variability of y;
2. Fuzzify the histogram by calculating its convolution with a Gauss-like function;
3. Use the minima of the fuzzy histogram as boundaries of subintervals of the range of y;
4. Divide all the training pairs into clusters. Pairs which y belongs to the same subinterval form the same cluster W_I.

– Exact clustering according to x and y in which the seed growing approach known in the digital picture processing is developed. The following algorithm is executed for each input cluster W_I created as the result of the rough clustering.
1. Create new empty output cluster W_O.
2. Select seed pair p_s in W_I and move it to W_O.
3. Assume that $c(p_s, p_s) = 1$.
4. Find the maximum proximity pair p in W_I.
5. Calculate $c(p, p_s)$.
6. If the connectedness $c(p, p_s) \geq \alpha$ then move p from W_I to W_O and go to 3.
7. If there are any pairs in W_I then go to 1.
8. Stop the algorithm.

For the better understanding the algorithm we recall some definitions
1. similarity function is defined on the basis of a distance in the domain Y:

$$\chi(p_1, p_2) = \frac{1}{1 + |y_1 - y_2|} \in (0, 1)$$

2. neighbourhood function is defined on the basis of a distance in the domain X:

$$\nu(p_1, p_2) = \frac{1}{1 + \|x_1 - x_2\|} \in (0, 1)$$

where $\| \cdot \|$ denotes the Euclidean norm in $\mathbb{R}^n$;

3. proximity function is the combination of similarity and neighbourhood functions:

$$\rho(p, p_s) = \Phi(\beta, \chi(p, p_s), \max_{z \in W_O} \nu(p, z)) \in (0, 1)$$

where W_O is the currently created output cluster, and $\Phi(\beta, u, v) \in \langle 0, 1 \rangle$ is an influence function. For each β the function Φ should be nondecreasing with respect to the second and third arguments. For example: $\Phi(\beta, u, v) = \beta v + (1 - \beta)u$ or $\Phi(\beta, u, v) = u(1 + \beta(v - 1))$. The parameter β measures the level of influence of v on u, and moreover β, u, v should be taken from the interval $\langle 0, 1 \rangle$.

4. connectedness is defined by the following recurrence formula basing on the proximity function instead of the similarity one:

$$c(p, p_s) = \min \left[\rho(p, p_s), \min_{z \in W_O} c(z, p_s) \right] \in (0, 1). \tag{13}$$

6.2 Evolutionary splitting procedure

We have proposed in [13, 14] a procedure which implements an evolutionary algorithm for extracting the knowledge from the data base by splitting it into clusters. Here we sketch only its main features. We use the following notation:

- cluster $\mathcal{K} = \{p_k : k = 1, ..., N\}$ is a subset of the training set;
- covering $\mathcal{C}$ is a union of clusters, that covers the whole set TRE;
- if p_k is the kth training pair from the cluster $\mathcal{K}$, N – number of pairs in the cluster, then we define p^s as a centroid of the cluster $\mathcal{K}$ in similar way to (3).

In the present approach we are distinguishing two types of evolutions, that appear as a result of a single genetic operation: external, at the level of clusters, and internal, at the level of training pairs.

We are introducing three type of relations:

- between each two training pairs in the set TRE as a distance function, based on the Euclidean or Mahalanobis (cf. [14]) metrics,
- between each two clusters in a covering $\mathcal{C}$ as a distance function between their centroids,
- between two different coverings $\mathcal{C}$ and $\mathcal{C}'$: as a set-theoretical relation: may have common part, or be disjoint.

We are introducing: three types of selection operators for the population of clusters: roulette, tournament selection, proximity selection, as a combination of two others, and three types of genetic operators (cf.[13]) for the population of clusters:

- union (in [13] we called it a unification operator) that acts on a pair of clusters and produces a new cluster as a union of both parents,
- crossover operator that exchanges parts of two clusters,
- split (in [13] we called it a separation operator) that produces two other clusters by splitting a cluster into two.

To evaluate each off-spring cluster in the population we are using different local fitness functions: it could be the maximal distance of the training pair of the cluster from the cluster centroid, or a mean variation of all training pairs in the cluster. Basing on the local fitness function of cluster one can build a global fitness function for the whole covering as a sum of local fitness function of its clusters.

We are using a variable probability of genetic operators: the action of each operator is evaluated (judged) at the end of the evolution step: if due its action the resulting covering is better then the operator probability is increasing in the next evolution step.

In the first stage we have decided to perform m independent evolutionary processes of m coverings by creating m initial coverings. It is done with the help of the histogram of the variability of y by a particular choice of discrete values from the range of y's. Then we fuzzify the histogram by calculating its convolution with m Gauss–like functions $\phi(y, \sigma_i) = c_i \exp(\frac{-y^2}{\sigma_i^2})$, with different σ_i, where c_i are normalization constants for $1 \leq i \leq m$. For each fuzzy histogram, its minima determine subintervals in the space $\mathcal{Y}$. For each fuzzy histogram, we create a covering in which clusters consist of all training pairs with the last (i.e., y) coordinate belonging to the same subinterval. During the evolutionary process, at each generation we apply one of the genetic operators that acts on clusters (or pairs of clusters). Then we use one of the selection procedure to the whole population. In this way a new covering is constructed that forms the population for the next generation. After a fixed number of generations we are applying a global fitness (evaluation) function to each covering. Then a population of all coverings is formed.

Different selection methods are applied to coverings. One of them is called a roofing, in which a number of partial coverings of the set TRE is constructed, and if in different coverings some clusters have a common points we are exchanging clusters between them or performing external genetic operators on them. Such an action can be called a distributed evolutionary algorithm (see [10] for a numerical implementation).

Acknowledgements

This material is based upon work supported by the grant 8 T11C 011 12 from the Polish State Committee for Scientific Research (KBN).

References

1. Anderberg, M.R.: Cluster Analysis for Applications. Probability and Mathematical Statistics, Academic Press, New York (1973)
2. Buckley, J. J. and Hayashi, Y.: Numerical relationships between neural networks, continuous functions, and fuzzy systems. Fuzzy Sets and Systems, 60 (1993) 1–8
3. Cybenko, G.: Approximation by superpositions of sigmoidal function. Mathematics of Control, Signals, and Systems, 2 (1989) 303–314
4. Funahashi, K.: On the approximate realization of continuous mapping by neural networks. Neural Networks, 2 (1989) 183–192
5. Hand, D.J.: Discrimination and Classification. Wiley, Chichester (1981)
6. Gołąbek, B., Kosiński, W. and Weigl, M.: Adaptation of learning rate via adaptation of weight vector in modified M-Delta networks. In Computational Intelligence and Applications, P.S. Szczepaniak (Ed.), (Studies in Fuzziness and Soft Computing, Vol. 23), Physica-Verlag, c//o Springer-Verlag, (1999) 156–163
7. Hornik, K.: Approximation capabilities of multilayer feedforward networks. Neural Networks, 4 (1991) 251–257
8. Jang, J. and Sun, T-C.: Functional equivalence between radial basis function networks and fuzzy inference system. IEEE Transaction on Neural Networks, 4 (1), (1993) 156–159
9. Kieś, P., Kosiński, W. and Weigl, M.: Seed growing approach in clustering analysis. In Intelligent Information Systems. Proceedings of the VI-th International Workshop on Intelligent Information Systems, Zakopane, 9-13 June, 1997, Instytut Podstaw Informatyki, PAN, Warszawa (1997) 7–15

10. Koleśnik, R., Koleśnik, L. and Kosiński, W.: Genetic operators for clustering analysis, In Intelligent Information Systems VIII, Proc. of the Workshop held in Ustroń, Poland, 14-18 June, 1999, Instytut Podstaw Informatyki, PAN, Warszawa (1999) 203–208

11. Kosiński, W., Weigl, M. and Kieś, P.: Fuzzy domain covering of an inference system. In Proceedings of the Third National Conference on Neural Networks and Their Applications, October 14 - 18, 1997, Kule n. Częstochowa, PNNS, Częstochowa (1997) 310–321

12. Kosiński, W. and Weigl, M.: General mapping approximation problems solving by neural networks and fuzzy inference systems. Systems Analysis Modelling Simulation, 30 (1), (1998) 11–28

13. Kosiński, W., Weigl, M., and Michalewicz Z.: Evolutionary domain covering of an inference system for function approximation. In Evolutionary Programming VII V. W. Port, N. Saravanam, D. Waagen and A. E. Eiben (Eds.), Proceedings of the 7th International Conference, EP'98, San Diego, California, USA, March 25-27, 1998, Lecture Notes in Computer Science, Vol. 1447, New York (1998) 167–180

14. Kosiński W. Weigl M., Michalewicz Z. and Koleśnik R. Genetic algorithms for preprocessing of data for universal approximators. In Intelligent Information Systems VII, Proceedings of the Workshop held in Malbork, Poland, 15-19 June, 1998, Instytut Podstaw Informatyki, PAN, Warszawa (1998) 320–331

15. Kowalczyk, D.: An adaptive fuzzy inference system as a fuzzy controller (Rozmyty adaptacyjny system wnioskujący jako sterownik rozmyty). Master Thesis, (in Polish), WSP w Bydgoszczy, Wydział Matematyki, Techniki i Nauk Przyrodniczych, Specjalność: Technika Komputerowa, Bydgoszcz, czerwiec (1999)

16. Mhaskar, H. and Micchelli, C.: Approximation by superposition of a sigmoidal function and radial basis functions. Advances Applied Mathematics, 13 (1992) 350–373

17. Mhaskar, H. and Micchelli, C.: Dimension - independent bounds on the degree of approximation by neural network. IBM Journal of Research and Development, 38 (3) (1994) 277–284

18. Scarselli, F. and Tsoi, A. Ch.: Universal approximation using feedforward neural networks: A survey of some existing methods, and some new results. Neural Networks, 11 (1) (1998) 15–37

19. Takagi, H. and Sugeno, M.: Fuzzy identification of systems and its applications to modeling and control. IEEE Trans. on Systems, Man. and Cybernetics, 15 (1985) 116–132

20. Weigl, M. and Kosiński, W.: Fuzzy inference system and modified back–propagation network in approximation problems. Proceedings of the III-rd International Symposium on Intelligent Information Systems, Wigry n. Suwałki, June 1994, Instytut Podstaw Informatyki, PAN, Warszawa (1994) 427–442

21. Weigl, M. Neural networks and fuzzy inference systems in approximation problems (Sieci neuronowe i rozmyte systemy wnioskujące w problemach aproksymacji). Ph. D. Thesis, (in Polish), IPPT PAN, Warszawa, czerwiec (1995)

22. Weigl, M. and Kosiński, W.: Fuzzy reasoning in adaptive expert systems for approximation problems. In Proceedings of the 3-th Zittau Fuzzy - Colloquy Zittau, September 5-6, 1995, Wissenschaftliche Berichte, Heft 41, (1995) 163–174

23. Weigl, M. and Kosiński, W.: Approximation of multivariate functions by generalized adaptive fuzzy inference network. In Proceedings of the 9-th International Symposium on Methodologies for Intelligent Systems ISMIS'96, Zakopane, June, 1996, IPI PAN, Warszawa (1996) 120-133

FGCounts of Fuzzy Sets with Triangular Norms

Maciej Wygralak, Daniel Pilarski

Adam Mickiewicz University, Faculty of Mathematics and Computer Science
Matejki 48/49, 60-769 Poznań, Poland
{wygralak, pilarski}@math.amu.edu.pl

Abstract. The "fuzzy" approach to the question of cardinality of a fuzzy set offers a very adequate and complete cardinal information in the form of a convex fuzzy set of ordinary cardinal numbers (of nonnegative integers, in the finite case). The existing studies of that approach, however, are restricted to cardinalities of fuzzy sets with the classical min and max operations. In this paper, we like to present a generalization of FGCounts to triangular norm-based fuzzy sets. Some remarks about an analogous generalization of FLCounts and FECounts will be given, too.

1 Introduction

There are solid motivations, both theoretical and applicational, for considering the question of cardinality for fuzzy sets. Very natural examples of queries about cardinalities of fuzzy sets or comparisons of such cardinalities can be found in many areas of applications. Let us mention communication with data bases, modeling the meaning of imprecise quantifiers in natural language statements, computing with words, computational theory of perceptions, decision-making in a fuzzy environment, etc. A typical instance is the query "How many individuals in the data base are middle-aged and about 180 cm tall?", which requires counting how many elements are in the intersection of two fuzzy sets. The graduation of membership values makes counting and cardinal calculus under fuzziness a task which is much more complicated and advanced than in the case of ordinary sets. The literature offers us two main approaches to that problem: the scalar approach and the "fuzzy" one in which cardinalities of fuzzy sets are defined as some convex fuzzy sets of usual cardinal numbers (see reviews and further references in [1, 4, 8, 9, 11]).

In this paper, we focus our attention on the "fuzzy" approach which, in comparison with the scalar one, gives us a considerably more complete and adequate cardinal information about a fuzzy set. Moreover, we restrict our discussion to finite fuzzy sets, i.e. to fuzzy sets with finite supports, which play the central role in applications. Their cardinalities are then some convex fuzzy sets in the set $\mathbb{N}$ of all nonnegative integers. In other words, they are some weighted families of nonnegative integers.

Throughout, fuzzy sets in a universe $\mathbf{M}$, finite or not, will be denoted by single capitals in italic, whereas single bold capitals will denote sets. In particular, $T := 1_{\varnothing}$ and $M := 1_{\mathbf{M}}$, where 1_D symbolizes the characteristic function of a set $D \subset M$; as usual, := stands for "equals by definition". FFS (FCS, respectively) symbolizes the family of all finite fuzzy sets (finite sets, respectively) in **M**. Moreover, let $A_t := \{ x \in M: A(x) \geq t \}$ with $t \in (0, 1]$ (*t-cut set of A*) and $A^t := \{ x \in M: A(x) > t \}$ with $t \in [0, 1)$ (*sharp t-cut set*), i.e. $\mathrm{core}(A) = A_1$ and $\mathrm{supp}(A) = A^0$ for a fuzzy set $A: M \to [0, 1]$.

We shall consider fuzzy sets with triangular operations. More precisely, we shall use the *sum* $A \cup_s B$ of $A, B \in$ FFS induced by a t-conorm **s** with $(A \cup_s B)(x) := A(x) \, \mathbf{s} \, B(x)$, the *intersection* $A \cap_t B$ and the *cartesian product* $A \times_t B$ induced by a t-norm **t** with $(A \cap_t B)(x) := A(x) \, \mathbf{t} \, B(x)$ and $(A \times_t B)(x, y) := A(x) \, \mathbf{t} \, B(y)$. If $\mathbf{s} = \vee := \max$ or/and $\mathbf{t} = \wedge := \min$, we simply write $A \cup B$ or/and $A \cap B$ and $A \times B$. On the other hand, the inclusion and equality between two fuzzy sets will be defined in the standard way by $A \subset B \leftrightarrow \forall x \in M: A(x) \leq B(x)$ and $A = B \leftrightarrow \forall x \in M: A(x) = B(x)$, no matter which t-operations are chosen.

The subject literature offers a general scalar approach to cardinalities of fuzzy sets with triangular operations (see [11]). On the other hand, the hitherto existing studies of the "fuzzy" approach concern only the case of the triangular operations $\wedge$ and $\vee$. The main aim of this paper is to outline a generalization of the well-known and already classical concept of the FGCount to (finite) fuzzy sets with tringular norm-based operations. Moreover, some remarks about an analogous generalization of the related concepts of the FLCount and the FECount will be given, too. A more extensive study of these generalizations will be presented in [12].

For the sake of self-containment and since different terminological conventions are used in the literature, let us recall some basic notions and facts from the theory of triangular operations which will be useful in the main part of the paper (see [2, 5-7]).

A binary operation $\mathbf{t}: [0, 1] \times [0, 1] \to [0, 1]$ which is commutative (i), associative (ii), nondecreasing in each argument (iii), and has 1 as unit is called a *triangular norm* (*t-norm*, in short). If a binary operation $\mathbf{s}: [0, 1] \times [0, 1] \to [0, 1]$ does satisfy (i)-(iii) and has 0 as unit, it is called a *triangular conorm* (*t-conorm*, in short). Triangular norms and conorms together will be called *triangular operations* (*t-operations*, in short). If

$$a \, \mathbf{s} \, b = 1 - (1 - a) \, \mathbf{t} \, (1 - b)$$

for each a and b, one says that **t** and **s** are *associated* t-operations, and one writes $\mathbf{s} = \mathbf{t}^*$. For each t-norm **t** and t-conorm **s**, we have

$$\mathbf{t}_d \leq \mathbf{t} \leq \wedge \leq \vee \leq \mathbf{s} \leq \mathbf{s}_d,$$

where the ordering relation $\leq$ is understood pointwise, the so-called *drastic t-norm* $\mathbf{t}_d$ is defined as $a \mathbf{t}_d b := (a \wedge b$ if $a \vee b = 1$, else 0), and $a \mathbf{s}_d b := (a \vee b$ if $a \wedge b = 0$, else 1) defines the so-called *drastic t-conorm* $\mathbf{s}_d$.

Our special interest will be in archimedean t-operations, including $\wedge$ and $\vee$. A t-norm **t** (t-conorm **s**, respectively) is called *archimedean* if it is continuous and $a \mathbf{t} a < a$

($as a > a$, respectively) for each $a \in (0, 1)$. An archimedean t-operation is called *strict* if it is strictly increasing in $(0, 1) \times (0, 1)$. The family of all archimedean t-norms (t-conorms, respectively) will be denoted by *Arch* (*ArchS*, respectively), whereas *Str* will denote the class of all strict t-norms. Moreover, $Arch^\wedge := Arch \cup \{\wedge\}$, $ArchS^\vee := ArchS \cup \{\vee\}$, and $Str^\wedge := Str \cup \{\wedge\}$. Prototypical examples of strict t-operations are the *algebraic t-norm* $\mathbf{t}_a$ and the associated *algebraic t-conorm* $\mathbf{s}_a$, where $a\mathbf{t}_a b := ab$ and $a\mathbf{s}_a b := a + b - ab$. The *Łukasiewicz t-operations* $\mathbf{t}_{\!L}$ and $\mathbf{s}_{\!L}$, where $a\mathbf{t}_{\!L}b := 0 \vee (a+b-1)$ and $a\mathbf{s}_{\!L}b := 1 \wedge (a+b)$, are prototypical examples of non-strict archimedean t-operations. Let us recall that each strict t-norm $\mathbf{t}$ satisfies the *cancellation law*: if $a\mathbf{t}b = a\mathbf{t}c$ with $a > 0$, then $b = c$.

Theorem 1.1 (Ling, [3]). *Let* $\mathbf{t}, \mathbf{s}: [0,1] \times [0,1] \to [0,1]$.
(a) $\mathbf{t} \in Arch$ *iff there exists a strictly decreasing and continuous generator* $g: [0,1] \to [0, \infty]$ *with* $g(1) = 0$ *such that* $a\mathbf{t}b = g^{-1}(g(0) \wedge (g(a) + g(b)))$ *for each* $a, b \in [0,1]$. *Moreover,* $\mathbf{t}$ *is strict iff* $g(0) = \infty$.
(b) $\mathbf{s} \in ArchS$ *iff there exists a strictly increasing and continuous generator* $h: [0,1] \to [0, \infty]$ *with* $h(0) = 0$ *such that* $a\mathbf{s}b = h^{-1}(h(1) \wedge (h(a) + h(b)))$ *for each* $a, b \in [0,1]$. *Finally,* $\mathbf{s}$ *is strict iff* $h(1) = \infty$.

Consequently, for $\mathbf{t} \in Arch$ with generator g, we get

$$a_1 \mathbf{t} a_2 \mathbf{t} \ldots \mathbf{t} a_k = g^{-1}(g(0) \wedge \Sigma i, \mathbf{k}_{=1} g(a_i))$$

for $a_1, a_2, \ldots a_k \in [0,1]$ with $k \in \mathbb{N}$; in particular, $a_1 \mathbf{t} a_2 \mathbf{t} \ldots \mathbf{t} a_k = 1$ for $k = 0$. Hence

$$a_1 \mathbf{t} a_2 \mathbf{t} \ldots \mathbf{t} a_k > 0 \text{ iff } \Sigma i, \mathbf{k}_{=1} g(a_i) < g(0)$$

and

$$a_1 \mathbf{t} a_2 \mathbf{t} \ldots \mathbf{t} a_k \geq t \text{ iff } \Sigma i, \mathbf{k}_{=1} g(a_i) \leq g(t), \ t \in (0, 1].$$

Each nonincreasing function $v: [0,1] \to [0,1]$ with $v(0) = 1$ and $v(1) = 0$ is called a *negation*. The negation v_* such that $v_*(a) = 0$ for $a > 0$ is the smallest possible negation, whereas v^* such that $v^*(a) = 1$ for $a < 1$ is the largest possible one. If v is strictly decreasing and continuous, it is said to be a *strict negation*. An involutive strict negation ($v(v(a)) = a$ for each a) is called a *strong negation*.

2 FGCounts with Triangular Norms

Let $A \in FFS$ and $[A]_k := \vee\{t : |Y_t| \geq k\}$ for $k \in \mathbb{N}$. Throughout, $n := |\text{supp}(A)|$ and $m := |\text{core}(A)|$. We see that $[A]_k = 1$ for each $k \leq m$, $[A]_k = 0$ for each $i > n$, and $0 < [A]_k < 1$ for each $m < k \leq n$. Each $[A]_k$ with $0 < k \leq n$ is the k-th element in the nonincreasingly ordered sequence of all positive values $A(x)$, including their possible repetitions. Clearly,

$A \subset B$ implies $[A]_k \leq [B]_k$ for each $k \in \mathbb{N}$.

As one knows, FGCount: FFS $\to [0,1]^{\mathbb{N}}$ is a mapping such that

$$FGCount(A)(k) := [A]_k.$$

The following properties are then satisfied for each $A, B \in FFS$:

$$A \in FCS \;\Rightarrow\; FGCount(A) = 1_{\{0,1,\ldots,n\}},$$

$$A \subset B \;\Rightarrow\; FGCount(A) \subset FGCount(B),$$

$$FGcount(A) + FGCount(B) = FGCount(A \cap B) + FGCount(A \cup B),$$

$$FGCount(A) \cdot FGCount(B) = FGCount(A \times B),$$

where the operations $+$ and $\cdot$ are defined by means of the ordinary extension principle, i.e. $(P*Q)(k) := \vee\{P(i) \wedge Q(j): i*j \geq k\}$ for nonincreasing $P, Q: \mathbb{N} \to [0,1]$ and $* \in \{+, \cdot\}$; if $* = \cdot$, the condition "$i*j \geq k$" can be replaced by "$i*j = k$". The FGCount of A is chronologically the first well-defined "fuzzy" approach to cardinalities of fuzzy sets with the ordinary min and max operations (see [13-16]). As we mentioned above, the valuation property and the the cartesian product rule are then preserved. From the semantical viewpoint, $FGCount(A)(k)$ can be interpreted as a degree to which A contains at least k elements.

One has the impression that the FGCount is well-defined for fuzzy sets with arbitrary t-operations, not only for those with $\mathbf{t} = \wedge$ and $\mathbf{s} = \vee$. But $A \in FFS$ is a sum of disjoint singletons, no matter which t-norm is used to execute the summation of the singletons. Consequently, $FGCount(A)$ should be equal to the sum of the FGCounts of the singletons composing A, i.e. we should get

$$FGCount(A) = FGCount([A]_1/x_1) +_t FGCount([A]_2/x_2) +_t \ldots +_t FGCount([A]_n/x_n)$$

with the addition $+_t$ defined via the t-norm-based extension principle $(P +_t Q)(k) := \vee\{P(i) \mathbf{t} Q(j): i+j = k\}$. However, it is easy to check that the sum on the right side of the equality is a fuzzy set α such that $\alpha(k) = [A]_1 \mathbf{t} [A]_2 \mathbf{t} \ldots \mathbf{t} [A]_k$ which generally differs from $FGCount(A)$, unless $\mathbf{t} = \wedge$. Thus, $FGCount_t: FFS \to [0,1]^{\mathbb{N}}$ such that

$$FGCount_t(A)(k) := [A]_1 \mathbf{t} [A]_2 \mathbf{t} \ldots \mathbf{t} [A]_k$$

for each $k \in \mathbb{N}$ should be taken as an appropriate generalization of FGCount to fuzzy sets with t-norms. So, we have $FGCount_t(A)(0) = 1$ and

$$FGCount_t(A)(k) = FGCount_t(A)(k-1) \mathbf{t} [A]_k$$

for each t-norm **t**, $A \in$ FFS and $k \geq 1$. Obviously, $\text{FGCount}_\wedge(A) = \text{FGCount}(A)$. Each $\text{FGCount}_t(A)$ with a t-norm **t** will be called a *generalized FGCount of A* or a *t-norm-based FGCount of A*. In the language of many-valued sentential calculus, $\text{FGCount}_t(A)(k)$ is the truth value of a sentence saying that A contains (at least) i elements for each $i \leq k$; it is generally equal to the truth value of "A contains at least k elements" only if **t** $= \wedge$. Let us consider a few simple examples with

$$A = 0.9/x_1 + 1/x_2 + 0.6/x_3 + 0.3/x_4 + 1/x_5 + 0.2/x_6 + 0.4/x_7.$$

Then

$$\text{FGCount}_\wedge(A) = (1, 1, 1, 0.9, 0.6, 0.4, 0.3, 0.2),$$

$$\text{FGCount}_t(A) = (1, 1, 1, 0.9, 0.54, 0.216, 0.648, 0.1296) \text{ for } \mathbf{t} = \mathbf{t}_a,$$

$$\text{FGCount}_t(A) = (1, 1, 1, 0.9, 0.5) \text{ for } \mathbf{t} = \mathbf{t}_L.$$

Generalized FGCounts of fuzzy sets are nonincreasing functions $f: \mathbb{N} \to [0,1]$ with $f(0) = 1$ and $f(k) = 0$ for some $k \in \mathbb{N}$. Conversely, if **t** $= \wedge$, then each such a function represents the generalized FGCount of a fuzzy set. This is no longer valid for **t** $=, /\wedge$. In the following, we present other basic properties of generalized FGCounts.

Corollary 2.1. *For each t-norm **t** and $A \in$ FFS, we have:*

(a) $\text{FGCount}_t(A)$ *is normal and convex.*

(b) $\text{FGCount}_t(A) = 1_{\{0,1,...,n\}}$ *whenever $A \in$ FCS.*

(c) $\text{FGCount}_t(A)(k) = 1$ *for $k \leq m$, $\text{FGCount}_t(A)(k) = 0$ for $k > n$.*

(d) $\text{FGCount}_t(A) \subset \text{FGCount}_u(A)$ *whenever* **t** $\leq$ **u**.

Hence $\text{FGCount}_w(A) \subset \text{FGCount}_t(A) \subset \text{FGCount}_\wedge(A)$ *with* $\mathbf{w} = \mathbf{t}_d$.

Proof. It follows from the definition of FGCount_t and properties of t-norms. $\square$

Our further discussion of generalized FGCounts will be restricted to t-norms from $Arch^\wedge$. We easily notice that if $\mathbf{t} \in Arch$ is non-strict, then $\text{FGCount}_t(A)(k) = 0$ possibly for $k < n$. Consequently, for $\mathbf{t} \in Arch^\wedge$ and $A \in$ FFS, we define

$$e(A) := \vee\{k \in \mathbb{N}: [A]_1 \mathbf{t}[A]_2 \mathbf{t}...\mathbf{t}[A]_k > 0\}.$$

Clearly, $e(A) = \vee\{k \in \mathbb{N}: \Sigma i, k_{=1} g([A]_i) < g(0)\}$ for $\mathbf{t} \in Arch$ (see Section 1). Moreover, e is nondecreasing with respect to A and **t**, $e(A) = 0$ iff $A = T$, and $e(A) = n$ for $\mathbf{t} \in Str^\wedge$. Let & denote the conjunction connective and

$$d_{A,B} := [A]_{e(A)+1} \vee [B]_{e(B)+1}.$$

Theorem 2.2. *Let* $t \in Arch^\wedge$ *and* $A, B \in FFS$. *The following statements are pairwise equivalent*:

(a) $FGCount_t(A) = FGCount_t(B)$,

(b) $e(A) = e(B)$ & $\forall k \le e(A)$: $[A]_k = [B]_k$,

(c) $e(A) = e(B)$ & $[A]_{e(A)} = [B]_{e(B)}$ & $\forall t \in (d_{A,B}, 1]$: $|A_t| = |B_t|$,

(d) $e(A) = e(B)$ & $[A]_{e(A)} = [B]_{e(B)}$ & $\forall t \in [d_{A,B}, 1)$: $|A'| = |B'|$.

Proof. (a)$\leftrightarrow$(b) is obvious for $t = \wedge$. If $t = ,/\wedge$, it follows from Theorem 1.1 and its consequences. The proof of (b)$\rightarrow$(c)$\rightarrow$(d)$\rightarrow$(b) is a routine task. $\square$

Corollary 2.3. *If* $t \in Str^\wedge$, *then the following conditions are pairwise equivalent*:

(a) $FGCount_t(A) = FGCount_t(B)$,

(b) $\forall i \in \mathbb{N}$: $[A]_k = [B]_k$,

(c) $\forall t \in (0, 1]$: $|A_t| = |B_t|$,

(d) $\forall t \in [0, 1)$: $|A'| = |B'|$.

Proof. Indeed, $t \in Str^\wedge$ implies $e(A) = n$ and $d_{A,B} = 0$. $\square$

The above theorem leads to the following definition of equipotency for fuzzy sets, i.e. of having the same generalized FGCount.

Definition 2.4. Let $t \in Arch^\wedge$ and $A, B \in FFS$. We say that A and B are *equipotent* iff the condition (b) in Theorem 2.2 is satisfied. We then write $A \sim B$ or $A \sim_t B$ if we like to emphasize which t-norm is used.

So, $A \sim B$ means that A and B are identical up to the permutation of their membership values, including possible repetitions. This identity is restricted to some top membership values in A and B if $t\_,/Str^\wedge$. Theorem 2.2(c, d) and Corollary 2.3(c, d) allow to express $A \sim B$ in terms of equipotencies of *t*-cuts. Further, $\sim$ is an equivalence relation which collapses to the ordinary equipotency of two (finite) sets whenever $A, B \in FCS$. It is easy to notice that a counterpart of the classical Cantor-Bernstein theorem does work for $t \in Arch^\wedge$ and $A, B \in FFS$:

$$A \subset B \subset C \ \& \ A \sim C \ \Rightarrow \ A \sim B \sim C.$$

The following properties are other simple consequences of Definition 2.4:

$$A \sim B \ \Rightarrow \ |core(A)| = |core(B)| \ \& \ |supp(A)| = |supp(B)| \ \text{ if } t \in Str^\wedge,$$

$$A \sim B \ \Rightarrow \ |core(A)| = |core(B)| \ \text{ if } t \in Arch \text{ is non-strict.}$$

By Corollary 2.3(b), all t-norms in $Str^\wedge$ are equivalent with respect to equipotency: if $A \sim_t B$ for a t-norm $t \in Str^\wedge$, then $A \sim_u B$ for each $u \in Arch^\wedge$. If t is archimedean and non-strict, we have the following:

$$A \sim_t B \ \& \ u \le t \ \Rightarrow \ A \sim_u B.$$

A partial ordering between generalized FGCounts (even a lattice one) can be introduced in the classical-like manner:

$$FGCount_t(A) \le FGCount_t(B) \ \Leftrightarrow \ \exists B^* \subset B: A \sim_t B^*.$$

Hence

$$A \subset B \ \Rightarrow \ FGCount_t(A) \le FGCount_t(B).$$

It is easy to point out that (see Definition 2.4)

$$FGCount_t(A) \le FGCount_t(B) \ \Leftrightarrow \ FGCount_t(A) \subset FGCount_t(B) \text{ if } t = \wedge,$$

$$FGCount_t(A) \le FGCount_t(B) \ \Rightarrow \ FGCount_t(A) \subset FGCount_t(B) \text{ if } t \in Arch.$$

Replacing in Theorem 2.2(b-d) each symbol $=$ by $\le$, we get three useful characterizations of the relationship $FGCount_t(A) \le FGCount_t(B)$ for $t \in Arch^\wedge$. The relation $FGCount_t(A) < FGCount_t(B)$ can be defined as $FGCount_t(A) \le FGCount_t(B)$ with $FGCount_t(A) \ne FGCount_t(B)$.

Addition of generalized FGCounts of fuzzy sets can be realized by means of the t-norm-based extension principle.

Theorem 2.5. *Let* $t \in Arch^\wedge$. *If* $A, B \in FFS$ *are such that* $A \cap B = T$, *then*

$$FGCount_t(A) +_t FGCount_t(B) = FGCount_t(A \cup B).$$

Proof. The thesis for $t = \wedge$ is well-known. For $t = /\wedge$ with generator g, it principally follows from the construction of $FGCount_t$. An analytical proof consists in showing that the corresponding t-cuts of both sides in the thesis are always identical. This can be done by applying Theorem 1.1 and its consequences. $\qquad\square$

The commutative and associative addition $+_t$ of generalized FGCounts is thus always well-defined and has a classical-like interpretation. Moreover, for each $t \in Arch^\wedge$ and $A, B \in FFS$ with $0 := FGCount_t(T)$, we have:

$$FGCount_t(A) +_t 0 = FGCount_t(A),$$

$$FGCount_t(A) +_t FGCount_t(B) = 0 \ \Leftrightarrow \ FGCount_t(A) = FGCount_t(B) = 0,$$

$$\text{FGCount}_t(A) \le \text{FGCount}_t(A) +_t \text{FGCount}_t(B).$$

Theorem 2.6. *Let* $t \in Arch^\wedge$ *and* $s \in ArchS^\vee$. *The valuation property*

$$\forall A, B \in FFS: \text{FGCount}_t(A) +_t \text{FGCount}_t(B) = \text{FGCount}_t(A \cap_t B) +_t \text{FGCount}_t(A \cup_s B)$$

is satisfied iff $t = \wedge$ *and* $s = \vee$.

Proof. If the valuation property were fulfilled by t-operations **t** and **s**, we would have $\text{FGCount}_t(A^*) +_t \text{FGCount}_t(B^*) = \text{FGCount}_t(A^* \cap_t B^*) +_t \text{FGCount}_t(A^* \cup_s B^*)$, where $A^* = B^* := a/x$ with $a \in (0, 1)$ and $x \in M$. But

$$(\text{FGCount}_t(A^*) +_t \text{FGCount}_t(B^*))(1) = a,$$

$$(\text{FGCount}_t(A^*) +_t \text{FGCount}_t(B^*))(2) = a\,t\,a,$$

$$(\text{FGCount}_t(A^* \cap_t B^*) +_t \text{FGCount}_t(A^* \cup_s B^*))(1) = a\,s\,a,$$

$$(\text{FGCount}_t(A^* \cap_t B^*) +_t \text{FGCount}_t(A^* \cup_s B^*))(2) = a\,t\,a\,t(a\,s\,a).$$

This implies $t = \wedge$ and $s = \vee$. On the other hand, it is well-known that $t = \wedge$ and $s = \vee$ do fulfil the valuation property. $\square$

Theorem 2.7. *Let* $t \in Arch^\wedge$ *and* $s \in ArchS^\vee$. *The subadditivity property*

$$\forall A, B \in FFS: \text{FGCount}_t(A \cup_s B) \le \text{FGCount}_t(A) +_t \text{FGCount}_t(B)$$

holds true iff $s = \vee$.

Proof. Cf. the proof of Theorem 2.6. $\square$

Let $t \in Arch^\wedge$, $\alpha := \text{FGCount}_t(A)$, $\beta := \text{FGCount}_t(B)$, $\gamma := \text{FGCount}_t(C)$, and $\delta := \text{FGCount}_t(D)$ with $A, B, C, D \in FFS$. Another consequence of Theorem 2.5 is the following *law of side-by-side addition of inequalities*:

$$\alpha \le \beta \ \& \ \gamma \le \delta \ \Rightarrow \ \alpha +_t \gamma \le \beta +_t \delta.$$

Hence

$$\alpha \le \beta \ \Rightarrow \ \alpha +_t \gamma \le \beta +_t \gamma.$$

If $t \in Str^\wedge$, one proves the *cancellation law* and the *strict version* of the law of side-by-side addition of inequalities:

$$\alpha +_t \beta = \alpha +_t \gamma \ \Rightarrow \ \beta = \gamma,$$

$$\alpha < \beta \ \& \ \gamma \leq \delta \ \Rightarrow \ \alpha +_t \gamma < \beta +_t \delta,$$

which implies that

$$\beta > 0 \ \Rightarrow \ \alpha < \alpha +_t \beta.$$

Details can be found in [12]. As we see, there are many analogies between the arithmetic of nonnegative integers and the arithmetic of generalized FGCounts. Those analogies are more numerous and stronger if t and s are strict and, especially, if $t = \wedge$ and $s = \vee$. On the other hand, an essential difference between those arithmetics is the lack of the compensation property: $FGCount_t(A) < FGCount_t(C)$ with $t \in Arch^\wedge$ does not generally imply the existence of a fuzzy set $B \in FFS$ such that $FGCount_t(A) +_t FGCount_t(B) = FGCount_t(C)$ (see [10, 12]).

Multiplication of generalized FGCounts with $t \in Arch^\wedge$ can be defined via the triangular norm-based extension principle:

$$(FGCount_t(A) \cdot_t FGCount_t(B))(k) := \vee \{FGCount_t(A)(i) \, t \, FGCount_t(B)(j): ij \geq k\}.$$

One shows that the rule

$$\forall A, B \in FFS: FGCount_t(A) \cdot_t FGCount_t(B) = FGCount_t(A \times_t B)$$

holds true iff $t = \wedge$. If $t =,/\wedge$, $FGCount_t(A) \cdot_t FGCount_t(B)$ does not generally form the generalized FGCount of any fuzzy set.

3 Generalized FLCounts and FECounts

The dual "fuzzy" approach to cardinality of a fuzzy set $A \in FFS$ is its FLCount. More precisely, $FLCount: FFS \rightarrow [0,1]^N$ is a mapping such that ([14])

$$FLCount(A)(k) := v_{\text{L}}([A]_{k+1})$$

with the *Łukasiewicz negation* $v_{\text{L}}(a) := 1 - a$. $FLCount(A)(k)$ is a degree to which A has at most k elements. The next related approach is the FECount of A ([8, 14]), where

$$FECount(A) := FGCount(A) \cap FLCount(A),$$

i.e.

$$FECount(A)(k) = [A]_k \wedge v_{\text{L}}([A]_{k+1}).$$

This time, $FECount(A)(k)$ expresses a degree to which A contains exactly k elements. A comprehensive study of FECounts is placed in [10].

Using argumentation similar to that given in Section 2 for $FGCount(A)$, one shows that the ideas of $FLCount(A)$ and $FECount(A)$ are suitable only for fuzzy sets with the operations $\wedge$ and $\vee$. Their generalizations to fuzzy sets with arbitrary t-operations do have the following forms for $k \in \mathbb{N}$, a t-norm t and a negation v:

$$\text{FLCount}_{t,v}(A)(k) := v([A]_{k+1}) \, t \, v([A]_{k+2}) \, t \ldots t \, v([A]_n),$$

$$\text{FECount}_{t,v}(A) := \text{FGCount}_t(A) \cap_t \text{FLCount}_{t,v}(A).$$

Hence

$$\text{FECount}_{t,v}(A)(k) = [A]_1 \, t \ldots t \, [A]_k \, t \, v([A]_{k+1}) \, t \ldots t \, v([A]_n).$$

$\text{FLCount}_{t,v}(A)(k)$ is a degree to which A has at most i elements for each $i \geq k$. So, $\text{FECount}_{t,v}(A)(k)$ is a degree to which A has at least i elements for each $i \leq k$ and, simultaneously, A has at most j elements for each $j \geq k$. Clearly, $\text{FLCount}_{t,v}(A)$ is convex. A less trivial fact is that $\text{FECount}_{t,v}(A)$ is always convex, too. $\text{FLCount}_{t,v}(A) = 1_{\{n,n+1,\ldots\}}$ and $\text{FECount}_{t,v}(A) = 1_{\{n\}}$ whenever $A \in \text{FCS}$. If $t = \wedge$ and $v = v_{\text{Ł}}$, we get $\text{FLCount}_{t,v}(A) = \text{FLCount}(A)$ and $\text{FECount}_{t,v}(A) = \text{FECount}(A)$. For $t = \wedge$ and $v = v^*$, $\text{FECount}_{t,v}(A)$ becomes the cardinality of A due to Dubois ([1]).

Non-strict archimedean t-norms do have zero divisors. If t is such a t-norm, it is possible that $\text{FECount}_{t,v}(A)(k) = 0$ for each k, which means that A is totally indescribable (indefinite) with respect to cardinality. This astonishing result suggests that generalized FECounts are suitable rather for strict t-norms, including $\wedge$.

4 Concluding Remarks

The subject of this paper were cardinalities of fuzzy sets with triangular operations. We proposed and investigated in many respects a generalization of the classical concept of the FGCount to such fuzzy sets. Also, we outlined an analogous generalization of FLCounts and FECounts. One should emphasize that leaving the world of min and max operations we lose some classical-like operational features of cardinalities of fuzzy sets.

References

1. Dubois, D., Prade, H.: Fuzzy Cardinality and the Modeling of Imprecise Quantification. Fuzzy Sets and Systems 16 (1985) 199-230
2. Gottwald, S.: Many-Valued Logic and Fuzzy Set Theory. In: Höhle, U., Rodabaugh, S.E. (eds.): Mathematics of Fuzzy Sets. Logic, Topology, and Measure Theory. Kluwer Acad. Publ., Boston Dordrecht London (1999) 5-89
3. Ling, C.H.: Representation of Associative Functions, Publ. Math. Debrecen 12 (1965) 189-212
4. Liu, Y., Kerre, E.E.: An Overview of Fuzzy Quantifiers. Part I: Interpretation. Fuzzy Sets and Systems 95 (1998) 1-21
5. Lowen, R.: Fuzzy Set Theory. Basic Concepts, Techniques and Bibliography. Kluwer Acad. Publ., Dordrecht Boston London (1996)

6. Nguyen, H.T., Walker, E.A.: A First Course in Fuzzy Logic. CRC Press, Boca Raton (1997)
7. Weber, S.: A General Concept of Fuzzy Connectives, Negations and Implications Based on t-Norms and t-Conorms. Fuzzy Sets and Systems 11 (1983) 115-134
8. Wygralak, M.: Fuzzy Cardinals Based on the Generalized Equality of Fuzzy Subsets. Fuzzy Sets and Systems 18 (1986) 143-158
9. Wygralak, M.: Vaguely Defined Objects. Representations, Fuzzy Sets and Nonclassical Cardinality Theory. Kluwer Acad. Publ., Dordrecht Boston London (1996)
10. Wygralak, M.: Questions of Cardinality of Finite Fuzzy Sets. Fuzzy Sets and Systems 102 (1999) 185-210
11. Wygralak, M.: Triangular Operations, Negations, and Scalar Cardinality of a Fuzzy Set. In: Zadeh, L.A., Kacprzyk, J. (eds.): Computing with Words in Information/Intelligent Systems 1. Foundations. Physica-Verlag, Heidelberg New York (1999) 326-341
12. Wygralak, M.: Fuzzy Sets with Triangular Norms and their Cardinality Theory. Fuzzy Sets and Systems, submitted
13. Zadeh, L.A.: A Theory of Approximate Reasoning. In: Hayes, J.E., Michie, D., Mikulich, L.I. (eds.): Machine Intelligence 9. Wiley, New York (1979) 149-184
14. Zadeh, L.A.: A Computational Approach to Fuzzy Quantifiers in Natural Languages. Comput. and Math. with Appl. 9 (1983) 149-184
15. Zadeh, L.A.: From Computing with Numbers to Computing with Words - From Manipulation of Measurements to Manipulation of Perceptions. IEEE Trans. on Circuits and Systems - I: Fundamental Theory and Appl. 45 (1999) 105-119
16. Zadeh, L.A.: Fuzzy Logic = Computing with Words. In: Zadeh, L.A., Kacprzyk, J. (eds.): Computing with Words in Information/Intelligent Systems 1. Foundations. Physica-Verlag, Heidelberg New York (1999) 3-23

On the Approximate Solution of Fuzzy Equation Systems

Michael Wagenknecht, Volker Schneider and Rainer Hampel

University of Applied Sciences
Zittau/Görlitz
IPM
Theodor-Körner-Allee 16
02763 Zittau, Germany
{m.wagenknecht,v.schneider,r.hampel}@htw-zittau.de

Abstract. We consider the solution of fuzzy equation systems within the framework of (L,R)-numbers. Therefore, we have to define the notion "solution" and we have to determine arithmetic operations approximations by the above class of fuzzy numbers.

1 Introduction

In chemical engineering quantitative models, parameters, data etc. are vague and can often be described by fuzzy sets. One of the main tasks in simulation and optimization is the handling of crisp mathematical models with fuzzy parameters particularly leading to the solution of fuzzy equation systems. Unlike the crisp case one is faced with at least three problems. First, how can we model fuzzy parameters? Second, how shall we perform arithmetic operations transparently and effectively? And finally, what actually is a solution of the system? Whereas the first question can be answered satifactorily by using (L,R)-numbers [2,4], already simple arithmetic operations (e.g. multiplication) cannot be carried out within this class of fuzzy numbers. On the other hand, the latter are of great advantage because of their simple structure and interpretability, so that especially in applications one aims to stay within this class. But then we have to determine (L,R)-number approximations of arithmetic operations.

In what follows we will consider fuzzy linear systems. We will demonstrate how the approximations mentioned can be involved in an applicable notion of solution and how they can be computed effectively.

2 Basics

Definition 1 [4]. A continuous triangular norm (t-norm) T is called *Archimedean* if $T(a,a) < a$ for all $a \in (0,1)$.

Remark 1. Denote $T_M = min$ and

$$T_W(a,b) = \begin{cases} \min(a,b) & \text{for } \max(a,b) = 1, \\ 0 & \text{otherwise.} \end{cases} \tag{1}$$

Then one sees that for any t-norm T

$$T_W(a,b) \le T(a,b) \le T_M(a,b) \tag{2}$$

In this sense we can call T_M the strongest and T_W the weakest t-norms (note that T_W is *not* continuous).

Proposition 1 [4]. *The norm T is Archimedean iff it has the following representation*

$$T(a,b) = f^{(-1)}\big(f(a) + f(b)\big) \cdot \tag{3}$$

Here the *generator function* $f: [0,1] \to \mathbf{R}^+$ is continuous, strictly decreasing, $f(1) = 0$, and the *pseudoinverse* of f is given by

$$f^{(-1)}(x) = f^{-1}\big(\min(f(0),x)\big) \cdot$$

Remark 2. The norm T_M is *not* Archimedean and has non representation similar to (3).

Definition 2 (Extension principle [2-4]). Let $F: \mathbf{R}^n \to \mathbf{R}$ be any n-ary function. Moreover, let $A_1, ..., A_n$ be fuzzy sets over the real axis. Then we define for a given norm T fuzzy set B by

$$\mu_B(t) = \begin{cases} \sup_{t = F(x_1,...,x_n)} T\big(\mu_{A_1}(x_1),...,\mu_{A_n}(x_n)\big) & \text{if the inverse image of } F \text{ is not empty;} \\ 0 & \text{otherwise.} \end{cases} \tag{4}$$

Thus we are able to compute functions of fuzzy sets.

Remark 3. The extension principle is not restricted to Euclidean spaces. However, since we are dealing with arithmetic operations we do not need more general forms.

Remark 4. Let $n = 2$, i.e. F is a binary operation (e.g. addition, multiplication, etc.) which we denote by "$*$". Then (4) has the special shape

$$\mu_B(t) = \sup_{t = x_1 * x_2} T\big(\mu_{A_1}(x_1), \mu_{A_2}(x_2)\big) \tag{5}$$

for the non-zero case.

For a fuzzy set A, we denote by A_α its α-level set for $\alpha \in (0,1]$, i.e.

$A_\alpha = \{x \in \mathbf{R} : \mu_A(x) \ge \alpha\}$. If A_α is a closed interval we will use the notation $A_\alpha = [\underline{A}_\alpha, \overline{A}_\alpha]$, as well. It is well-known that A can equivalently be represented by

$$\mu_A(x) = \sup_{x \in A_\alpha} \alpha \tag{6}$$

(note that the *support* of A is defined by $\operatorname{supp} A = \bigcup_{\alpha \in (0,1]} A_\alpha$).

On the other hand, the somehow unwieldy extension principle may essentially simplify using the fact that is possibles a relationship between B_α and the $A_{i\alpha}$ (e.g. for T_M it turns to interval computations). Therefore, we need the notion of a fuzzy number.

Definition 3 [1,2,4].A fuzzy set A over the real axis is called a *fuzzy number* (*FN*) if it is *unimodal*, *normal*, *compactly supported*, with *upper semi-continuous* (usc) and *quasi-concave* membership function.

A *FN* A is called *nonnegative* if $\operatorname{supp} A \subset \mathbf{R}^+$. It is nonpositive if $-A$ is nonnegative.

Theorem 1 [3]. *Let F and A_i be as in Definition 2. Additionally, suppose F to be continuous over the Cartesian product of $cl\,\operatorname{supp} A_i$ whereby the A_i are fuzzy numbers. Moreover, let T be usc. Then*

$$B_\alpha = \left[F(A_1, \dots, A_n) \right]_\alpha = \bigcup_{T(\alpha_1, \dots, \alpha_n) \ge \alpha} F\left(A_{1\alpha_1}, \dots, A_{n\alpha_n} \right) \tag{7}$$

for all $\alpha \in (0,1]$.

Remark 5. For T_M one gets the well-known result

$$B_\alpha = F\left(A_{1\alpha}, \dots, A_{n\alpha} \right) . \tag{8}$$

Now, let be unique $\overline{a}_i$ with $\mu_{A_i}(\overline{a}_i) = 1$. Then for T_W we obtain

$$B_\alpha = \bigcup_{i=1}^{n} F\left(\overline{a}_1, \dots, A_{i\alpha}, \dots, \overline{a}_n \right) . \tag{9}$$

One may ask, if B is also a fuzzy number (e.g., the sum of fuzzy numbers is expected to be a fuzzy number again). The answer is positive.

Proposition 2. *Under the assumptions of Theorem 1, fuzzy set B is a fuzzy number.*

Definition 4. Let $L, R: [0,1] \to [0, \infty)$ be two continuous, decreasing functions fulfilling $L(0) = R(0) = 1$, $L(1) = R(1) = 0$. Moreover, let $\overline{a}$ be any real number, and suppose a_l, a_r to be positive numbers. Then fuzzy set A is an (*L,R*)-number if

$$\mu_A(t) = \begin{cases} L\left(\dfrac{\overline{a}-t}{a_l}\right) & \text{for } t \le \overline{a}; \\[3mm] R\left(\dfrac{t-\overline{a}}{a_r}\right) & \text{for } t > \overline{a}. \end{cases} \tag{10}$$

Symbolically, we will write $A = \left(\overline{a}, a_l, a_r\right)_{L.R}$.

Remark 6. Obviously, every (*L,R*)-number is also a fuzzy number where the support is included in $[\overline{a} - a_l, \overline{a} + a_r]$. Crisp real numbers can *formally* be represented as $A = \left(\overline{a}, 0, 0\right)_{L.R}$. With this convention we may allow the spreads to be 0.

Suppose L and R to be invertible over [0,1]. Then

$$A_\alpha = \left[\overline{a} - L^{-1}(\alpha) \cdot a_l, \overline{a} + R^{-1}(\alpha) \cdot a_r\right]. \tag{11}$$

3 Solution of Fuzzy Equation Systems

Consider the following system of equations

$$\sum_{j=1}^{n} A_{ij} \cdot X_j = B_i \, ; i = 1, \dots, m. \tag{12}$$

The left-hand side in (12) is denoted by LH_i.

Suppose that there a metric between fuzzy sets is given. For instance

a) $\rho_l(A, B) = \rho_u(A, B) + \rho_o(A, B)$ with

$$\rho_u(A, B) = \left(\int_0^1 \left|\underline{A}_\alpha - \underline{B}_\alpha\right|^p d\alpha\right)^{\frac{1}{p}}, \quad \rho_o(A, B) = \left(\int_0^1 \left|\overline{A}_\alpha - \overline{B}_\alpha\right|^p d\alpha\right)^{\frac{1}{p}} \tag{13}$$

and $p \ge 1$.

b) For fuzzy numbers of (*L,R*)-type, i.e. $A = \left(\overline{a}, a_l, a_r\right)$, $B = \left(\overline{b}, b_l, b_r\right)$ the following metric is senseful

$$\rho_{LR}(A,B) = \left(\left| \bar{a} - \bar{b} \right|^p + \left| a_l - b_l \right|^p + \left| a_r - b_r \right|^p \right)^{\frac{1}{p}}. \tag{14}$$

One easily shows the equivalence of both in the sense of convergency. If the kind of metrics is regardless we simly use ρ.

A solution of (12) is given as a solution of the optimization task

$$\sum_{i=1}^{m} \rho(LH_i, B_i) \xrightarrow[x_1,\ldots x_n]{} \min \tag{15}$$

As a rule, LH_i is not known exactly, or is not of (L,R)-type. Suppose that there are (L,R)-inclusions, i.e.

$$C_i^u \subseteq LH_i \subseteq C_i^o. \tag{16}$$

Now we get an "approximating left-hand side" $C_i(\lambda)$ by the convex combination of C_i^u and C_i^o

$$C_i(\lambda) = \left(c_i, \lambda \cdot \underline{c}_i^1 + (1-\lambda) \cdot \bar{c}_i^1, \lambda \cdot \underline{c}_i^2 + (1-\lambda) \cdot \bar{c}_i^2 \right) \tag{17}$$

with $\lambda \in [0,1]$. Obviously,

$$C_i^u \subseteq C_i(\lambda) \subseteq C_i^o \tag{18}$$

and we consider the following substitute for (12)

$$C_i(\lambda) = B_i ; i = 1,\ldots,m. \tag{19}$$

The corresponding optimization problem in analogy to (15) is

$$\sum_{i=1}^{m} \rho(C_i(\lambda), B_i) \xrightarrow[x_1,\ldots x_n]{} \min \tag{20}$$

$$x_j^l, x_j^r \geq 0, \ j = 1,\ldots,n$$

where λ may be optimized, as well (each equation may have its individual λ).

One gets for the metrics ρ_l and ρ_{LR} an error estimation as follows

$$\rho(LH_i, B_i) \leq \rho(LH_i, C_i(\lambda)) + \rho(C_i(\lambda), B_i) \leq \rho(C_i^u, C_i^o) + \rho(C_i(\lambda), B_i). \tag{21}$$

The first summand in the right-hand side of (21) is the approximation error by including the exact LH_i. The second summand describes the error of the approximative solution of (20). And both are known! It seems to be senseful to minimize not only the latter one, but to consider the following polyoptimization task

$$
\begin{pmatrix}
\sum\limits_{i=1}^{m} \rho(C_i^u, C_i^o) \\
\sum\limits_{i=1}^{m} \rho(C_i(\lambda), B_i)
\end{pmatrix}
\xrightarrow[x_1,...,x_n]{} \min
\tag{22}
$$

thus keeping the inclusion error within reasonable bounds, as well.

4 Fuzzy Arithmetics

For modelling equation systems of type (12) we need the basic operations of fuzzy addition and multiplication. Regarding the former we refer to [1,5-7]. On the other hand, since the product of two (L,R)-number is not a number of this class (unless one factor is crisp or we apply T_W) we aim at getting (L,R)-approximations of multiplication. Throughout this section we assume L and R to be invertible on $[0,1]$.

Let A,B be sign-determined FN, e.g. suppose both to be nonnegative. Then for $C = A \cdot B$ we get from (4)

$$
\underline{C}_\alpha = \inf_{T(\xi,\eta) \geq \alpha} \underline{A}_\xi \cdot \underline{B}_\eta , \quad \overline{C}_\alpha = \sup_{T(\xi,\eta) \geq \alpha} \overline{A}_\xi \cdot \overline{B}_\eta
\tag{23}
$$

With respect to the t-norm applied we will distinguish several cases.

1. $T = T_M$

Here we simply get

$$
\underline{C}_\alpha = \underline{A}_\alpha \cdot \underline{B}_\alpha , \quad \overline{C}_\alpha = \overline{A}_\alpha \cdot \overline{B}_\alpha .
\tag{24}
$$

Assuming $A = (\overline{a}, a_l, a_r)_{L,R}$, $B = (\overline{b}, b_l, b_r)_{L,R}$ we see that

$$
\underline{C}_\alpha = \overline{a} \cdot \overline{b} - (\overline{a} \cdot b_l + \overline{b} \cdot a_l) \cdot L^{-1}(\alpha) + a_l \cdot b_l \cdot (L^{-1}(\alpha))^2 ,
$$

$$
\overline{C}_\alpha = \overline{a} \cdot \overline{b} + (\overline{a} \cdot b_r + \overline{b} \cdot a_r) \cdot R^{-1}(\alpha) + a_r \cdot b_r \cdot (R^{-1}(\alpha))^2 .
\tag{25}
$$

As mentioned above, multiplication leads out of the (L,R)-class. Under the assumption that $a_l \cdot b_l, a_r \cdot b_r$ are small in comparison with the main values we proceed to the well-known approximation

$$A \cdot B \approx \left(\overline{a} \cdot \overline{b}, \overline{a} \cdot b_l + \overline{b} \cdot a_l, \overline{a} \cdot b_r + \overline{b} \cdot a_r \right)_{L,R} \tag{26}$$

However, multiple application of (26) can lead to large errors. Thus we will show how to get estimations enabling an error control. One immediately sees that

$$\overline{a} \cdot \overline{b} - \left(\overline{a} \cdot b_l + \overline{b} \cdot a_l \right) \cdot L^{-1}(\alpha) \leq \underline{C}_\alpha \leq \overline{a} \cdot \overline{b} - \left(\overline{a} \cdot b_l + \overline{b} \cdot a_l \right) \cdot L^{-1}(\alpha) + a_l \cdot b_l \cdot L^{-1}(\alpha) \ ,$$

$$\overline{a} \cdot \overline{b} + \left(\overline{a} \cdot b_r + \overline{b} \cdot a_r \right) \cdot R^{-1}(\alpha) \leq \overline{C}_\alpha \leq \overline{a} \cdot \overline{b} + \left(\overline{a} \cdot b_r + \overline{b} \cdot a_r \right) \cdot R^{-1}(\alpha) + a_r \cdot b_r \cdot R^{-1}(\alpha).$$

$$\tag{27}$$

Since the condition

$$\frac{\overline{a}}{a_l} + \frac{\overline{b}}{b_l} \geq 1$$

is fulfilled we can introduce two FN

$$C^{lo} = \left(\overline{a} \cdot \overline{b}, \overline{a} \cdot b_l + \overline{b} \cdot a_l - a_l \cdot b_l, \overline{a} \cdot b_r + \overline{b} \cdot a_r \right)_{L,R} \ ,$$

$$C^{up} = \left(\overline{a} \cdot \overline{b}, \overline{a} \cdot b_l + \overline{b} \cdot a_l, \overline{a} \cdot b_r + \overline{b} \cdot a_r + a_r \cdot b_r \right)_{L,R} \tag{28}$$

and obviously $C^{lo} \subseteq C \subseteq C^{up}$ in the sense of fuzzy inclusion. Hence, we can enclose the true curve by (L,R)-numbers.

2. $T = T_W$

From (4) one immediately obtains

$$A \cdot B = \left(\overline{a} \cdot \overline{b}, max \left(\overline{a} \cdot b_l, \overline{b} \cdot a_l \right), max \left(\overline{a} \cdot b_r, \overline{b} \cdot a_r \right) \right)_{L,R} \tag{29}$$

3. T = general Archimedean t-norm

Now (7) specifies to

$$\underline{C}_\alpha = \inf_{f^{(-1)}(f(\xi)+f(\eta))\geq\alpha} \underline{A}_\xi \cdot \underline{B}_\eta \ , \ \overline{C}_\alpha = \sup_{f^{(-1)}(f(\xi)+f(\eta))\geq\alpha} \overline{A}_\xi \cdot \overline{B}_\eta \ .$$

After simple computations we get

$$\underline{C}_\alpha = \inf_{\xi\in[\alpha,1]} \underline{A}_\xi \cdot \underline{B}_\eta \ , \ \overline{C}_\alpha = \sup_{\xi\in[\alpha,1]} \overline{A}_\xi \cdot \overline{B}_\eta \tag{30}$$

under the condition $\eta = f^{-1}(f(\alpha) - f(\xi))$.

On the other hand,

$$\underline{A}_\xi \cdot \underline{B}_\eta = \overline{a}\cdot\overline{b} - \overline{b}\cdot a_l \cdot L^{-1}(\xi) - \overline{a}\cdot b_l \cdot L^{-1}(\eta) + a_l \cdot b_l \cdot L^{-1}(\xi) \cdot L^{-1}(\eta)$$
$$= \overline{a}\cdot\overline{b} - \overline{b}\cdot a_l \cdot L^{-1}(\xi) - \overline{a}\cdot b_l \cdot L^{-1} \circ f^{-1}(f(\alpha) - f(\xi))$$
$$+ a_l \cdot b_l \cdot L^{-1}(\xi) \cdot \left[L^{-1} \circ f^{-1}(f(\alpha) - f(\xi)) \right],$$

$$\overline{A}_\xi \cdot \overline{B}_\eta = \overline{a}\cdot\overline{b} + \overline{b}\cdot a_r \cdot R^{-1}(\xi) + \overline{a}\cdot b_r \cdot R^{-1}(\eta) + a_r \cdot b_r \cdot R^{-1}(\xi) \cdot R^{-1}(\eta)$$
$$= \overline{a}\cdot\overline{b} + \overline{b}\cdot a_r \cdot R^{-1}(\xi) + \overline{a}\cdot b_r \cdot R^{-1} \circ f^{-1}(f(\alpha) - f(\xi))$$
$$+ a_r \cdot b_r \cdot R^{-1}(\xi) \cdot \left[R^{-1} \circ f^{-1}(f(\alpha) - f(\xi)) \right].$$

To proceed further let us suppose T to have normed generator f and

$$L^{-1} = f^p, \ R^{-1} = f^q \tag{31}$$

with certain positive p, q. Hence, we are led to

$$\underline{A}_\xi \cdot \underline{B}_\eta = \overline{a}\cdot\overline{b} - \overline{b}\cdot a_l \cdot \tau^p - \overline{a}\cdot b_l \cdot (f(\alpha) - \tau)^p + a_l \cdot b_l \cdot \tau^p \cdot (f(\alpha) - \tau)^p, \tag{32}$$

$$\overline{A}_\xi \cdot \overline{B}_\eta = \overline{a}\cdot\overline{b} + \overline{b}\cdot a_r \cdot \tau^q + \overline{a}\cdot b_r \cdot (f(\alpha) - \tau)^q + a_r \cdot b_r \cdot \tau^q \cdot (f(\alpha) - \tau)^q, \tag{33}$$

where we used $\tau = f(\xi)$. With the denotations

$$h(\tau) = -\overline{b}\cdot a_l \cdot \tau^p - \overline{a}\cdot b_l \cdot (f(\alpha) - \tau)^p, \ g(\tau) = \overline{b}\cdot a_r \cdot \tau^q + \overline{a}\cdot b_r \cdot (f(\alpha) - \tau)^q \tag{34}$$

we obtain from (30),(32) and (33)

$$\bar{a}\cdot\bar{b} + \min_{\tau\in[0,f(\alpha)]} h(\tau) \leq \underline{C}_\alpha \leq \bar{a}\cdot\bar{b} + \min_{\tau\in[0,f(\alpha)]} h(\tau) + a_l\cdot b_l \cdot\frac{(f(\alpha))^p}{4^p} \qquad (35)$$

$$\bar{a}\cdot\bar{b} + \max_{\tau\in[0,f(\alpha)]} g(\tau) \leq \overline{C}_\alpha \leq \bar{a}\cdot\bar{b} + \max_{\tau\in[0,f(\alpha)]} g(\tau) + a_r\cdot b_r \cdot\frac{(f(\alpha))^q}{4^q} \qquad (36)$$

Hence we have to determine the extremal values of g and h on $[0,f(\alpha)]$. Therefore we yet introduce $\mu_* = \dfrac{a_*}{\bar{a}}$, $\nu_* = \dfrac{b_*}{\bar{b}}$ with $* = l,r$.

L1. $p \geq 1$

Then h is concave and one has

$$\min_{\tau\in[0,f(\alpha)]} h(\tau) = \min\left[h(0), h(f(\alpha))\right] = -(f(\alpha))^p \cdot \max(\bar{b}\cdot a_l, \bar{a}\cdot b_l) \qquad (37)$$

and

$$\bar{a}\cdot\bar{b} - \max(\bar{b}\cdot a_l, \bar{a}\cdot b_l)\cdot L^{-1}(\alpha) \leq \underline{C}_\alpha \leq \bar{a}\cdot\bar{b} - \left(\max(\bar{b}\cdot a_l, \bar{a}\cdot b_l) - \frac{a_l\cdot b_l}{4^p}\right)\cdot L^{-1}(\alpha). \qquad (38)$$

L2. $p \leq 1$

Now h is convex, hence for getting the minimum of h we determine the root of h' leading to $\min_{\tau\in[0,f(\alpha)]} h(\tau) = h(\tilde{\tau})$ with

$$\tilde{\tau} = \frac{v_l^{\frac{1}{p-1}}}{\mu_l^{\frac{1}{p-1}} + v_l^{\frac{1}{p-1}}}\cdot f(\alpha)$$

since obviously $\tilde{\tau}\in[0,f(\alpha)]$. This leads to

$$\min_{\tau\in[0,f(\alpha)]} h(\tau) = h(\tilde{\tau}) = -a_l\cdot b_l\cdot\left(\mu_l^{\frac{1}{p-1}} + v_l^{\frac{1}{p-1}}\right)^{1-p}\cdot(f(\alpha))^p \qquad (39)$$

and finally

$$\bar{a}\cdot\bar{b} - a_l\cdot b_l\cdot\left(\mu_l^{\frac{1}{p-1}} + v_l^{\frac{1}{p-1}}\right)^{1-p}\cdot L^{-1}(\alpha) \leq \underline{C}_\alpha \leq \bar{a}\cdot\bar{b} - a_l\cdot b_l\cdot\left(\left(\mu_l^{\frac{1}{p-1}} + v_l^{\frac{1}{p-1}}\right)^{1-p} - \frac{1}{4^p}\right)\cdot L^{-1}(\alpha). \qquad (40)$$

Corresponding inclusions for $\overline{C}_\alpha$ we get in analogy to the above cases

U1. $q \geq 1$

Now g is convex and attains its maximum in one of the border points. That is,

$$\overline{a} \cdot \overline{b} + \max(\overline{b} \cdot a_r, \overline{a} \cdot b_r) \cdot R^{-1}(\alpha) \leq \overline{C}_\alpha \leq \overline{a} \cdot \overline{b} + \left(\max(\overline{b} \cdot a_r, \overline{a} \cdot b_r) + \frac{a_r \cdot b_r}{4^q} \right) \cdot R^{-1}(\alpha).$$

(41)

U2. $q < 1$

The concavity of g demands the determination of the root of g' (see $\tilde{\tau}$ above) yielding

$$\overline{a} \cdot \overline{b} + a_r \cdot b_r \cdot \left(\mu_r^{\frac{1}{q-1}} + \nu_r^{\frac{1}{q-1}} \right)^{1-q} \cdot R^{-1}(\alpha) \leq \overline{C}_\alpha \leq \overline{a} \cdot \overline{b} + a_r \cdot b_r \cdot \left(\left(\mu_r^{\frac{1}{q-1}} + \nu_r^{\frac{1}{q-1}} \right)^{1-q} + \frac{1}{4^q} \right) \cdot R^{-1}(\alpha).$$

(42)

Now the constitution of inclusions causes no difficulties.

References

1. De Baets, B., Markova-Stupnanova, A.: Analytical Expressions for the Addition of Fuzzy Intervals. Fuzzy Sets and Systems 91 (1997) 203-213
2. Dubois, D., Prade, H.: Fuzzy Sets and Systems. Academic Press, N.J. (1980)
3. Fuller, R., Keresztfalvi, T.: On Generalization of Nguyen's Theorem. Fuzzy Sets and Systems 41 (1991) 371-374
4. Kruse, R. et al.: Foundations of Fuzzy Systems. J. Wiley, Chichester (1994)
5. Markova-Stupnanova, A.: A Note to the Addition of Fuzzy Numbers Based on a Continuous Archimedean t-Norm. Fuzzy Sets and Systems 91 (1997) 251-256
6. Mesiar, R.: A Note to the T-Sum of L-R-Numbers. Fuzzy Sets and Systems 79 (1996) 259-261
7. Mesiar, R.: Shape Preserving Additions of Fuzzy Intervals. Fuzzy Sets and Systems 86 (1997) 73-78

Using Fuzzy Parasets in Problem-Solving Under Uncertainty

Jozef Šajda

Institute of Control Theory and Robotics
Slovak Academy of Sciences,
Dúbravská 9, 84237 Bratislava, Slovakia
utrrsajd@nic.savba.sk

Abstract . An interpretation of problem-solving with uncertainty supported by fuzzy paraset methodology is presented. Using the concept of uncertainty by Shafer, Zadeh and Dubois-Prade, a rough algorithm for general problem-solving with uncertainty is proposed. To measure uncertainty for subsets of a basic space, the known functions of plausibility and belief are used, as well as the possibility and necessity measures. The algorithm is outlined not only for a finite, but also for an infinite basic space. To improve the effect of the procedure proposed, specific methodological tools are designed: an evidence paraset and an uncertainty paraset.

1 Introduction

The problem of uncertainty in logic is still very actual. In everyday life we learn that many of real logical systems immanently contain a truth uncertainty which is principally unavoidable by the use of any logical procedure. The uncertainty in logic is something like the fenomenon of nonlinearity in mathematics or physics. Probably, we are now at the beginning only to rule solving of uncertainty problem in general. In spite of the fundamental works by Dempster [2], Shafer [1], Zadeh [3], Dubois-Prade [4], Yager [6], and many others who have the problem of uncertainty addressed and solved, the most of work is still to be done by their followers.

This paper pretends to be at least a small positive contribution to the general effort to rule the problem of logical uncertainty, especially in the field of problem-solving. We have attempted to embed uncertainty in the framework of paraset methodology, which enables us to involve into consideration a time ordering as a basic element of the concept of (fuzzy) paraset.

2 Some Information on (General Abstract) Parasets

The concept of a *general abstract paraset* (shortly a *paraset*) has been first defined and used in the papers [7,8] in the following way.

Let $X \neq \varnothing$ be a *basic space* of elements x, T be a linear set of *time points* t (shortly a *time set*) with the ordering $\leq$, and A be a (multivalued) *mapping* $A : T \to P(X)$, of the time set T into the power-set $P(X)$ of the space X. Then the triple $\langle X, T, A \rangle$ is called a *(general abstract) paraset*, A, in the space X, symbolically $A = \langle X, T, A \rangle$. The set $A(t)$, (shortly A_t) in $P(X)$, which is the map of the element $t \in T$ in the mapping A, is called a *point-representation* of the paraset A (at the point t). The set of all point-representations A_t ordered by the relation $\preceq$ in the sense $A_t \preceq A_s$ iff $t \leq s$, is called a *history* of the paraset A. Thus, for a given paraset A, the triple $\langle X, T, A \rangle$ is its formal representation, while the history of A is its physical representation. Hence,

$$A = \langle X, T, A \rangle = \{A_t \mid A_t \in P(X), \ A_t \preceq A_s, \ s, t \in T\}.$$

The history of a paraset provides one a whole picture of the paraset. In general, this picture shows the paraset as a "living" thing which is able to change both its shape and the position in the space X.

For practical reason, the point-representation A_t belonging to the time point t_k will be written in the simplified form A_k; e.g. $A_1, A_2, ..., A_n$ means n point-representations A_t at time points $t_1, t_2, ..., t_n$, respectively.

A paraset $A = \langle X, T, A \rangle$ of which the time set T is finite is said to be a *finite paraset*, otherwise A is *infinite*. Typically, for any infinite paraset A there exists a non-empty point-representation, $A_n \neq \varnothing$, for any $n = 1, 2, ...$

The paraset $A' = \langle X, T', A' \rangle$ in the space X, for which $T' \subseteq T$ and $A_t' = A_t$ for all $t \in T'$, is called a *subparaset* of the paraset $A = \langle X, T, A \rangle$, symbolically $A' \subseteq A$. If for all $t, t' \in T$ with the property $t < t'$ holds $A_t \subseteq A_{t'}$ or $A_t \supseteq A_{t'}$ then A is said to be *monotonously rising*, or *monotonously decreasing;* at a strong inclusion – *rising*, or *decreasing* paraset, respectively. If for all $t \in T$ holds $A_t = C$, where $C \in P(X)$, then A is called a *constant* paraset. Hence, classical sets may be viewed as special cases of constant parasets. Another special case of constant paraset is *empty* paraset, denoted $\varnothing$, of which all point-representations are empty sets, $\varnothing$, and the *universal* paraset, X, of which all point-representations are identical with the whole space X.

If the time set T of an infinite paraset A is defined by a sequence of discrete values $t_1 < t_2 < ... < t_n < ...$ aiming to a value $t_0 \in T$ and the corresponding sequence of point-representations $A_1, A_2, ..., A_n, ...$ aims to a set A_0, $A_0 \in P(X)$, then A is said to be a *convergent* paraset.

In the paper [7] the operations of *union* ($\cup$), *intersection* ($\cap$) and *complementation* ($\overline{\ \ }$) over parasets are defined, resulting into parasets again. In that way, a basis for operational calcul is prepared to calculate with parasets in practice. It is also shown, that the paraset operations above satisfy the known laws for classical set theory, as they are reducing themselves in real computations to the common set rules. This enables us to form new, composed parasets by means of the operations $\cup, \cap, \overline{\ \ }$. Thus, given parasets $A, B, C, ...$ all defined in the space X, some new parasets composed from them are, e.g.

$$A \cup B, \quad (A \cup B) \cap C, \quad (\overline{A} \cap B) \cup \overline{C}, \quad (A \cup B) \cap (\overline{A} \cap \overline{C}). \quad (1)$$

3 Basic Definitions and Properties of Fuzzy Parasets

Let $\Phi(X)$ be the power-set of all fuzzy subsets in X. Recall that a fuzzy set $F \in \Phi(X)$ is characterized by the mapping $F: X \to [0,1]$, where the value $F(x)$ expresses the degree of membership of the element $x \in X$ in the fuzzy set F. A fuzzy set F is *normalized* if $F(x) = 1$ for at least one $x \in X$. Given $F, G \in \Phi(X)$, F is a fuzzy *subset* of G, symbolically $F \subseteq G$, iff $F(x) \leq G(x)$ for all $x \in X$. It is obvious that the relation $\subseteq$ of inclusion defines a partial ordering in the set $\Phi(X)$: for any $F, G \in \Phi(X)$

$$F \subseteq G \quad iff \quad (\forall x \in X)(F(x) \leq G(x)).$$

It is easy to show that the structure $(\Phi(X), \subseteq)$ is a complete lattice [10] with the minimal element $\emptyset$ and the maximal element X. Recall that for any fuzzy set $F \in \Phi(X)$ the set *supp F* (from "support") defined by

$$supp \, F = \{x \in X \mid F(x) > 0\}, \tag{2}$$

can be established.

An important class of sets useful for various applications, are *cut sets* of a fuzzy set. Given a fuzzy set F and a number $\alpha \in [0,1]$, the α - *cut* of F, F_α, is the crisp set

$$F_\alpha = \{x \in X \mid F(x) \geq \alpha\}. \tag{3}$$

The concept of cut set is interesting due to the following obvious property: for any fuzzy set F and numbers $\alpha, \beta \in [0,1]$,

$$\alpha > \beta \Rightarrow F_\alpha \subseteq F_\beta. \tag{4}$$

Now, analogically to the definition of (general abstract) paraset, we define the concept of fuzzy paraset as follows.

Definition. Let X be a *basic space* of elements x, T be a *time set* of time points t, and $F: T \to \Phi(X)$ be a fuzzy multifunction [9] which to every value $t \in T$ assigns a non-empty fuzzy set from $\Phi(X)$, and to any value $t \notin T$ assigns empty set $\emptyset$, i.e.

$$F(t) = F_t, \, F_t \neq \emptyset, \, t \in T$$
$$= \emptyset, \quad t \notin T.$$

Then we call:
(a) the triple $F = \langle X, T, F \rangle$ a *fuzzy paraset F* in X,
(b) the subset $F_t \in \Phi(X)$ of elements x, for $t \in T$, a *point-representation* of the fuzzy paraset F (at the point t),
(c) the set of all point-representations ordered by $\preceq$, a *history* of the fuzzy paraset F.

Thus, a fuzzy paraset F may be written in two equivalent representations,

$$F = \langle X, T, F \rangle = \{F_t \mid F_t \in \Phi(X), \, F_t \neq \emptyset, \, \forall t \in T\}, \tag{5}$$

where the second right-hand expression represents the history of the fuzzy paraset F.

Analogically to crisp parasets, we will use the simplified indexing $t_k = k$ also for fuzzy parasets, e.g. a point-representation F_t at $t = t_k$ will be denoted F_k, etc.

Operations over fuzzy parasets can be formally defined in the same way like those over crisp parasets [9]. Thus, given two arbitrary fuzzy parasets, $F = \langle X, T_F, F \rangle$ and $G = \langle X, T_G, G \rangle$, we can define the following new fuzzy parasets:

(i)   the *complement*, $\overline{F}$, of F by

$$\overline{F} = \langle X, T_F, \overline{F} \rangle,$$

where the time set remains unchanged, and the point-representations, $\overline{F}_t$, are defined by

$$\overline{F}_t = X - F_t, \quad \forall t \in T_F; \tag{6}$$

(ii)   the *union*, $F \cup G$, of F and G by

$$F \cup G = \langle X, T_{F \cup G}, F \cup G \rangle,$$

where the time set $T_{F \cup G} = T_F \cup T_G$ is union of the time sets T_F, T_G and is ordered by $\leq$, and the point-representations are defined by

$$(F \cup G)(t) = (F \cup G)_t = F_t \cup G_t, \quad \forall t \in (T_F \cup T_G); \tag{7}$$

(iii)   the *intersection*, $F \cap G$, of F and G by

$$F \cap G = \langle X, T_{F \cap G}, F \cap G \rangle,$$

where the time set $T_{F \cap G} = T_F \cap T_G$ is intersection of the time sets T_F, T_G and is ordered by $\leq$, and the point-representations are defined by

$$(F \cap G)(t) = (F \cap G)_t = F_t \cap G_t, \quad \forall t \in (T_F \cap T_G). \tag{8}$$

Analogically to crisp parasets, if $F, G, H, ...$ are arbitrary fuzzy parasets in the space X, we can create new fuzzy parasets from them, using the operations of union, intersection and complementation, e.g.

$$F \cap G, \quad (F \cup G) \cap H, \quad (\overline{F} \cap G) \cup \overline{H}, \quad (F \cup \overline{G}) \cap (G \cup \overline{H}). \tag{9}$$

4   An Interpretation of Uncertain Problem-Solving

To represent problem with uncertainty, either logical propositions or equivalent set-theory expressions are frequently used. It is assumed that an amount of uncertainty involved can be reduced by a relevant information, acquired as a result of an action, e.g. finding a new fact, receiving an important message, discovering a historical record, learning a result of an experiment, etc. The more information is obtained, the greater is decreasing uncertainty involved. We accept the concept of uncertainty like described and searched in [1-6].

In general, let $X \neq \varnothing$ be a set of elements x representing *all possibilities*, from whose only one (for simplicity) corresponds to the solution of a given problem. Accepting Shafer [1], X is a *frame of discernment* containing this (unknown, in general) specific element. It is clear that such a conception represents only one of the possible interpretations of general problem-solving under uncertainty. Nevertheless, an appropriate procedure is expected to determine the specific element.

To describe the situation mathematically, let m be Shafer *basic probability assignment* defined in [1] by a set function $m: P(X) \rightarrow [0,1]$, with the property

$$m(\varnothing) = 0, \qquad \sum_{A \subseteq X} m(A) = 1, \tag{10}$$

where the quantity $m(A)$ expresses the degree of one's subjective belief, based on some true evidence available, that the set A contains the specific element. The purpose of m is to make the first qualified assessment of the probability for such a belief.

A subset $A \subset X$ for which $m(A) > 0$ is called a *focal element* of the belief m; the union of all focal elements forms a subset $Z \subseteq X$ called a *core* of the belief. Given a set F^* of all focal elements, the pair (F^*, m) is called a *body of evidence*.

To express a measure of uncertainty that a typical element is in A, a *belief function Bel*, based on the basic probability assignment m, is defined in X by

$$Bel(A) = \sum_{B \subset A} m(B), \qquad \forall A \in P(X). \tag{11}$$

To describe the whole situation connected with the uncertainty involved, it is reasonable to define in X also a *plausibility function, Pl,* by

$$Pl(A) = \sum_{B \cap A \neq \varnothing} m(B), \qquad \forall A \in P(X). \tag{12}$$

From the definitions of *Bel* and *Pl* it follows

$$0 \leq Bel(A) \leq Pl(A) \leq 1, \qquad \forall A \in P(X). \tag{13}$$

Both of these uncertainty functions should satisfy the so-called *modal duality* equation,

$$Pl(A) = 1 - Bel(\bar{A}), \qquad \forall A \in P(X) \tag{14}$$

which is a consequent of modal logic consideration. The duality implies mutual exchangebility of *Bel* and *Pl* in the expression (14).

The number interval *[Bel(A), Pl(A)]* expresses the range of possible uncertainty for the subset A, where the lower bound, *Bel(A)*, may be regarded as a degree of certainty (or necessity) that a specific element is in A, while the upper bound, *Pl(A)*, means a degree of plausibility (or possibility) of that proposition.

The most interesting case for subsets A's is becomming when $A_1, A_2,..., A_n$ form a *nested* sequence of focal elements,

$$A_1 \supset A_2 \supset ... \supset A_n ; \tag{15}$$

then the functions *Bel* and *Pl* can be reduced [3-5] to necessity and possibility measures.

Zadeh [3] introduced a *possibility distribution*, π, as a mapping $\pi : X \to [0,1]$, by means of which the *possibility measure*, Π, and the *necessity measure*, N, can be defined [4] by

$$\Pi(A) = sup\{\ \pi(x)\ |\ x \in A\ \}, \qquad \forall A \in P(X) \tag{16}$$

$$N(A) = inf\{\ 1 - \pi(x)\ |\ x \in \overline{A}\ \}, \qquad \forall A \in P(X). \tag{17}$$

It is easy to check that for the nested focal elements (15) the measures, Π and N, satisfy the following conditions

$$\Pi(A_i \cup A_j) = max\{\ \Pi(A_i),\ \Pi(A_j)\}, \qquad i,j = 1,2,...,n \tag{18}$$

$$N(A_i \cap A_j) = min\{\ N(A_i), N(A_j)\}, \qquad i,j = 1,2,...,n. \tag{19}$$

as well as the modal duality equation (14),

$$\Pi(A_i) = 1 - N(\overline{A_i}), \qquad i = 1,2,...,n. \tag{20}$$

It is very important that the possibility distribution, π, may be regarded as the membership function, μ_F, of a fuzzy set in X. Consequently,

$$\pi(x) = \mu_F(x), \qquad \forall x \in X \tag{21}$$

and, since there exists (according to the basic presupposition) a specific element $x \in X$ for which $\mu_F(x) = 1$, the fuzzy set F is normalized.

Let F_i be α_i – cuts of the fuzzy set F, $i=1,2,...,n$. Then, according to (3) and (4), the sets F_i form a nested structure

$$F_1 \supset F_2 \supset ... \supset F_n \tag{22}$$

for the corresponding ordered α_i – cut numbers

$$\alpha_1 < \alpha_2 < ... < \alpha_n . \tag{23}$$

Hence, at suitable α's the structure (22) can represent the nested sequence (15) of focal elements of the possibility measure Π. Denoting $\pi_i = inf\{\pi(x)\ |\ x \in A_i\}$ for $i = 0,1,...,n$ we obtain $\pi_i = \alpha_i$, where $\pi_0 = \alpha_0 = 0$ and $A_0 = supp\ F$. Obviously, (23) implies

$$0 = \pi_0 < \pi_1 < \pi_2 < \dots \pi_n \le 1. \tag{24}$$

It can be shown [4,5] that the basic probability assignment m determining the possibility measure Π is completely defined in terms of possibility distribution,

because for every focal subset A_i the basic probability assignment, $m(A_i)$, can be computed as the difference

$$m(A_i) = \pi_i - \pi_{i-1}, \quad i = 1,2,\dots,n \tag{25}$$

of two neighbouring values of the possibility distribution π.

5 Fuzzy Paraset Representation of Solving Uncertain Problems

The process of problem-solving with uncertainty, roughly described in the previous section, is based on the idea of a gradual decreasing uncertainty by new information acquired, and/or by realizing some actions providing a new relevant evidence. As a rule, such a process, if completely realized in practice, requires a time period. Therefore, it seems to be reasonable to represent the process by (fuzzy) paraset methodology, which has been described in [7-9]. Bringing a time order into the process, parasets can be very useful tool at creating focal elements and reducing the uncertainty involved.

5.1 The Evidence Paraset

Let Y be a set of *elementary* (or *atomic*) propositions concerning both elements $x \in X$ and their properties expressing mutual relations. Let $Y^* = (Y, \wedge, \vee, \neg)$ be the Boolean lattice over Y of compound propositions p, q, ... generated from atomic propositions by logical connectives of disjunction ($\vee$), conjunction ($\wedge$) and negation ($\neg$), as well as by implication ($\rightarrow$) and equivalence ($\equiv$), derived from $\vee$, $\wedge$, $\neg$ in the common way. It is easy to see that the minimal element of the lattice Y^*, denoted 0, is any contradiction, while the maximal element, denoted 1, is every tautology, and that for any $p \in Y^*$ hold $p \wedge \neg p = 0$ and $p \vee \neg p = 1$. Further, for any $p, q \in Y^*$, p is said to be *supporting* q iff $p \rightarrow q$. The set of all p supporting q is denoted $S(q)$.

Analogically to the previous section, we can define in the Boolean lattice Y^* the Shafer's *basic probability assignment, m^**, with the property

$$m^*(0) = 0, \qquad \sum_{\forall p \in Y^*} m^*(p) = 1.$$

In the same way, we define a belief function Bel and a plausibility function Pl, both depending on m^*, by

$$Bel(q) = \sum_{p \in S(q)} m^*(p), \qquad Pl(q) = \sum_{p \in SC} m^*(p) \qquad (26)$$

for all $q \in Y^*$, where $SC = S(\neg q)^c - \{0\}$ and c is the operation of complementation [4].

The quantity $m^*(p)$ expresses one's personal belief that the proposition p concerning the specific x is true. Analogically, the functions $Bel(q)$ and $Pl(q)$, $q \in Y^*$, express degrees of belief and plausibility, respectively, that q is true because of the specific element. Of course, they also satisfy the condition (13),

$$0 \le Bel(p) \le Pl(p) \le 1 \qquad \forall p \in Y^*$$

as well as the modal duality equation (14), i.e.

$$Pl(p) = 1 - Bel(\neg p), \qquad \forall p \in Y^*. \qquad (27)$$

The utilities of uncertainty functions, Bel and Pl, for propositions in Y^* consist in their abilities to measure the truth of propositions related to the specific element. In this way, a relevant knowledge about the specific element can be obtained.

From basic properties of Bel and Pl follow some interesting results usable at forming focal subsets in X. For example, if propositions $p_i \in Y^*$ are such that p_{i+1} supports p_i for every $i=1,2,...,n-1$, i.e.

$$p_n \to p_{n-1} \to ... \to p_2 \to p_1 \qquad (28)$$

then for any two $p, q \in \{p_1,..., p_n\}$ the following assertions hold [4],

$$Bel(p \wedge q) = min \{Bel(p), Bel(q)\},$$

$$Pl(p \vee q) = max\{Pl(p), Pl(q)\},$$

$$min \{Bel(p), Bel(\neg p)\} = 0, \qquad (29)$$

$$max\{Pl(p), Pl(\neg p)\} = 1,$$

$$Bel(p) > 0 \to Pl(p) = 1,$$

etc.

Now, we are going to form in a constructive way a special paraset,

$$E = \langle Y^*, \{t\}, E \rangle, \qquad (30)$$

where the basic space, Y^*, is the Boolean lattice above of propositions on elements $x \in X$ including the specific one; the time set, $\{t\}$, is a sequence $t_1, t_2, ...$ of discrete real time points t ordered by the relation $\le$, covering the whole time period of the our problem being solved; the point-representations are generated by the mapping

$E: \{ t \} \rightarrow P(Y^*)$, which to a time point t_i assigns the non-empty set $E(t_i) = E_i$ in $P(Y^*)$ of the propositions, concerning the specific element, true at the time point t_i; and the history of E,

$$\{ E_i / E_i \in P(Y^*), \ E_i \preceq E_j, \ t_i, t_j \in \{ t \} \}, \tag{31}$$

is generated in a gradual way, immediately after creating the point-representation E_i, $i=1,2,...$

Since the sets E_i are for creating evidences of fundamental importance, we will call E_i shortly an *evidence* (at the time point t_i). Recall that acording to the common practice in logical systems, instead of a (finite) set of several true propositions the conjunction of them can be used. Consequently, this simplifying step can be used also here, by considering evidences E_i as singletons in $P(Y^*)$ formed by propositions p_i. That is the case in (28), so that we can simply write $E_i = p_i$, $i=1,2,...$

Taking into consideration all the properties or knowledge like (27) and (29), the point-representations E_i include all the relevant truth about the situation with the specific element in X at every present time t_i. This information should be used to derive the corresponding evidence needed for gaining the focal set A_i.

Since the goal of the paraset E is to produce evidences on the elements $x \in X$ related to the specific element, in order to obtain more and more knowledge for creating nested sequence of focal subsets A_i, $i=1,2,...$, we call the triple (30) an *evidence paraset*.

5.2 The Ucertainty Paraset

Analogically to evidence paraset E, we define the so-called *uncertainty paraset* U,

$$U = \langle X, \{ t \}, U \rangle \tag{32}$$

of which the main purpose is to decrease uncertainty involved by creating new focal subsets. Similarly to the evidence paraset, E, the uncertainty paraset U will be also generated in a constructive way. The basic space X of U is the Shafer's frame of discernment of the problem being solved; the time set, $\{ t \}$, is the same one like in E; the point-representations $U(t_i) = U_i$ are produced by the mapping $U: \{ t \} \rightarrow P(X)$, which to time points $t_i \in \{ t \}$ assigns non-empty subsets $U_i \subset X$; and the history of U,

$$\{ U_i / U_i \in P(X), \ U_i \preceq U_j, \ t_i, t_j \in \{ t \} \} \tag{33}$$

is generated analogically to (31) from the point-representations U_i, $i=1,2,...$

Assume that after the choice of the basic probability assignment, m, due to the first true evidence E_1 available as the first point-representation of the evidence paraset E, we are able to indicate the first focal subset $A_1 \subset X$ at the time t_1 as the point-representation U_1 of the uncertainty paraset U.

Note that, according to the definition of both possibility distribution, π, and the set *supp* by (2), as the first focal set, A_1, may be accepted

$$A_1 = supp\ Z = supp\ \pi,$$

where Z is union of all focal sets based on the actual m. Next steps are roughly described by the following procedure for both obtaining new evidences and related focal elements. New evidences are represented by propositions in Y^*, and focal elements by nested subsets in X.

The procedure

1. At the time $t = t_1$, according to initial information and evidence, select in Y^* the best proposition p_1 and propositions p_{1i} with the property $p_{1i} \rightarrow p_1$; construct the point-representations $E_1 = p_1$ and $U_1 = A_1$.
2. At the time $t = t_2$, from the propositions p_{1i} select the best one, $p2$, and find propositions p_{2j} supporting p_2, according to new information or action; construct the evidence $E_2 = p_2$ and the focal set $U_2 = A_2$, $A_2 \subset A_1$.
.... etc. ....
n. At the time $t = t_n$, from the propositions p_{nk} supporting p_n, select the best one, according to new information or action; construct the corresponding point-representations $E_n = p_n$ of the paraset E, and $U_n = A_n$ of the paraset U.
.... etc. ....

Going in this way on, we can constructively build up the whole sequence (15) of nested focal subsets $A_i = U_i$, $i=1,2,...,n$ for arbitrary n.

If the space X is finite, the parasets E and U are finite too, and obviously then there exists $n = N$ such that A_N is a singleton which is $A_N = \{x_0\}$, where x_0 is the specific element mentioned above as a solution of the initial problem.

If the space X is not finite, the problem solved requires both parasets, E and U, to be infinite. Then the time set $\{t\}$ is represented by the sequence

$$t_1 < t_2 < ... < t_n < ... \tag{34}$$

and the corresponding point representations are $E_1 = p_1, ..., E_n = p_n,...$ for the paraset E and $U_1 = A_1, ..., U_n = A_n, ...$ for the paraset U.

We can choose an infinite sequence of α_n– cut numbers related to the fuzzy set F mentioned above, to be aiming to 1. This implies that a sequence of the corresponding values of the possibility disribution, π, is aiming to 1, because F is normalized. Consequently, the sequence

$$F_1 \supset F_2 \supset ... \supset F_n \supset ... \tag{35}$$

of nested α_n – cut sets of the fuzzy set F aims to a singleton. Thus, the uncertainty paraset U is convergent.

On the other hand, if the sequence (35) is *consonant*, i.e. it is formed in agreement with the above procedure of gaining evidences and decreasing uncertainty, then all F_n are focal. Therefore, the limit of the sequence (35) is focal. Thus, the limit singleton is necessary $\{x_0\}$.

Perhaps the following consequence of the result above is important for the practical application. Given an arbitrarily small number $\varepsilon > 0$, there exist numbers $\delta > 0$ and $n = N$ such that for any focal set $A_k \cong F_k$, $k \geq N$, the expression

$$(1 - \pi_k) < \delta \rightarrow |x - x'| < \varepsilon \tag{36}$$

holds for all $x, x' \in A_k$. This means that the right-hand expression in (36) holds also for the specific element, $x' = x_0$. Thus, the inequality $|x - x_0| < \varepsilon$ holds for any $x \in A_k$, implying that the specific element x_0, as the solution of the initial problem, can be achieved by the procedure above with an *arbitrary precision*. Thus, using the procedure on, the greater is n, the better is aproximation of A_n by F_n.

6 Conclusion

To solve a problem with truth uncertainty, by the methodology roughly described above, assumes a mutual cooperation of steps accomplished at the same time in both of the evidence and the uncertainty parasets. These parasets are generated con- structively and gradually and provide a suitable methodological aid at solving the problem. To increase possibility of using knowledge from various information resources in the process of acquiring true evidences, the evidence paraset can be composed of several component evidence parasets of different types, like in (1). The effect of such a combination of various (fuzzy) parasets as information resources will be more interesting, provided that the component parasets have more complex basic spaces.

References

1. Shafer, G.: A Mathematical Theory of Evidence. Univ. Press, Princeton (1976)
2. Dempster, A. P.: Upper and Lower Probabilities Induced a Multivalued Mapping. Annals of Math. Stat. 38 (1967) 325-339
3. Zadeh, L.A.: Fuzzy Sets as a Basis for a Theory of Possibility. Fuzzy Sets and Systems, 1 (1978) 3-28
4. Dubois, D., Prade, H.: Representation and Combination of Uncertainty with Belief Functions and Possibility Measures. Univ. P. Sabatier, 31062 Toulouse Cédex, France, (1978) 244-264
5. Klir, G.J.: Uncertainty in the Dempster-Shafer Theory: A Critical Re-Examination. Int. J. General Systems 18 (1990) 155-166
6. Yager, R.R.: On the Normalization of Fuzzy Belief Structures. Int. J. of Approximate Reasoning 14 (1996) 127-153
7. Šajda, J.: Introduction to the Intuitive Theory of General Abstract Parasets. Proc. of the 7th World Congress IFSA'97, Prague (1997) 67-72
8. Šajda, J.: Elliptic Parasets. Proc. of the 7th Int. Conf. on AI and Inf.-Control Systems of Robots'97, Smolenice Castle-Bratislava (1997) 333-346
9. Šajda, J.: A Fuzzification of Paraset Methodology. Proc. of the 6th Fuzzy Colloquium, Zittau, Germany (1998) 44-49
10. De Beats, B., Cooman, G., Kerre, E.: The Construction of Possibility Measures from Samples on T-semi-partitions. J. of Information Sciences 106 (1998) 3-22

11. Harmanec, D., Klir, G.J., Wang, Z.: Modal Logic Interpretation of Dempster-Shafer Theory: An Infinite Case. J. of Approximate Reasoning 14 (1996) 81-93
12. Churn-Jung, L.: Possibilistic Residuated Implication Logics with Applications. Int. J. of Uncertainty, Fuzziness and Knowledge-Based Systems 6 (1998) 365-385
13. Goodman, I.R., Nguyen, H.T.: Uncertainty Models for Knowledge-Based Systems. North-Holland, Amsterdam (1985)

On the Change of the Distribution Shape of Randomized Fuzzy Variables by Filtering over Compatibility Degrees

Jiri Georg Sustal

BTU Cottbus - Brandenburg Technical University of Cottbus,
Postfach 10 13 44, D-03013 Cottbus, Germany
sustal@math.tu-cottbus.de

Abstract. A fuzzy output of a fuzzy controller is usually determined by a *deterministic* procedure e.g. by the center-of-gravity method. A fuzzy controller is a special case of a fuzzy-rule based system. If a fuzzy-rule based system should be used to *simulate* a real system, some *indeterminacy and uncertainty* about the system output may be desirable. The output can be generated as follows: first a λ-level set is randomly generated according to some rules, then an element of the generated λ-level set is chosen according to the uniform distribution over the chosen λ-level set. The result of this procedure is a probability distribution over the output set. We can influence this process in various ways e.g. we can "filter out" some undesirable compatibility degrees λ. We study the impact of such a "filtering" on the shape of the resulting probability density function over the output space.

Keywords: probabilistic filtering, possibilistic filtering, λ-level sets, fuzzy sets, modelling of fuzzy-rule based systems, identification of fuzzy sets

1 Introduction

A fuzzy output of a fuzzy controller is usually determined by a *deterministic* procedure e.g. by the center-of-gravity method. A fuzzy controller is a special case of a fuzzy-rule based system. If a fuzzy-rule based system should be used to *simulate* a real system, some *indeterminacy and uncertainty* about the system output may be desirable because this feature is typical for real systems which are modeled by fuzzy models. The output can be generated as follows: first a λ-level set is randomly generated according to some rules, then an element of the generated λ-level set is chosen according to the uniform distribution over the chosen λ-level set. The result of this procedure is a probability distribution over the output set. We can influence this process in various ways e.g. we can "filter out" some undesirable compatibility degrees λ. We study the impact of such a "filtering" on the shape of the resulting probability density function over the output space. Why should we study such a filtering at all? Because there is no clear way how to determine the output if one wants to simulate it and a possibility distribution of the output is given. It reveals that some possibility distributions are ruther insensitive to various filterings, on the other hand some are very sensitive.

To simplify the notation we do not exactly distinguish between a fuzzy set, its membership function and the associated possibility distribution e.g. we speak about λ-level sets of a possibility distribution. To better compare various shapes of functions, the domain X of a normalized possibility distribution $\pi, \pi : X \to L, L = [0,1]$, will be supposed to be the interval $[0,1] \in \mathbb{R}$. The discrete case can be treated separately. The following symbols are used: f, f_X, f_L etc. for probability density functions, π, π_X, π_L etc. for possibility distributions. Because of the special importance of these functions over L, we shall alternatively use g for f_L (i.e., for a density function over L) and ρ for π_L (i.e., for a possibility distribution over L). A λ-level set of π, $[\pi]_\lambda$, is $[\pi]_\lambda = \{x \in X : \pi(x) \geq \lambda\}$.

2 Filtering over Compatibility Degrees

A *probabilistic filtering* of a possibility distribution π over X is given by a probability distribution over the set L, i.e., over the degrees of compatibility, in our continuous case we prefer to use a density function g. A *possibilistic filtering* of a possibility distribution π over X is given by a possibility distribution ρ over L.

The procedure of the probabilistic filtering can be described as follows [2].

Step1. Generate a value λ over $[0,1]$ according to the probability distribution with a density g (in the case of no additional information it will be the uniform distribution over L).

Step 2. Generate a value $x, x \in X$, by the uniform distribution over the level set $[\pi]_\lambda$ where λ has already chosen by the step 1.

The generation of λ-level sets have been studied in detail in connection with random sets cf. [8],[3]. Similar steps 1 and 2 have been also used by [1],[7] in the mass assignment theory The steps 1 and 2 are also implicitly contained in the formula for the U-uncertainty cf. [4] which is given for normalized fuzzy sets as $U(\mu) = \int_0^1 \log_2 |[\mu]_\lambda| d\lambda$. This formula can be derived from the entropy of the joint density $f_{X,L}$, $f_{X,L}(x,\lambda) = |[\pi]_\lambda|^{-1}$ over all (x,λ). The joint density $f_{X,L}(x,\lambda) = |[\pi]_\lambda|^{-1}$ is a special case of the joint density (2) for $g(\lambda) = 1$, hence the steps 1 and 2 are implicitly involved.

In the case of the possibilistic filtering we have two possibility distributions: $\pi, \pi : X \to L_1, L_1 = [0,1]$ and $\rho, \rho : L_1 \to L_2, L_2 = [0,1]$. L_1 and L_2 mean the same set of compatibility degrees. The possibilistic filtering consists of two stages. In the first stage a $\lambda_1 \in L_1$ is chosen according to the steps 1 and 2 for the distribution ρ. In the second stage a $x \in X$ is chosen for a given λ_1 according to the step 2 and the level set $[\pi]_{\lambda_1}$.

If some preconditions are fulfilled, each possibility distribution can be converted by the above steps to a probability distribution, see the next section. Hence also the possibility distribution ρ over L can be (in theory) converted to a probability distribution g over L ,i.e., the possibilistic filtering can be converted to the probabilistic filtering.

The probability distribution with the density g or equivalently a possibility distribution ρ over L can be thought of as a kind of filter which "filters out" not desirable compatibility degrees.

In the case of the possibilistic filtering there is a strong analogy to the truth-qualification, which is also given by two fuzzy sets, the first μ over X, the second τ over L where L is the set of truth values. It generates a new fuzzy set $\tau \circ \mu$ over X by concatenating μ and τ cf.[4].

We prefer not to use the terms *truth-qualified* and *probability-qualified* fuzzy propositions because we are using these qualifications in a different way as it is common in the literature.

There is of course a question how to choose the right g or ρ over L. This has been discussed e.g. in [2], [3], [7]. Here are some comments. We understand the density g mainly as the designer's tool. I.e., the system designer should first understand how the filtering g or ρ over L effects the shape of π_X and only after that he chooses the right form. The interplay between π_X and g or ρ will be studied further bellow. Secondly, in the sequel we consider only simple forms of π_X - triangular and trapezoidal forms, which simplifies all considerations significantly. Nevertheless also these simple forms give us some intuitive hints regarding more complicated shapes of π_X.

3 Generated Probability Distribution

The above steps generate under some preconditions a probability distribution f_X over X. A formula for f_X can be given as follows. Let us denote the length of a λ-level set as $|[\pi]_\lambda|$. Then $|[\pi]_\lambda|^{-1}$ for $x \in [\pi]_\lambda$ and 0 otherwise is a density function over X for a given λ. It is natural to define $f_{X,L}$, an assumed joint density over all (x, λ), as

$$f_{X,L}(x, \lambda) = |[\pi]_\lambda|^{-1} g(\lambda) \ . \tag{1}$$

We suppose that $f_{X,L}$ is integrable and that $\int_D f_{X,L}(x, \lambda) dx d\lambda = 1$ where D is the domain of all (x, λ), $D = \{(x, \lambda) : x \in X, \lambda \in [0, 1], \lambda \leq \pi_X(x)\}$ Integrating over L we get the marginal density over X

$$f_X(x) = \int_0^{\pi(x)} |[\pi]_\lambda|^{-1} g(\lambda) d\lambda \text{ for } x \in X \ . \tag{2}$$

We see that the above steps 1 and 2 generate a probability distribution over X. Hence we could also generate x directly by this distribution without using the above steps. However the calculation of f_X need not be simple if it is feasible at all. On the other hand, the above steps can be easily implemented provided the calculation of the λ-level sets is not too difficult.

4 Special Cases

In the sequel various shapes of π and g will be discussed.

Case 1. π_X triangular in the middle of $[0,1]$, g uniform over L.

$$\pi_X(x) = \begin{cases} 2x & if\ 0 \le x \le 0.5 \\ -2x + 2\ if\ 0.5 < x \le 1 \end{cases}$$

For the length of a λ-level set $|[\pi]_\lambda|$ we have $|[\pi]_\lambda| = 1 - \lambda$. The density f_X can be calculated as

$$f_X(x) = \begin{cases} -\ln(1 - 2x)\ if\ 0 \le x \le 0.5 \\ -\ln(2x - 1)\ if\ 0.5 < x \le 1 \end{cases}$$

The results of the filtering are depicted in Fig. 1 and 2.

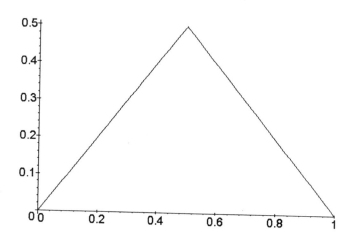

Fig. 1. triangular π_X

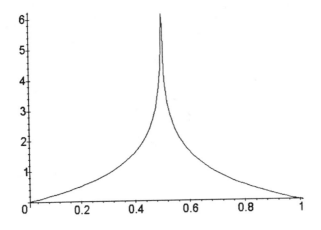

Fig. 2. Density f_X

Case 2. π_X as in the case 1 , g triangular at the upper end of $[0,1]$, i.e., the lower compatibility degrees are filtered out.

$$g(\lambda) = \begin{cases} 0 & if\ 0 \leq \lambda \leq 0.5 \\ 8\lambda - 4\ if\ 0.5 < \lambda \leq 1 \end{cases}$$

The density f_X can be calculated as

$$f_X(x) = \begin{cases} 0 & if\ 0 \leq x \leq 0.25 \\ -16x - 4\ln(1 - 2x) + 1.227 & if\ 0.25 < x \leq 0.5 \\ 16x - 4\ln(2x - 1) - 14.772 & if\ 0.5 < x \leq 0.75 \\ 0 & if\ 0.75 < x \leq 1 \end{cases}$$

The results of the filtering are depicted in Fig. 3 and 4.

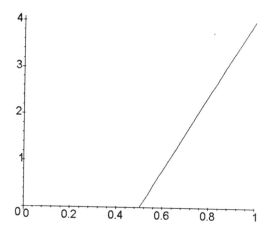

Fig. 3. high λ levels

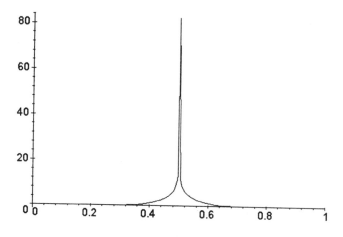

Fig. 4. Density f_X

Case 3. π_X as in the case 1 , g triangular at the lower end of $[0, 1]$, i.e., the upper compatibility degrees are filtered out.

$$g(\lambda) = \begin{cases} -8\lambda + 4 \; if \; 0 \le \lambda \le 0.5 \\ 0 \quad if \; 0.5 < \lambda \le 1 \end{cases}$$

The density f_X can be calculated as

$$f_X(x) = \begin{cases} 16x + 4\ln(1 - 2x) & if \quad 0 \le x \le 0.25 \\ 1.22741 & if \ 0.25 < x \le 0.75 \\ 16 - 16x + 4\ln(2x - 1) & if \quad 0.75 < x \le 1 \end{cases}$$

The results of the filtering are depicted in Fig. 5 and 6.

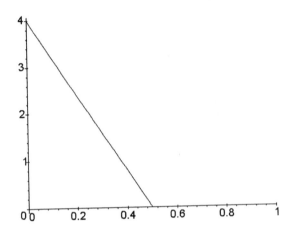

Fig. 5. low λ levels

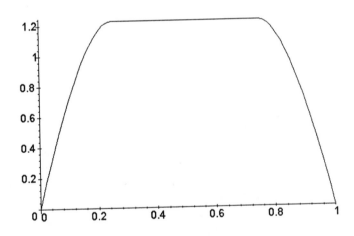

Fig. 6. Density f_X

Case 4. π_X trapezoidal on $[0,1]$, g uniform over $[0,1]$.

$$\pi_4(x) = \begin{cases} 4x & if \quad 0 \leq x \leq 0.25 \\ 1 & if \ 0.25 < x \leq 0.75 \\ -4x + 4 & if \quad 0.75 < x \leq 1 \end{cases}$$

The density f_X can be calculated as

$$f_X(x) = \begin{cases} -2\ln(1 - 2x) & if \quad 0 \leq x \leq 0.25 \\ 2\ln 2 & if \ 0.25 < x \leq 0.75 \\ -2\ln(2x - 1) & if \quad 0.75 < x \leq 1 \end{cases}$$

The results of the filtering are depicted in Fig. 7 and 8.

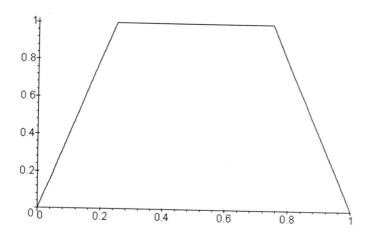

Fig. 7. trapezoidal π_X

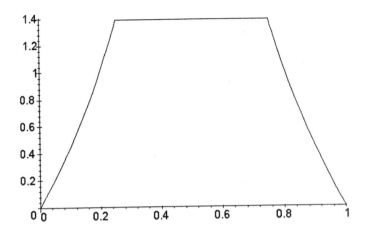

Fig. 8. Density f_X

Case 5. (Possibilistic filtering by ρ over L.) π_X triangular, ρ triangular, both defined as in the case 1.

In the case 1 we have seen that a triangular possibility distribution without additional information leads to a probability density f_1 of the form as in the case 1 (only the domain is now L and not X). Hence we have actually the case of the probabilistic filtering with the density $g(\lambda) = f_1(\lambda)$ and we can generate an $x \in X$ according the steps 1 and 2.

For $0 \leq x \leq 0.25$ we get the following density f_X^1.

$$f_X^1(x) = \int_0^{2x} \frac{1}{1-y}(-\ln(1-2y))dy$$
$$= \mathrm{dilog}(2-4x) + \ln(1-4x)\ln 2 + \ln(1-4x)\ln(1-2x) + \frac{1}{12}\pi^2$$

where

$$\mathrm{dilog}\,(x) = \int_1^x \frac{\ln t}{1-t}dt.$$

For $0.25 \leq x \leq 0.5$ we get the following density f_X^2.

$$f_X^2(x) = \int_0^{0.5} \frac{1}{1-y}(-\ln(1-2y))dy + \int_{0.5}^{2x} \frac{1}{1-y}(-\ln(2y-1))dy$$
$$= \frac{1}{4}\pi^2 - \mathrm{dilog}(4x-1).$$

On the whole interval $0 \leq x \leq 1$ we get the density f_X

$$f_X(x) = \begin{cases} f_X^1(x) & if \ 0 \le x < 0,25 \\ f_X^2(x) & if \ 0,25 \le x < 0.5 \\ f_X(1-x) \ if & 0.5 \le x \le 1 \end{cases}$$

The results are depicted in Fig.9 and 10.

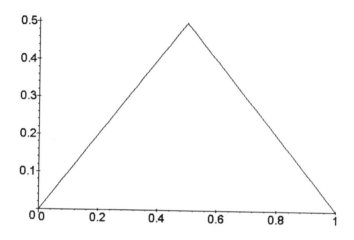

Fig. 9. triangular π_X

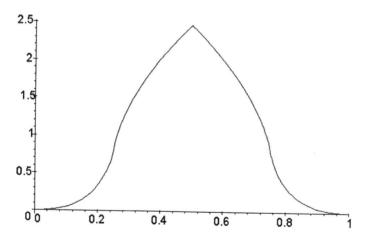

Fig. 10. Density f_X

5 Identification of Fuzzy Sets

We can also reverse the situation and ask: given a probability density f_X what are the corresponding possibility distribution π and the filtering g or ρ? The answer is not generally unique as can be seen for the pairs

$$(\pi_1 = 1, \quad g(\lambda) = 2\lambda) \text{ and } (\pi_1 = 1, \quad g(\lambda) = 2(1 - \lambda)).$$

Both pairs gives the same uniform distribution f_X over X. If we put more restrictions on possible candidates for π and g, the answer may turn to be unique. E.g., if π can be only triangular as in the cases 1 to 3 and if g can have only one of the forms 1 to 3 then the answer is unique because the density functions f_X differ in their shape as has been shown above.

Let us consider the following situation. A random sample $S = \{x_1, \ldots, x_n\}$, $x_i \in X, X$ an interval in $\mathbb{R}$ is given. On the basis of this evidence the designer wishes to determine a suitable triangular π if the probabilistic filtering g like in the case 3 is supposed. According to Fig. 5 and 6 only the center c of the triangle base and the length $2l$ of the triangle base should be estimated. This can be done by putting $c = \bar{x}$ and $\frac{l^2}{6} = S^2$ where $\bar{x}$ is the sample mean and S^2 is the sample variance. Other filtering forms of a triangular shape of π_X can be treated similarly.

6 Multidimensional Case

The case $X = X_1 \times \ldots \times X_n$ can be treated similarly. It is a common practice to use in X prototypes of the form $A = A'_1 \wedge \ldots \wedge A'_n$ where A'_i are cylindrical extions of one-dimensional prototypes having e.g. triangular ore trapezoidal forms and $\wedge$ stands for a triangular norm. Then $A_\lambda = A_{1_\lambda} \times \ldots \times A_{n_\lambda}$ and a uniform generation of an element $x = (x_1, \ldots, x_n)$ of A_λ means an independent uniform generation of elements x_i of $A_i, i = 1, \ldots, n$. The joint density f over X is given as $\prod_i f_i$ where f_i is the density generated in space f_i. Conversely if we want to estimate A from statistical data, we can estimate it as $\hat{A} = \hat{A}_1 \times \ldots \times \hat{A}_n$ where $\hat{A}_i$ is an estimate in the space A_i.

7 Summary

We have studied the effects of probabilistic and possibilistic filtering on a generated probability distribution and its shape. We see that the effect of the filtering on the triangular shape of π is rather strong - cases 1 to 3. Here, a probabilistic filtering can be used as the designer's tool to influence the shape of the generated probabilistic distribution. On the other hand the trapezoidal shape does not change its form too much (case 4). The possibility qualification has been studied only shortly in the case 5.

References

1. Baldwin, J.F.: Mass assignments and fuzzy sets for fuzzy databases. In: Fedrizzi M., Kacprzyk J., Yager R. R. (eds.): Advances in the Shaffer Dampster Theory of Evidence, John Wiley, (1992)
2. Chanas, S., Nowakowski M.: From fuzzy data to a single action - a simulation approach. In: Kacprzyk J., Fedrizzi M. (eds.): Combining fuzzy imprecision with probabilistic uncertainty in decision making. Springer Verlag (1988) 331–341
3. Goodman, I.R.: Fuzzy sets as equivalence class of random sets. In: Yager R.T.(ed.): Recent Developments in Fuzzy Set and Possibility Theory. Pergamon Press, (1981)
4. Klir, G.J., Yuan, B.: Fuzzy sets and fuzzy logic. Prentice Hall, (1995)
5. Li, Q.D.: The random set and the cutting of random fuzzy sets. Fuzzy Sets and Systems 86 (1997) 223-234
6. Sustal, J.G.: On fuzzy confidence intervals and modelling of random variables. In: Zimmermann, H.J.(ed.): Proceedings of EUFIT 97, Vol. 1 (1997) 350–352
7. Weber, K.: Fuzzy rules generation and fuzzy inference based on frquency distributions. In: Brewka, G. et. al. (eds.): Fuzzy-Neuro Systems 99, Leipziger Universitätsverlag, (1999)
8. Wang, P.Z.: Random sets and fuzzy sets. In: International Encyclopedia on Systems and Control, Pergamon Press, (1983)

On Optimization with Fuzzy Constraint Conditions

Jun Liu[1], Da Ruan[1], Zhenming Song[2], and Yang Xu[2]

[1] Belgian Nuclear Research Centre (SCK•CEN), Boeretang 200,
B-2400 Mol, Belgium
{jliu, druan}@sckcen.be
[2] Department of Applied Mathematics, Southwest Jiaotong University
Chengdu 610031, Sichuan, P. R. China
yxu@home.swjtu.edu.cn

Abstract. In general, it is a difficult problem to solve an optimal criterion set with fuzzy constraint conditions. In this paper, the difficult problem is changed into a simple problem to solve a conditional extreme value problem in mathematics. Thus it becomes rather simpler to solve the problems for fuzzy linear programming and fuzzy nonlinear programming. The paper provides with detailed case studies on optimization with single fuzzy constraint condition as well as on optimization with multi-fuzzy constraint conditions and multi-objective functions. Finally, this paper presents related theories on determination of the optimal criterion set.

1 On Optimization with Single Fuzzy Constraint Condition

Case 1: The universe of discourse of fuzzy constraints is the same as the domain of definition of objective function

Let X be the universe of discourse, $\underset{\sim}{C} \in F(X)$ be a fuzzy constraint condition, $f : X \to R$ be a real-value objective function. $\forall \lambda \in (0, 1]$, Set

$$M_\lambda = \left\{ x' \mid f(x') = \vee_{x \in C_\lambda} f(x),\ x' \in C_\lambda \right\} \tag{1}$$

$$M = \bigcup_{\lambda \in (0,1]} M_\lambda \tag{2}$$

$$\overline{M} = \bigcup_{\lambda \in [0,1]} M_\lambda \tag{3}$$

Definition 1.1

$$\underset{\sim}{C}_f = \bigcup_{\lambda \in [0,1]} \lambda M_\lambda \tag{4}$$

is called a fuzzy superior set of f under $\underset{\sim}{C}$. x^* is called an optimal point under $\underset{\sim}{C}$ if it satisfies the following equation

$$\underset{\sim}{C}_f (x^*) = \vee_{x \in X} \underset{\sim}{C}_f (x) \tag{5}$$

$$D = \left\{ x^* \mid C_{\underset{\sim}{f}}(x^*) = \vee_{x \in X} C_{\underset{\sim}{f}}(x), x^* \in X \right\} \tag{6}$$

is called the optimal criterion set under fuzzy constraint condition $C_{\sim}$.

In fact, the determination of D is just the decision problem under fuzzy constraint conditions

Case 2: The universe of discourse of fuzzy constraints differs from the domain of definition of objective function

Let fuzzy constraint conditions $C_{\sim} \in F(X)$, $G: Y \to R$ be a real-value function, $X \neq Y$.

If there exist a function $F : X \to Y$, set $f = G \circ F$, then $f : X \to R$, where "$\circ$" is a compound operation of function. So this case can be translated into Case 1.

2 On Optimization with Multi-Fuzzy Constraint Conditions and Multi-Objective Functions

Let X be the universe of discourse, $C_{\underset{\sim}{i}} \in F(X)$ be the fuzzy constraint conditions, $f_i : X \to R$ be the real-value objective functions, where $i=1, 2, \ldots, m; j=1, 2, \ldots, n$. Set

$$F_1 : \underbrace{F(X) \times F(X) \times \ldots \times F(X)}_{m} \to F(X)$$

We write $f_j : X \to R$ as $f_j \in (X \to R)$. And set

$$F_2 : \underbrace{(X \to R) \times (X \to R) \times \cdots \times (X \to R)}_{n} \to (X \to R)$$

Then we have $C_{\sim} = F_1(C_1, C_2, \cdots, C_m) \in F(X)$, $f = F_2(f_1, f_2, \cdots, f_n) \in (X \to R)$.

So the problem is translated into the optimal problem with single fuzzy constraint condition. Hence, we only focus on the optimal problem with single fuzzy constraint condition.

Theorem 1.1[1, 2]

$$C_{\underset{\sim}{f}} = C \cap M = C \cap \overline{M} \tag{7}$$

We can notice from Theorem 1.1 that D can be obtained upon M or $\overline{M}$ which has been solved. However, it also can be seen from the equations (2) and (3) that M or $\overline{M}$ generally is difficult to be solved (Certainly, it is easy to solve $\overline{M}$ for some special cases, e.g., $C_{\sim}$ and f are MESA-type functions [2]). So, can D be directly determined without determining M and $\overline{M}$? The answer is affirmative. It is also the purpose of this paper.

3 Determination of the Optimal Criterion Set

In the following, we always suppose $C \neq \emptyset$, $\vee_{x \in X} C(x) = \alpha(>0)$, $M_\alpha \neq \emptyset$. Then

Lemma 2.1[2] Let $\lambda_1, \lambda_2 \in [0, 1]$, $\lambda_1 < \lambda_2$. If $C_{\lambda_1} \cap M_{\lambda_2} \neq \emptyset$, then

$$M_{\lambda_2} = C_{\lambda_2} \cap M_{\lambda_1} \tag{8}$$

Definition 2.1 Set

$$\vee_{x \in X} f(x) = T, \quad \wedge_{x \in X} f(x) = t \tag{9}$$

(1) If $t \neq -\infty$, and f is not a constant function, then set

$$\underset{\sim}{f}(x) = \frac{f(x) - t}{T - t} \tag{10}$$

(2) If f is a constant function, then set

$$\underset{\sim}{f}(x) = \begin{cases} 0, & f(x) \equiv 0 \\ 1, & f(x) \equiv c(\neq 0) \end{cases} \tag{11}$$

(3) If $t = -\infty$, then set

$$\underset{\sim}{f}(x) = \begin{cases} \dfrac{f(x) - m}{T - m}, & f(x) \geq m \\ 0, & f(x) < m \end{cases} \tag{12}$$

where $-\infty < m < \vee_{x \in C_\alpha} f(x)$.

Obviously, $\underset{\sim}{f} \in F(X)$.

Set

$$D' = \left\{ x^* \mid C_{\underset{\sim}{f}}(x^*) = \vee_{x \in X} C_{\underset{\sim}{f}}(x), \ x^* \in X \right\} \tag{13}$$

$$M_\alpha' = \left\{ x' \mid \underset{\sim}{f}(x') = \vee_{x \in C_\alpha} \underset{\sim}{f}(x), \ x' \in C_\alpha \right\} \tag{14}$$

$$D^* = \left\{ x^* \mid \underset{\sim}{f}(x^*) = \text{hgt}(\underset{\sim}{f} \cap C_\alpha), \ x^* \in X \right\} \tag{15}$$

Theorem 2.1

$$D = M_\alpha = M_\alpha' = D^* = D' \tag{16}$$

Proof. It only needs to prove:

(1) $x^* \in D \Leftrightarrow$ (2) $x^* \in M_\alpha \Leftrightarrow$ (3) $x^* \in M_\alpha' \Leftrightarrow$ (4) $x^* \in D^* \Leftrightarrow$ (5) $x^* \in D'$.

(1)$\Rightarrow$(2) Assume that $x^* \notin M_\alpha$.

If $x^* \in \overline{M}$ and $x^* \in C_\alpha$, then $\exists \beta \in [0, \alpha)$, $x^* \in M_\beta$. Hence, $x^* \in C_\alpha \cap M_\beta$. It follows from Lemma 2.1 that $x^* \in M_\alpha$, which is a contradiction.

If $x^* \notin \overline{M}$, then

$$\vee_{x \in X} [C \wedge \overline{M}(x)] = \mathop{C}_{\sim f}(x^*) = \mathop{C}_{\sim}(x^*) \wedge \overline{M}(x^*) = 0.$$

It follows from $M_\alpha \neq \varnothing$ that $\exists y \in C_\alpha, y \in M_\alpha \subset \overline{M}$. Hence,

$$\alpha = \mathop{C}_{\sim}(y) \wedge \overline{M}(y) \leq \vee_{x \in X} [\mathop{C}_{\sim}(x) \wedge \overline{M}(x)] = 0$$

which implies that $\alpha = 0$, it is a contradiction. Consequently, $x^* \in \overline{M}$.

(c) If $x^* \notin C_\alpha$, then

$$\vee_{x \in X} [\mathop{C}_{\sim}(x) \wedge \overline{M}(x)] = \mathop{C}_{\sim f}(x^*) = \mathop{C}_{\sim}(x^*) \wedge \overline{M}(x) = \mathop{C}_{\sim}(x^*) < \alpha.$$

Analogous to (b), $\exists y \in C_\alpha$ such that

$$\alpha = \mathop{C}_{\sim}(y) \wedge \overline{M}(y) \leq \vee_{x \in X} [\mathop{C}_{\sim}(x) \wedge \overline{M}(x)] < \alpha,$$

which is a contradiction.

Consequently, from (a), (b) and (c), $x^* \in M_\alpha$.

(2) $\Rightarrow$ (1) $x^* \in M_\alpha \subset \overline{M}$ implies that $\forall x \in X$,

$$\vee_{y \in X} [\mathop{C}_{\sim}(y) \wedge \overline{M}(y)] = \vee_{y \in X} \mathop{C}_{\sim f}(y)$$

$$\geq \mathop{C}_{\sim f}(x^*) = \mathop{C}_{\sim}(x^*) \wedge \overline{M}(x^*) = \mathop{C}_{\sim}(x^*) = \alpha$$

$$\geq \mathop{C}_{\sim}(x) \geq \mathop{C}_{\sim}(x) \wedge \overline{M}(x)$$

Moreover, since x is arbitrary, we have

$$\vee_{y \in X} [\mathop{C}_{\sim}(y) \wedge \overline{M}(y)] \geq \mathop{C}_{\sim f}(x^*) \geq \vee_{x \in X} [\mathop{C}_{\sim}(x) \wedge \overline{M}(x)]$$

Hence, $\mathop{C}_{\sim f}(x^*) = \vee_{x \in X} [\mathop{C}_{\sim}(x) \wedge \overline{M}(x)] = \vee_{x \in X} \mathop{C}_{\sim f}(x)$, i.e., $x^* \in D$.

(2) $\Leftrightarrow$ (3), that is,

$$\mathop{f}_{\sim}(x^*) = \vee_{x \in C_\alpha} \mathop{f}_{\sim}(x) \Leftrightarrow f(x^*) = \vee_{x \in C_\alpha} f(x).$$

If we take $\mathop{f}_{\sim}$ as the equation (10), then

$$(\Rightarrow) \qquad \mathop{f}_{\sim}(x^*) = \frac{f(x^*) - t}{T - t} = \frac{\vee_{x \in C_\alpha} f(x) - t}{T - t}$$

$$= \vee_{x \in C_\alpha} \frac{f(x) - t}{T - t} = \vee_{x \in C_\alpha} \mathop{f}_{\sim}(x)$$

$(\Leftarrow)$ Since $f(x) = \mathop{f}_{\sim}(x)(T - t) + t$, so

$$f(x^*) = \mathop{f}_{\sim}(x^*)(T - t) + t = \vee_{x \in C_\alpha} \mathop{f}_{\sim}(x)(T - t) + t$$

$$= \vee_{x \in C_\alpha} [f(x)(T-t)+t]$$

$$= \vee_{x \in C_\alpha} f(x)$$

If we take $\underset{\sim}{f}$ as the equation (11), then

(a) If $f(x) \equiv 0$, then we have

$(\Rightarrow)$ $\underset{\sim}{f}(x^*) = 0 = \vee_{x \in C_\alpha} \underset{\sim}{f}(x)$; $(\Leftarrow)$ $\underset{\sim}{f}(x^*) = 0 = \vee_{x \in C_\alpha} f(x)$

(b) If $f(x) \equiv c(\neq 0)$ then we have

$(\Rightarrow)$ $\underset{\sim}{f}(x^*) = 1 = \vee_{x \in C_\alpha} \underset{\sim}{f}(x)$; $(\Leftarrow)$ $\underset{\sim}{f}(x^*) = c = \vee_{x \in C_\alpha} f(x)$

If we take $\underset{\sim}{f}$ as the equation (12), then

$(\Rightarrow)$ (c_{11}) If $f(x^*) < m$, then $\forall x \in C_\alpha$, $f(x) < m$. So $\underset{\sim}{f}(x^*) = 0 = \vee_{x \in C_\alpha} \underset{\sim}{f}(x)$

(c_{12}) If $f(x^*) \geq m$, i.e., $\vee_{x \in C_\alpha} f(x) \geq m$. If $\vee_{x \in C_\alpha} f(x) > m$, we set

$$S = \{x \in C_\alpha \mid f(x) \geq m\},$$

then $x^* \in S$ and

$$\vee_{x \in C_\alpha} f(x) = \vee_{x \in S} f(x)$$

$$\underset{\sim}{f}(x^*) = \frac{f(x^*) - m}{T-m} = \frac{\vee_{x \in C_\alpha} f(x) - m}{T-m}$$

$$= \frac{\vee_{x \in S} f(x) - m}{T-m} = \vee_{x \in S} \frac{f(x) - m}{T-m} \quad (\geq 0)$$

$$= (\vee_{x \in S} \frac{f(x) - m}{T-m}) \vee (\vee_{x \in C_d - S} 0)$$

$$= \vee_{x \in C_\alpha} \underset{\sim}{f}(x).$$

If $\vee_{x \in C_\alpha} f(x) = m$, set $A = \{x \in C_\alpha \mid f(x) = m\}$. If $A \neq \varnothing$, then $x^* \in A$ and

$$\vee_{x \in C_\alpha} f(x) = \vee_{x \in A} f(x)$$

$$\underset{\sim}{f}(x^*) = \frac{f(x^*) - m}{T-m} = \vee_{x \in A} \frac{f(x) - m}{T-m} = 0$$

$$= (\vee_{x \in A} \frac{f(x) - m}{T-m}) \vee (\vee_{x \in C_\alpha - A} 0)$$

$$= \vee_{x \in C_\alpha} \underset{\sim}{f}(x)$$

If $A = \varnothing$, then $\forall x \in C_\alpha$, $f(x) < m$. So $\forall x \in C_\alpha$, $\underset{\sim}{f}(x) = 0$. Hence

$$f(x^*) = 0 = \vee_{x \in C_\alpha} f(x).$$

Consequently, from (c_{11}) and (c_{12}) as above, we have $f(x^*) = \vee_{x \in C_\alpha} f(x)$.

$(\Leftarrow)$ (c_{21}) We first prove that $f(x^*) \neq 0$. If $f(x^*) = \vee_{x \in C_\alpha} f(x) = 0$, then $\forall x \in C_\alpha$, $f(x) = 0$. Hence, $\forall x \in C_\alpha$, $f(x) \leq m$. It follows that $\vee_{x \in C_\alpha} f(x) \leq m$, which is contradict to the selection of m. Consequently, $f(x^*) \neq 0$.

(c_{22}) If $f(x^*) = \vee_{x \in C_\alpha} f(x) \neq 0$, set $B = \{x \in C_\alpha \mid f(x) > 0\}$, then

$x^* \in B$ and $f(x^*) = \dfrac{f(x^*) - m}{T - m}$. It follows that $f(x^*) = f(x^*)(T - m) + m$.

Since $\forall x \in B$, $f(x) = \dfrac{f(x) - m}{T - m} > 0$ implies $f(x) \leq m$, so

$$f(x^*) = f(x^*)(T - m) + m$$
$$= \vee_{x \in B} [f(x)(T - m) + m] = \vee_{x \in B} f(x)$$
$$= (\vee_{x \in B} f(x)) \vee (\vee_{x \in C_\alpha - B} f(x))$$
$$= \vee_{x \in C_\alpha} f(x)$$

$(3) \Leftrightarrow (4)$ Note that $\forall x \in X$, $f(x) \in [0, 1]$. Hence

$x^* \in M'_\alpha$ if and only if

$$f(x^*) = \vee_{x \in C_\alpha} f(x) = \vee_{x \in X} [f(x) \wedge C_\alpha(x)] = \mathrm{hgt}(f \cap C_\alpha).$$

$(3) \Leftrightarrow (5)$ It only needs to replace the f in the proof of $(1) \Leftrightarrow (2)$ by f.

Note that C_α is a closed set which implies that $M_\alpha \neq \varnothing$, hence,

Corollary 2.1 If C_α is a closed set, then Theorem 2.1 holds.

Corollary 2.2 If C_α is a closed set and f is strictly monotone on C_α, then x is unique.

If C is a classic set, note that $\forall \alpha \in (0, 1]$, $C_\alpha = C_1 = C$, then the equation (15) is

$$D^* = \{x^* \mid f(x^*) = \mathrm{hgt}(f \cap C)\} \tag{17}$$

which shows that the optimal problem with fuzzy constraint conditions investigated in this paper is just the generalization of ordinary conditional extreme value problem There are other definitions of the optimal problem with fuzzy constraint condition as follows in some literatures[1, 2, 3].

Definition 2.2 Let C, $f \in F(X)$, C be the fuzzy constraint, f be the objective function. The optimal criterion set is defined as

$$\hat{D} = \{x^* \mid (f \cap C)(x^*) = \vee_{x \in X} [f(x) \wedge C(x)], \ x^* \in X\} \tag{18}$$

Now we give an example to show that $D \neq \hat{D}$, i.e., the two kinds of optimal criterion set are different in essence. In some literatures, the former is called the asymmetry-type optimal problem, the latter is called the symmetry-type optimal problem.

Example 2.1 Let $X = [0, +\infty)$, f, $C \in F(X)$, and

$$f(x) = x \exp(1 - x/5)/5$$

$$C(x) = \begin{cases} 1, & 0 \leq x \leq 1 \\ \dfrac{1}{1 + (x-1)^2}, & x > 1 \end{cases},$$

Then it is easy to solve that

$$D = \{1\}$$

$$\hat{D} = \{x^* \mid x^* \exp(1 - x^*/5)/5 = 1/[1 + (x^* - 1)^2]\}.$$

Obviously, $D \neq \hat{D}$.

4 Conclusion

In general, it is a difficult problem to solve the optimal criterion set under fuzzy constraint conditions. In this paper, the optimal problem with fuzzy constraint condition was translated into solving M_α or M_α', i.e., translated into solving a conditional extreme value problem in mathematics. It makes the problem simpler. And it also becomes simpler to solve the problems for fuzzy linear programming and for fuzzy nonlinear programming, which in fact is a special case of the optimal problem.

References

1. Peizhuang, W.: Fuzzy Set Theory and Their Application (in Chinese). Shanhai Science and Technology Press, Shanhai (1983)
2. Kaiqi, Z., Yang, X.: Fuzzy Systems and Expert Systems (in Chinese). Press of Southwest Jiaotong University, Jiaotong (1989)
3. Wangming, W. et al.: Application of Fuzzy Set Methods (in Chinese). Press of Beijing Normal University, Beijing (1985)
4. Yaohuang, G., Yang, X. et al.: Uncertainty Decision Making (in Chinese). Press of Southwest Jiaotong University, Jiaotong (1996)
5. Kacprzyk, J., Orlovski, S.A. (eds.): Optimization Models Using Fuzzy Sets and Possibility Theory. D. Reidel Publishing Company, Dordrecht (1987)
6. Zimmermann, H.-J.: Fuzzy Sets, Decision Making and Expert Systems. Kluwer Academic Publisher, Boston (1987)
7. Bellman, R.E., Zadeh, L.A.: Decision Making in a Fuzzy Environment. Management Science 17 (1970) 141-164

Fuzzy Control Theory

Inference Methods for Partially Redundant Rule Bases

Ralf Mikut, Jens Jäkel, and Lutz Gröll

Forschungszentrum Karlsruhe GmbH, Institute for Applied Computer Science (IAI)
P.O. Box 3640, D-76021 Karlsruhe, Germany
Phone: +49-7247-825731, Fax: +49-7247-825785
E-Mail: {mikut,jaekel,groell}@iai.fzk.de

Abstract. In this paper, a new inference strategy applicable to redundant or contradictory fuzzy rules is introduced. Both characteristics result mainly from a data-based generation of fuzzy systems where linguistic hedges are used to get an abstract description and where different rules' premises are overlapping. It is shown, that common fuzzy operators fail in these cases and that the newly introduced switching fuzzy operators solve these problems.

1 Introduction

This paper discusses inference strategies for fuzzy rule bases resulting from data-based automatic rule generation algorithms. Typical methods for rule generation are tree-oriented, statistical and evolutionary approaches. The aim of these data-based methods is the design of compact rule bases with a small number of interpretable rules which explain the learning data set and provide a sufficient statistical soundness. Powerful methods use linguistic hedges as *at least, approximately* or different abstraction levels of linguistic terms as *positive* for an abstract description or *positive small* and *positive large* for a more specialized description. In addition, many rule premises only specify some linguistic variables with linguistic terms.

The resulting rule bases often include partially redundant rules with identical conclusions or rules with overlapping premises and contradictory conclusions. Furthermore, rule premises can contain disjunctive combinations of linguistic terms.

Any fuzzy inference strategy should produce the results expected by human experts reading the rules and membership functions. Classical inference strategies and fuzzy operators partially give strange results if some of the specific characteristics of automatically generated rule bases occur. A first task is the formalization of these expected results into so-called semantic constraints. Secondly, modified inference approaches for more adequate results will be proposed.

The aims of this paper are

- to give a short overview about data-based methods for rule generation,
- to discuss problems of classical fuzzy inference strategies and
- to propose new operators and strategies to solve these problems.

2 Data-based rule generation

Inductive learning strategies for the generation of crisp rules from a set of examples have been studied for long time [1]. Today, heuristic approaches with a statistical point of view as the ROSA method [2, 3] and tree-oriented approaches [4, 5] play an important role.

The ROSA method generates single rules and evaluates these rules using statistical measures. The hypothesis generation uses different heuristics or evolutionary algorithms. Typically, the approach produces rules with an incomplete premise structure. It means, that only a few linguistic variables in the premise are specified by linguistic terms.

Tree-oriented approaches create a decision tree consisting of nodes and branches. A node indicates a linguistic term or class of the output and contains a test on an input variable $x_i = ?$ if it is a decision node. For each outcome of a test, a linguistic term $x_i = A_{ij}$ of the tested input, a branch starts from the decision node. A decision node receives the most frequent output term or class in the respective subset of examples. The construction algorithm consists of step-wise splits of the set of training examples using a test $x_i = ?$ in each step.

If the search algorithm terminates before specifying all linguistic variables, the same incomplete premise structure as for the ROSA method follows. The results of the subsequent pruning process depend on the strategy – firstly, pruning the tree by deleting subtrees [4, 5] or, secondly, pruning the (fuzzy) rules extracted from the tree deleting linguistic variables [4–6] or adding disjunctively linguistic terms [6]. The first approach guarantees a complete rule base with mutually exclusive rules. The latter is characterized by rules with overlapping premises but produces normally a more compact rule base with a better generalization ability. A further problem is the existence of local contradictory rules, which follows from small overlaps of different rules' premises. The disjunctive combination of linguistic terms is more transparent if new derived linguistic terms are created [6].

In the last step of rule search, only some significant rules will be chosen for the rule base [7]. The advantage is that the number of necessary rules is further reduced. But mostly the resulting rule base no longer covers the whole input space. Therefore, a default rule is introduced which complements the other rules to the whole input space. It is especially advantageous if a frequent output class (e. g. the class "normal") is spread on the input space.

As a consequence, inference strategies for data-based generated rule bases have to handle these characteristics in a satisfactory manner. In the next chapter, typical problems and solving strategies will be described. A more detailed description can be found in [8].

3 Inference strategies

As discussed above, the linguistic terms A_{ij} can be classified into *primary* and *derived* terms. Primary terms are the given terms in the classical sense, e.g. defined by an expert. Derived terms for ordered linguistic variables are automatically generated by modifications using linguistic hedges like *at least, rather, at most, not* or derived generalized terms as *positive* instead of the primary terms *positive small, positive big*. In contrast to [9] and [10], derived terms are built by the union of primary terms and not by modifying a primary term.

Linguistic terms and fuzzy sets should hold the following conditions (*semantic constraints*):

1. Fuzzy sets of the disjunctive primary terms are triangular or trapezoidal, normal, convex and have single overlap [11].
2. Fuzzy sets of primary terms cover the universe of discourse completely observing $\sum_{A_i \in T_{\text{prim}}} \mu_{A_i}(x) = 1$, $\forall x \in \mathcal{X}$ (T_{prim} set of primary terms).
3. For every pair of linguistic terms A, B, $\forall x \in \mathcal{X}$:

    if $A \subseteq B$ (a) $\mu_A(x) \leq \mu_B(x)$, (b) $\mu_{A \cup B}(x) = \mu_B(x)$, (c) $\mu_{A \cap B}(x) = \mu_A(x)$
    if $A \cap B = \emptyset$ (d) $\mu_{A \cup B} = \max\{1, \mu_A + \mu_B\}$, (e) $\mu_{A \cap B} = 0$

Commonly used operators partially fail in the construction of the fuzzy sets of derived terms according to conditions 3(a)–(e). For example, *min* does not hold 3(e), *(algebraic) product* 3(c,e), *max* 3(d), *sum* 3(b) and the *algebraic sum* 3(b,d).

If these conditions are not hold, unexpected membership functions result for the derived terms and for conjunctions and disjunctions using derived terms. One example is given in Fig. 1. Here, the derived term *positive* (POS) represented as disjunction of the primary terms *positive small* (PS) and *positive big* (PB) is considered. Following 3(d), the membership function of POS has to be as depicted in the upper left diagram of Fig. 1 (thick line). The *bounded sum* operator (Fig. 1 top, left) as well as the *sum* operator (Fig. 1 top, middle) produces in contrast to the *max* operator (Fig. 1 top, right) the correct result.

The disjunction of the derived term *positive* with a membership function as in in the upper left diagram of Fig. 1 (thick line) and the primary term *positive small* gives *positive*. Here, the *bounded sum* operator (Fig. 1 bottom, left) as well as the *sum* operator (Fig. 1 bottom, middle) fails, whereas the *max* operator (Fig. 1 bottom, right) gives the expected result.

The crisp set relations of different primary and derived terms are the key to solve these problems. If also in the crisp case, an intersection set exists as discussed above, a different approach has to be used. This idea leads to the

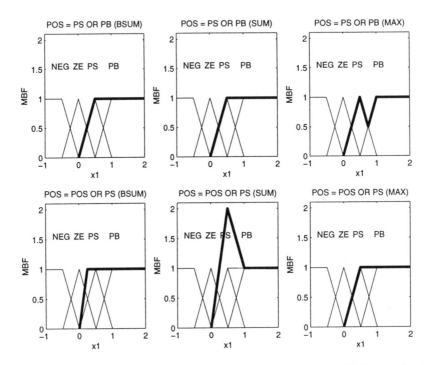

Fig. 1. Computing POS = PS OR PB (top) and POS = POS OR PS (bottom) using the operators BOUNDED SUM (left), SUM (middle) and MAX (right)

following fuzzy set operators:

$$\mu_{A \cap B}(x) = \begin{cases} 0 & \text{if } A \cap B = \emptyset \\ \min\{\mu_A(x), \mu_B(x)\} & \text{else,} \end{cases} \tag{1}$$

$$\mu_{A \cup B}(x) = \mu_A(x) + \mu_B(x) - \mu_{A \cap B}(x) = \begin{cases} \mu_A(x) + \mu_B(x) & \text{if } A \cap B = \emptyset \\ \max\{\mu_A, \mu_B\} & \text{else.} \end{cases} \tag{2}$$

These operators consider also the crisp set relations and fulfill the conditions 3(a)–(e). They are an operator pair according to DE MORGAN's law.

Rule premises consist of conjunctions of statements like "x_1 is *positive small*". Conclusions contain only elementary statements like "*y* is *small*". Moreover, rules are equally plausible. Rules which are partially redundant, i. e. which have overlapping premises, should not reinforce each other.

The idea of the inference strategy introduced in [8] is to consider the redundancy of rule premises. The composition of the results of the activated rules uses the multi-dimensional and multi-variable extension of the union operator

(2):

$$\mu_{P_1 \cup \cdots \cup P_q}(x) = \sum_{i=1}^{q} \mu_{P_i}(x) + (-1)^1 \sum_{i<j}^{q} \mu_{P_i \cap P_j}(x) + \cdots + (-1)^{q-1} \mu_{P_1 \cap \cdots \cap P_q}(x).$$

(3)

The membership value of the intersections $\mu_{P_i \cap P_j}(x)$, assuming P_i is $(x_1 = A_{1,i})$ AND $(x_2 = A_{2,i})$ AND ... AND $(x_m = A_{m,i})$ and P_j, respectively, is calculated as follows:

$$\mu_{P_i \cap P_j}(x) = \mu_{A_{1,i} \cap A_{1,j}}(x_1) \mu_{A_{2,i} \cap A_{2,j}}(x_2) \cdots \mu_{A_{m,i} \cap A_{m,j}}(x_m)$$

and $\mu_{A_{l,i} \cap A_{l,j}}(x_l)$ according to (1), analogically for intersections of more than two premises. It can be written in matrix notation as

$$\mu_y = C\big[\mu_{P_1} \ \cdots \ \mu_{P_q} \ \mu_{P_1 \cap P_2} \ \cdots \ \mu_{P_1 \cap \cdots \cap P_q}\big]^T = C\mu_{P'}.$$

(4)

The elements of C, c_{ij}, take the value 1 with a sign according to (3) if the rule(s) referred to in the j-th element of $\mu_{P'}$ have the i-th output term in the conclusion else the value 0. $\mu_{P'}$ is an extended premise vector including all intersections of different rule premises. If all intersections of premises are empty sets the inference is a *sum-prod* inference.

Further, it will prove advantageously to write the matrix C as product $C_1 C_2$. C_1 is of dimension (n, q) with elements $c_{1,ij} \in \{0, 1\}$ where $c_{1,ij} = 1$ if the conclusion of the j-th rule contains the i-th output term. This matrix connects the premises with the conclusions of the rules. With C_2, corrected premise activations concerning partially redundant rules with same or different conclusions are computed. Finally, the defuzzification transforms μ_y into a real value y. This inference scheme can be extended to the case of contradictory rules [8].

R_1 : IF x_1=PB	THEN y=ZE
R_2 : IF x_2=PS	THEN y=ZE
R_3 : IF x_1=POS $\wedge$ x_2=POS	THEN y=ZE
R_4 : IF x_1=ZE $\wedge$ x_2=NOT PS	THEN y=PS
R_5 : IF x_1=PS $\wedge$ x_2=ZE	THEN y=PB

Table 1. Rule base with simplified rules

		x_2		
		ZE	PS	PB
x_1	ZE	PS (4)	ZE (2)	PS (4)
	PS	PB (5)	ZE (2,3)	ZE (3)
	PB	ZE (1)	ZE (1,2,3)	ZE(1,3)

Table 2. Rule base with conclusions and the indices of rules from Table 1

The rule base in Table 1 shows an example of using partially redundant rules with two input variables and one output variable. The rule base is not contradictory. As a result of automatic rule generation, the premises of the rules are as generally as possible. This aim explains the use of the term *positive* (POS) instead of *positive small* (PS) or *positive big* (PB) and the unspecified linguistic variables x_2 in rule R_1 and x_1 in rule R_2. For the term *zero* (ZE), no modifications are introduced. The use of the negation NOT in rule R_4 eliminates another rule. As a consequence, a rule base with normally 9 rules (Table 2) is formulated with 5 rules. Table 2 displays the overlapping of these rules with the rule numbers in parenthesis. E.g., the premise x_1=PB AND x_2=PS activates 3 rules. For this example, C_1 and C_2 can be written as

$$C_1 = \begin{pmatrix} 1 & 1 & 1 & 0 & 0 \\ 0 & 0 & 0 & 1 & 0 \\ 0 & 0 & 0 & 0 & 1 \end{pmatrix}, \quad C_2 = \begin{pmatrix} 1 & 0 & 0 & 0 & 0 & -1/2 & -1/2 & 0 & 1/3 \\ 0 & 1 & 0 & 0 & 0 & -1/2 & 0 & -1/2 & 1/3 \\ 0 & 0 & 1 & 0 & 0 & 0 & -1/2 & -1/2 & 1/3 \\ 0 & 0 & 0 & 1 & 0 & 0 & 0 & 0 & 0 \\ 0 & 0 & 0 & 0 & 1 & 0 & 0 & 0 & 0 \end{pmatrix}.$$

In C_2, only columns with non-empty intersections are displayed. With this assumption, $\mu_{P'}$ has the structure

$$\boldsymbol{\mu}_{P'} = [\mu_{P_1}\ \mu_{P_2}\ \mu_{P_3}\ \mu_{P_4}\ \mu_{P_5}\ \mu_{P_1 \cap P_2}\ \mu_{P_1 \cap P_3}\ \mu_{P_2 \cap P_3}\ \mu_{P_1 \cap P_2 \cap P_3}]^T \quad (5)$$

and

$$C = \begin{pmatrix} 1 & 1 & 1 & 0 & 0 & -1 & -1 & -1 & 1 \\ 0 & 0 & 0 & 1 & 0 & 0 & 0 & 0 & 0 \\ 0 & 0 & 0 & 0 & 1 & 0 & 0 & 0 & 0 \end{pmatrix} \quad (6)$$

follows. Using this inference scheme, the input-output mappings (characteristic fields) of the reduced rule base in Table 1 (Fig. 2, top left) and the semantically equal full rule base in Table 2 (Fig. 2, top middle) do not differ (Fig. 2, top right). Otherwise, classical inference methods as the *sum-prod* method or the *min-max* method produce significant differences between the input-output mappings of these rule bases (Fig. 2, middle and bottom).

If the rule R_4 in Table 1 is changed to IF x_1=ZE THEN y=PS, a local contradiction arises for the premise x_1=ZE AND x_2=PS with the conclusions ZE (from rule R_2) and PS (from modified rule R_4). Here, C_1 remains the same and C_2 is changed to

$$C_2 = \begin{pmatrix} 1 & 0 & 0 & 0 & 0 & -1/2 & -1/2 & 0 & 0 & 1/3 \\ 0 & 1 & 0 & 0 & 0 & -1/2 & 0 & -1/2 & -1/2 & 1/3 \\ 0 & 0 & 1 & 0 & 0 & 0 & -1/2 & -1/2 & 0 & 1/3 \\ 0 & 0 & 0 & 1 & 0 & 0 & 0 & 0 & -1/2 & 0 \\ 0 & 0 & 0 & 0 & 1 & 0 & 0 & 0 & 0 & 0 \end{pmatrix} \quad (7)$$

with $\boldsymbol{\mu}_{P'} = [\mu_{P_1}\ \mu_{P_2}\ \mu_{P_3}\ \mu_{P_4}\ \mu_{P_5}\ \mu_{P_1 \cap P_2}\ \mu_{P_1 \cap P_3}\ \mu_{P_2 \cap P_3}\ \mu_{P_2 \cap P_4}\ \mu_{P_1 \cap P_2 \cap P_3}]^T$ and

$$C = \begin{pmatrix} 1 & 1 & 1 & 0 & 0 & -1 & -1 & -1 & -1/2 & 1 \\ 0 & 0 & 0 & 1 & 0 & 0 & 0 & 0 & -1/2 & 0 \\ 0 & 0 & 0 & 0 & 1 & 0 & 0 & 0 & 0 & 0 \end{pmatrix}. \quad (8)$$

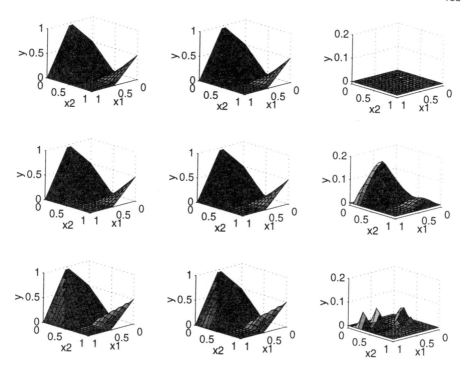

Fig. 2. Characteristic fields for the rule bases in Table 1 (left), Table 2 (middle) and its differences (right) - Inference methods: new inference method (top), *sum-prod* (middle), *min-max* (bottom)

As in the non-contradictory case, only the new introduced inference concept guarantees the same input-output behaviour for the modified rule base in Table 1 and the modified rule base in Table 2.

For rule generation and evaluation, a compact notation for the estimation of the output fuzzy sets $\hat{\mu}_{\tilde{y}}[k]$ will be used:

$$\underbrace{\left(\hat{\mu}_y[1] \ \ldots \ \hat{\mu}_y[k] \ \ldots \ \hat{\mu}_y[N]\right)}_{\hat{Y}} = C_1 \, C_2 \underbrace{\left(\mu_{P'}(x[1]) \ \ldots \ \mu_{P'}(x[k]) \ \ldots \mu_{P'}(x[N])\right)}_{P} .$$

$$(9)$$

From the constraints for fuzzy sets follows $1_n^T \hat{Y} = 1_N^T$ and from the completeness of the rule base $1_q^T P = 1_N^T$. Here, 1_n means a n-dimensional vector of ones. As mostly only a small number of rules have overlap most elements of $\mu_{P'}(x[k])$ are zero for all k. Therefore, in the implementation the output is calculated with reduced C or C_2, respectively, and $\mu_{P'}$.

The proposed inference concept provides additional advantages [6–8]:

- The formalization of semantic constraints into restrictions can be used to estimate the rule conclusions in a rule base. The chosen way is to formulate an optimization problem for the C_2-matrix.
- This optimization produces certainty factors of rules, which lead to higher approximation quality but reduced interpretability.
- The form with default rules enables an evaluation of single rules, group of rules or complete rule base according to different aims (approximation quality, clearness of rules, different levels of abstraction).
- These evaluation measures can be used to choose single rules for a rule base.
- The new inference scheme can be integrated in the search process of tree-oriented approaches. Then, evaluation of candidate rules and application of the chosen rule base employ the same inference scheme.

4   Conclusions

In this paper, a new inference method has been proposed. It is advantageous especially for data-based generated rule bases producing rules with derived terms and linguistic hedges, partially redundant premises and local contradictions. Common inference methods lead here to insufficient results. These effects have been demonstrated by means of an example.

The inference method contains new operators for conjunctive and disjunctive operations and bases on an analysis of qualitatively overlapping premises and terms. The use of these operators leads to a matrix formulation of the inference process. Semantic constraints which have to be held are formulated as restrictions for certain optimization problems. The inference allows also the processing of contradictory rule bases.

The application of the inference method in the design of a fuzzy system for a cardiologic diagnosis is described in [12].

References

1. E. B. Hunt, J. Marin, P. T. Stone: Experiments in Induction. Academic Press, New York, (1966)
2. A. Krone, H. Kiendl: Automatic generation of positive and negative rules for two-way fuzzy controllers. In Proc. 2nd Europ. Congr. on Intelligent Techniques and Soft Computing EUFIT'94, Aachen (1994) 438–442
3. M. Fritsch: Baumorientierte Regel-Induktionsstrategie für das ROSA-Verfahren zur Modellierung komplexer dynamischer Systeme. Fortschritt-Bericht VDI, Reihe 8, Nr. 565. VDI-Verlag, Düsseldorf, (1996)
4. L. Breiman, J. H. Friedman, R. A. Olshen, C. J. Stone: Classification and Regression Trees. Wadsworth, Belmont, Ca. (1984)
5. J. R. Quinlan: Induction of decision trees. Machine Learning 1 (1986) 81–106
6. J. Jäkel, L. Gröll, R. Mikut: Automatic generation and evaluation of interpretable rule bases for fuzzy systems. In Computational Intelligence for Modelling, Control and Automation CIMCA'99, IOS Press, Amsterdam (1999) 192–197
7. J. Jäkel, L. Gröll, R. Mikut: Tree-oriented hypothesis generation for interpretable fuzzy rules. In Proc. 7th Europ. Congr. on Intelligent Techniques and Soft Computing EUFIT'99, Aachen (1999)
8. J. Jäkel, L. Gröll, and R. Mikut: Bewertungsmaße zum Generieren von Fuzzy-Regeln unter Beachtung linguistisch motivierter Restriktionen. In Berichtsband 8. Workshop Fuzzy Control d. GMA-FA 5.22, Aachen (1998) 15–28
9. L. A. Zadeh: A fuzzy set theoretic interpretation of linguistic hedges. Journal of Cybernetics 2 (1972) 4–34
10. N. Sano, I. Koyauchi, H. Kadotani, R. Takahashi: Transformation of membership functions by hedge and primitive terms using the extension principle. In Proc. 5th Zittau Fuzzy Colloquium (1997) 13–16
11. G. J. Klir, B. Yuan: Fuzzy Sets and Fuzzy Logic: Theory and Applications. Prentice Hall, Upper Saddle River, NJ, (1995)
12. J. Jäkel, R. Mikut, H. Malberg, G. Bretthauer: Datenbasierte Regelsuche für Fuzzy-Systeme mittels baumorientierter Verfahren. In Berichtsband 9. Workshop Fuzzy Control d. GMA-FA 5.22, Dortmund (1999) 1–15

Analysis and Design of Fuzzy PID Controller Based on Classical PID Controller Approach

Petr Pivoňka

Brno University of Technology, Faculty of Electrical Engineering and Computer Science, Božetěchova 2, 61266 Brno, Czech Republic

Abstract. A fuzzy PID controllers are physically related to classical PID controller. The settings of classical controllers is based on deep common physical background. Fuzzy controller can embody better behaviour comparing with classical linear PID controller because of its non linear characteristics. Well tuned fuzzy controller can be also more stable and more robust for the time varying systems. On the other hand, when the fuzzy controller is tuned badly it can exhibit limit cycle which can decrease lifetime of the actuator. This phenomenon is critical especially when the actuator is valve. Knowing about these problems, more analytical methods of tuning fuzzy controllers can be found. The method with unified universe considerably simplifies the setting and realization of fuzzy controllers. This paper tries to analyze causes of oscillations and it outlines the possibilities how to reduce them. The paper also shows solution how to reduce time needed for computation of control signal by decreasing the number of membership functions and by changing defuzzification method.

1 Introduction

One of the main drawbacks of fuzzy controllers is big amount of parameters to be tuned. It is especially difficult to make initial approximate adjustment because there is no cookery book how to do it. Also it is very well known that good convergence of optimization method is strongly dependent on initial settings. In the literature are described many papers for solution of these problems. Till this time all the published papers can't ensure a setting of parameters fuzzy PID controller with the same physical meaning similar to classical PID controller, and can't solve the problems with time transformation.

The adjustment of fuzzy controllers may be considerably simplified when fuzzy controller with a unified universe is used. The parameters to be tune then have their physical meaning and fuzzy controller can be approximately adjusted using known rules for classical controllers. Probably the easiest way how to implement fuzzy PID controller is to create it as a parallel combination of basic fuzzy controllers P+I+D [4] or PI+D [2]. Suitable choice of inference method can ensure behaviour which is close to one of classical PID controller for both the tracking problem and the step disturbance rejection. The fuzzy sets are assumed to have initially symmetrical layout and the parameters of both regulators are tuned using for example by Ziegler - Nichols method.

To improve behaviour of such designed fuzzy controller it is necessary either to manually change the quantities of fuzzy controller or to use some optimization methods which does this operation. One which can be implied is genetic algorithms. Different quantities can be changed to reach the optimum values. These quantities are fuzzy set layout, number of fuzzy sets, rule base mapping, the parameters of basic fuzzy controllers and their various combinations. Note that all the optimizations must be always performed according to the chosen inference and defuzzification method. It is apparent that process of optimization can take a lot of time. Moreover this method is contingent on existence of accurate mathematical model of the process because in vast majority of the cases it is not possible to perform any kind of optimization directly on real process. The model usually does not correspond to real system which limits the success of optimization methods.

The prime idea of fuzzy control was to apply it at the place where there is no deep knowledge of transfer function of controlled system and where this knowledge can be hardly identified. These are often the cases where the fuzzy control leads to better performance comparing with classical approach. Also for this instance it seems to be advantageous to have physical connection between fuzzy controller and its classical counterpart because it can significantly simplify the adjustment of regulator for real process.

The heuristic optimization of parameters settings is also suitable for fuzzy PI controller with unified universe where the parameters are the same as the ones of classical PI controller. The parallel combination of fuzzy PI and PD controllers can be used for heuristic optimization of parameters settings but it should be noted that because of the presence of double proportional part in this regulator the adjusted parameters will differ from the ones of classical PID controller. But important thing is that the adjustment of this parameters is still in the same physical meaning. Note that for all previously mentioned controllers it is also possible to employ time transformation (sample time modification) without having to change the scope of universes.

2 Fuzzy PI Controller Design

A classical PI controller is described by equation (1) where K is the gain of PI controller, T_I is an integral constant, e(t) is an error signal, $e(t) = w(t) - y(t)$, $w(t)$ is the desired value, $y(t)$ is the output from process and $u(t)$ is the output from controller - the action value.

$$u(t) = K\left(e(t) + \frac{1}{T_I} \int_0^t e(\tau)\mathrm{d}(\tau)\right) \tag{1}$$

When we derive (1) we get

$$u'(t) = K\left(e'(t) + \frac{1}{T_I}e(t)\right) \qquad (2)$$

For a local extreme location we put

$$u'(t) = K\left(e'(t) + \frac{1}{T_I}e(t)\right) = 0 \qquad (3)$$

The solution of equation (3) is

$$e'(t) = -\frac{1}{T_I}e(t) \qquad (4)$$

because the PI controller gain $K > 0$. The equation (4) depends only on the PI controller integral time constant T_I! Its physical meaning lies in a fact that it determines a border where the action derivation changes a sign from positive to negative. By comparison we find out that the derivation of action value depends on the derivation of error magnitude. That is why we can map the rule base to the discrete state space - for example the shape which is in Fig. 1. We use the ZO elements on the diagonal for mapping of equation (4). If we translate the equation (3) to a discrete form, we get the equation for action value change of the discrete PI controller

$$\Delta u(k) = K\left(\Delta e(k) + \frac{1}{T_I}e(k)\right) \qquad (5)$$

where $\Delta u(k) = \big(u(k) - u(k-1)\big)/T$, $\Delta e(k) = \big(e(k) - e(k-1)\big)/T$, T is the sampling period, k is the step.

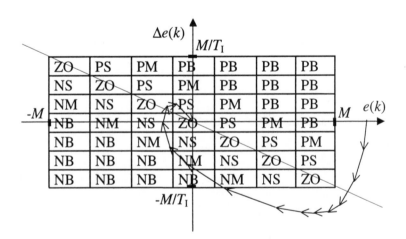

Fig. 1. The rule base of fuzzy PI controller mapped to the discrete state space

From [1] is obvious that the time constant T_I has relation to the change-in-error. Therefore we can modify the equation (5) for a fuzzy PI controller

$$\Delta u(k) = \frac{K}{T_I}\Big(T_I \Delta e(k) + e(k)\Big) \tag{6}$$

In the next step it is necessary to map the rule base to the discrete state space $\Delta e(k), e(k)$. We define the scale factor M for the universe range, $M > 0$. This scale factor sets the universe ranges for the error and its first difference (Fig. 1). We extend the equation (6) and get

$$\Delta u(k) = \frac{KM}{T_I}\left(\frac{T_I}{M}\Delta e(k) + \frac{1}{M}e(k)\right) \tag{7}$$

Input or output value is multiplied by a constant which indicates a real range of the universe (Fig. 2). If the universe range is multiplied by coefficient 5 before fuzzification, the real range of the universe is $< -0.2, 0.2 >$. For coefficient 0.1 the real range of the universe is $< -10, 10 >$. It is evident there's no conflict with commonest and this procedure leads to the significant simplicity for the fuzzy controller design as will be demonstrated.

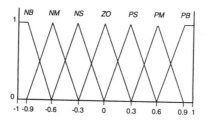

Fig. 2. Symmetrical membership function lay-out

We apply fuzzification to input variables and after defuzzification we get the equation

$$\Delta u(k) = \frac{KM}{T_I}D\left\{F\left\{\frac{T_I}{M}\Delta e(k) + \frac{1}{M}e(k)\right\}\right\} \tag{8}$$

where F is an operation for fuzzification and D for defuzzification. For $\Delta u(k)$ we put

$$\Delta u(k) = \frac{u(k) - u(k-1)}{T} = \frac{KM}{T_I}D\left\{F\left\{\frac{T_I}{M}\Delta e(k) + \frac{1}{M}e(k)\right\}\right\} \tag{9}$$

The control signal generated by fuzzy PI controller in step k is

$$u(k) = \frac{KMT}{T_I}D\left\{F\left\{\frac{T_I}{M}\Delta e(k) + \frac{1}{M}e(k)\right\}\right\} + u(k-1) \tag{10}$$

A realization of the fuzzy PI controller is in Fig. 3. A physical meaning of the parameters for the fuzzy PI controller remains the same like for the PI controller (the controller gain K and the time integral constant T_I). The saturation limits of the action value realizes antiwindup. Let us assume plant described by following transfer function to illustrate and to compare behaviour of fuzzy PI controller with classical continuous PI controller:

$$F(s) = \frac{2}{(10s + 1)(s + 1)^2} \tag{11}$$

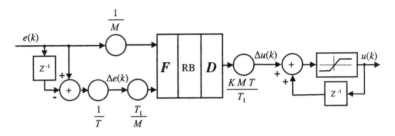

Fig. 3. Fuzzy PI controller structure with the normalized universe range with the antiwindup

Using the tuning method from [5] we obtain parameters $K = 2.14$, $T_I) = 5.8$ s. The responses of controlled system using this control algorithm are shown in Fig. 4. The disturbance acts on the input of the system. Fuzzy controller was realized according to Fig. 3 with rule base mapping according to Fig. 1. The membership functions were distributed as shown in Fig. 2. The similar settings of parameters $K = 2.14$, $T_I = 5$ s with respect to optimizations for different methods of inference and defuzzification methods was used. The scale was settled to $M = 10$ and sample period was set to $T = 0.1$ s. The time responses for different inference and defuzzification methods are shown in Fig. 5.

The following settings were tested - the inference method Min-Max and Prod-Max, defuzzification was done using COG method with singletons or triangles as an output membership functions. The simulations were launched with either three of seven fuzzy sets in all the normalized universes. Also singletons were realized using normalized universe. The singletons were located in vertexes of original fuzzy sets (see Fig. 2) Comparing the results obtained using classical PI and fuzzy PI controllers following discussion can take place. The output of the system has very small overshoot when it is controlled with fuzzy regulator. The disturbance rejection using fuzzy controller is comparable with disturbance rejection of classical PI controller.

In [5] in Fig. 5 there was shown that the step disturbance in the input of the system with amplitude 0.05 brings the system to the significant oscillations.

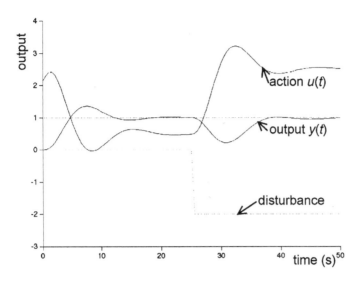

Fig. 4. Classical PI controller

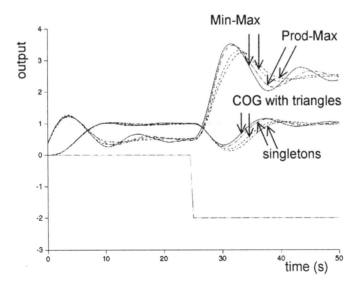

Fig. 5. Fuzzy PI controller

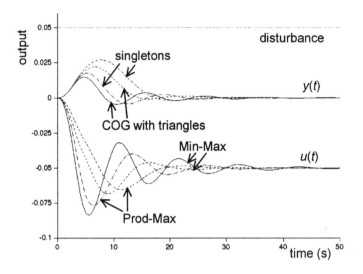

Fig. 6. Fuzzy PI controller with 7 membership functions

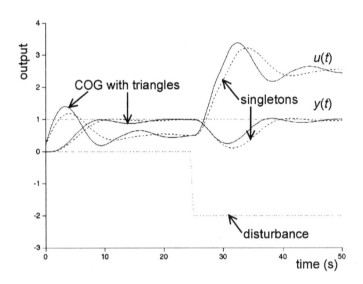

Fig. 7. Fuzzy PI controller with 3 membership (method Min-Max)

The same settings as in this article was used to obtain time responses. As it can be seen in Fig. 6, where there are corresponding time responses, there are no any oscillations visible. However, it was not possible to analyze potential causes of oscillations which were obtained in article [5] due to the lack of detailed settings of fuzzy controller. Significantly better results comparing with approximately adjusted fuzzy PI controller were not obtained even after the optimization using genetic algorithms. It is possible to shorten time required for computing the command by reducing the number of membership functions to three. Reduction of membership functions brings together small degradation of dynamical behaviour (Fig. 7). This degradation can be almost eliminated by fine tuning of the parameters. As a result of these simulations we can state that the assertions in paper [5] were not proved.

3 Fuzzy PD Controller Design

A classical ideal PD controller is described:

$$u(t) = K\big(e(t) + T_\mathrm{D}e'(t)\big) \tag{12}$$

We'd like to know when the action value is equal to zero, that is way we can write

$$e(t) + T_\mathrm{D}e'(t) = 0 \tag{13}$$

The solution of equation (13) is

$$e'(t) = -\frac{1}{T_\mathrm{D}}e(t) \tag{14}$$

The equation (14) depends only on the derivation time constant of the PD controller and its physical meaning is similar to equation (4) for the PI controller. If we transfer the equation (12) to the discrete form, we get an equation of the discrete PD controller

$$u(k) = K\big(e(k) + T_\mathrm{D}\Delta e(k)\big) \tag{15}$$

where $\Delta e(k) = \big(e(k) - e(k-1)\big)/T$ and T is the sample period. In the next step we map the rule base to the discrete state space $\Delta e(k)$, $e(k)$. We initiate the scale M for the universe range, $M > 0$. This scale sets ranges for the error and the change-in-error. After extending the equation (15) we get

$$u(k) = KM\Big(\frac{1}{M}e(k) + \frac{T_\mathrm{D}}{M}\Delta e(k)\Big) \tag{16}$$

We apply fuzzification to input variables and after defuzzification we get the equation

$$u(k) = KM\boldsymbol{D}\Big\{\boldsymbol{F}\Big\{\frac{1}{M}e(k) + \frac{T_\mathrm{D}}{M}\Delta e(k)\Big\}\Big\} \tag{17}$$

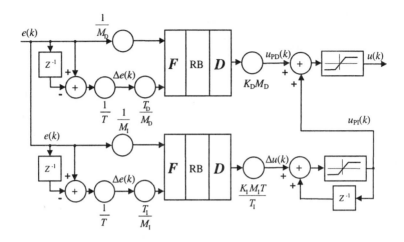

Fig. 8. Structure of the fuzzy PD+PI controller with the normalised universe range

4 Fuzzy PD+PI Controller Design

Let us create fuzzy controller as a parallel combination of fuzzy PI and PD controllers. Simulation results obtained using classical I-PD controller are shown in Fig. 9.

This controller was adjusted using Ziegler–Nichols method ($K = 7.28$, $T_I = 2.8$ s, $T_D = 0.7$ s). The initial adjustment of fuzzy PI+PD controller is taken up from reference [3] ($K_I = K_D = 4$, $T_I = 2.2$ s, $T_D = 2$ s, $M_I = M_D = 10$, $T = 0.1$ s), Min-Max inference method, COG method for defuzzification, 7 membership functions. Membership function layout and rule base for both PI and PD parts are shown in Fig. 2 and Fig. 1 respectively. The only exception is in increment of command which is realised as shown in Fig. 11. Simulation results can be seen in Fig. 10. Fig. 12 shows the time responses with three different settings. Solid line uses the same adjustment as previously described simulation but inference method is Prod-Max. Dash-dot-dot line represents the simulation with 3 membership functions, inference Min-Max and defuzzification using triangular membership functions with COG method. Dash-and-dot line shows simulation results with 3 membership functions, inference Min-Max and defuzzification using singletons. When the singletons are used they are placed into the vertex of original fuzzy membership functions. Fuzzy controller is generally inclined to oscillation with relatively small amplitude. The origin of this oscillations is not only incorrect tuning of the parameters but also the inference method Min-Max. The oscillations are considerably eliminated when the Prod-Max inference method is employed and singletons for defuzzification are used. Another potential source of oscillations is wrong implementation of inference engine. Shift of the vertex point of middle membership function just for 0.01 on the normalised universe (illustrated in Fig. 15) causes the limit

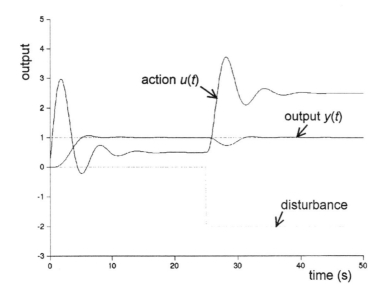

Fig. 9. Classical I-PD controller

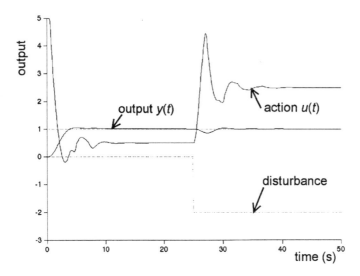

Fig. 10. Fuzzy PI+PD controller

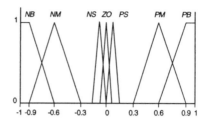

Fig. 11. Non-linear membership function lay-out for fuzzy PI controller (Δu)

cycles to appear (see results in Fig. 14) without any other external intervention. The initial deviation is caused by PD controller and the oscillations by PI controller. When the singleton membership functions are used instead of triangular ones the described phenomena of oscillations disappear. Following the results in Fig. 14 it can be seen that the singletons as an output membership functions give at the beginning higher oscillation but after the time they disappear (dashed line). Triangular membership functions with COG defuzzification give after short transient response steady limit cycle. Note the nonzero steady state error in both cases. As a consequence we can say that there is a difference between defuzzification using singletons and triangular membership functions even if the COG method is used.

5 Conclusion

For the fuzzy PID controller setting it is necessary to determine universe ranges and perform tens or hundreds simulation experiments until the acceptable values are found. A retrieval of optimal parameters is very difficult, because the setting is dependent on big amount of other parameters. Method with the unified universe range, stated in this article, considerably simplifies setting of fuzzy PI/PD/PID controllers. It allows approximate adjustment of controller's parameters according to well-known methods for PID controller synthesis. If the universe has non-linear membership function layout then the results can have better behaviour than the classical PID controller. The fuzzy PID controller can be programmed like a unified block in industrial controller and therefore work consumed on a implementation to the particular control system can be shortened.

Acknowledgement
This research was partially financially supported by the Czech Grant Agency GAČR under the grant number 102/98/0552 Research and Applications of Heterogeneous Models and by the MŠMT research plan CEZ: MSM 260000013 Automation of Technological and Manufacturing Processes.

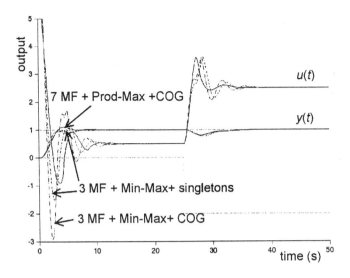

Fig. 12. Different settings of fuzzy PI+PD controller

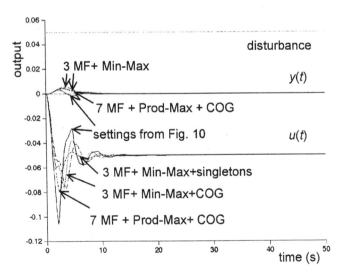

Fig. 13. Disturbance rejection of fuzzy PI+PD controller

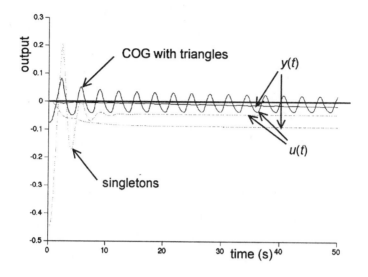

Fig. 14. Effect of shift fuzzy set

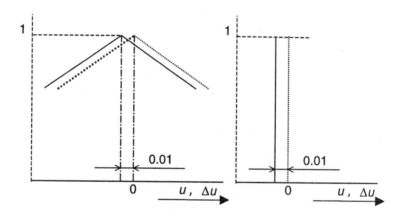

Fig. 15. Shift of middle fuzzy set

References

1. Pivoňka, P.: Physical Background of Fuzzy PI and PD Controller. In: The Proceedings of the Eighth International Fuzzy Systems Association World Congress. Taipei Taiwan (1999) 635–639
2. Pivoňka, P.: Fuzzy PI+D Controller with a Normalised Universe. In: The Proceedings of 6th European Congress EUFIT'98, Vol. **2**. Aachen Germany (1998) 890–894
3. Pivoňka, P., Šídlo, M.: Fuzzy PI+PD Controller. BUSEFAL, **74**. Toulouse France (1998) 93–97
4. Pivoňka, P., Findura, M.: The Alternative Realisation of Fuzzy Controllers. Automatizace, **41**. Czech Republic (1998), No. 10, 11, 12 P31–P38 in Czech
5. Rotach, V.: On Connection Between Traditional and Fuzzy PID Regulators. In: The Proceedings of 6th Fuzzy Colloquium. Zittau Germany (1998) 86–90

Expert Methods in the Theory of Automatic Control

V. Rotach

Moscow Power Institute (Technical University)
Krasnokazarmennaja 14, Moscow, 111250, Russia
rotach@acsnw.mpei.ac.ru

Abstract. We consider the basic rules of synthesis of automatic control systems with a large number of uncontrollable random disturbances including time-delay (for example, objects in thermal and atomic power stations, chemical manufactures, etc.). It is shown that in these situations, the methods of traditional formal theory cannot be successful, and they should be combined with heuristic expert methods, e.g. in formulating hypotheses. The initial check of hypotheses is carried out by computing the optimal parameters of the system applying reasonable methods and modeling. The final determination of the optimum is performed step-by-step through the application of adaptation algorithms using active actions on the system.

We shall consider the important and widespread class of objects with a certain degree of vagueness with the distributed parameters and time-delay. These objects are characterized by the following features:

- They are under influence random disturbances, which are partly uncontrollable.
- Their properties can vary in time, but this happens rather slowly.
- They are nonlinear, however, they permit linearization under small deviations that allows approximately to represent their properties by a family of linear models.

Examples of such kind of objects are given by thermal and atomic power stations, chemical, food, and other manufactures. In a more general environment, such objects are large complexes of manufactures, marketing campaigns, etc.

If the mathematical model of the object under control is known, the existing formal methods allow to find with sufficient accuracy the optimal algorithm of the control with respect to a given optimality criterion. In the present paper this point of view is exposed to doubt - it is argued that (at least for the specified class of objects) the synthesis of control systems is based on heuristic (expert) hypotheses rather, and formal computation is are necessary to accept or to reject hypotheses. It is natural that heuristic hypotheses can be described by fuzzy sets theory.

The random uncontrollable disturbances force to apply the closed structure of a control system. This is shown in Fig.1, where O, C - object and controller, y, x - controlled variable and its desirable meaning, u - control action, λ - random uncontrollable disturbances.

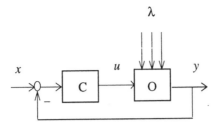

Fig. 1. The standard structure of a control system

After transforming into a linear model, all disturbances can be replaced by an equivalent one that directly acts on the controlled value. Thus, the system has only two input actions - the desired values of controlled value and equivalent disturbances. For minimizing the deviation of the controlled variable from its desirable value each of these influences should pass through an independent filter [1]. Practically, the structure in Fig. 2 is most convenient [2]. In this structure, from the controller the regulator *Reg* is allocated, whose task is to eliminate harmful actions of disturbances; this process we shall call *regulation*. The controller only carries out functions of desirable changes of controlled values; this process we call *control*. The closed structure of a subsystem of regulation can result in a loss of stability. Therefore, any criterion of system optimality should contain restrictions on the margins of stability.

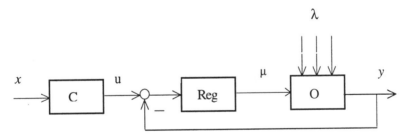

Fig. 2. Complete structure of control system

From the point of view of the purpose of object's functioning the optimality criteria can be divided into two classes:

- Statistical criteria describing the behavior of an object over long terms.
- Criteria that take into account rather large, but short-term deviations of the variable under control.

In both cases, the restriction on the margin of stability can be characterized by one parameter, for example by the size of frequency resonant peak of the closed loop characteristics.

The synthesis of algorithms for regulator actions with random character disturbances is usually made by methods of probability theory. It is natural that similar synthesis can be applied only if the quality of the system is estimated by the average over long time. Unfortunately, the so far developed methods of delayed control systems synthesis basically rest on optimum prediction theories and Wiener filters, and they do not take into account restrictions on stability margins of the closed loop. Thus the resulting optimal algorithm does not provide sufficient robustness for the control system.

The majority of technological objects require the absence of large short-term deviations of the controlled value. The applied statistical criteria (e.g. variance of controlled value) practically do not allow to estimate the size of such deviations. The only way out from this dilemma is the orientation at the worst-case scenario (largest random disturbances). The definition of such disturbances for linear systems is given through the principle of "accumulation of disturbances" by B. Bulgakov [3]. From this principle it follows that if the system has a sufficient margin of stability and inertness yielding the greatest deviation of the control variable, the usual step-wise variation can be accepted. In this case, optimality criteria are integral, as a rule.

Unfortunately, also in this case it was not possible to find a method of synthesizing an optimal regulator, which would account stability margins of the system. There can be disturbances inaccessible to control with unknown mathematical model of their actions.

Therefore, it is necessary to develop a procedure consisting of two stages:

- Acceptance of a hypothesis on the structure of the regulation algorithm.
- Computing optimal parameters of the accepted regulator structure (restricted synthesis) by a criterion, which gives the minimal deviation of the controlled value for any disturbance.

The hypothesis about the law of regulation can be accepted either on the basis of an optimiality criterion without restriction on stability margin, or with the help of experts. In particular, a PID algorithm can be chosen, what has been done in engineering during the past. Synthesis performed by minimizing the deviation variance criterion applying Wiener's theory of prediction (adding some features of real systems, e.g. the requirement that the deviation of the controlled value during automatic control should be essentially smaller than the deviation for no regulation) has shown that this algorithm is rather close to the optimal having properties of a prediction [4]. It puts under doubt the usefulness of numerous "advanced" algorithms (MIC regulators with internal model, Smith-type regulators, fuzzy-PID regulators, etc.).

The calculation of optimal parameters of a PID regulator should be carried out by methods strictly accounting all conditions that are valid for the accepted optimality criterion [4-6] (thus excluding Ziegler-Nichols methods). In particular, with the above given additional requirements on the quality of regulation (and a sufficiently large stock of stability), the minimum variance and short-term maximal deviation is reached by a maximum of the integral component of the PID regulator.

The analysis of the system quality is carried out after computing the optimal parameters of the regulator, and a PID algorithm may be taken into consideration. The demands for model validity should be maximal including random disturbances and parameter variations.

Let us notice that for the regulation within structures like in Fig.1, nonlinear algorithms that rest on system's reactions on step disturbances should be excluded. The application of such algorithms makes all closed loops nonlinear and the principle of accumulation of disturbances cannot be applied. However, all known nonlinear algorithms of regulation heavily base on step disturbances. Thus, there is no comparative analysis of behavior of such algorithms under real-life conditions with random disturbances. That is, the widespread approach of evaluating advantages of nonlinear algorithms over PID seems to be reconsidered. This refers to other "advanced" algorithms of regulation,as well. Nonlinear control algorithms can be used if (as it frequently happens in practice) the task is sufficiently determined in advance (for example, in maximal speed problems). See Fig. 2.

The increase of quality of regulation should be achieved by enhancing the information during "pure" regulation. For it, on the input of the regulator (besides the basic controlled variable) it is necessary to suppress signals from auxiliary adjustable variables or controllable disturbances [7]. In this case the system structure remains unchanged, only all variables should be considered as vectors (Fig.2).

Thus, the problem of synthesizing optimal regulation systems consists not only in optimal algorithms of regulation, but first of all in the search for optimal structures of regulator - object information realations. Unfortunately, the formal solution (except elementary cases like cascades) is not obvious. It is also necessary to solve this task with the help of experts, and the problem becomes rather difficult in situations with complex and several controlled variables.

We can conclude that the results obtained by the above approach must not be considered as final. This is due to the following reasons:

- The object model is afflicted with inevitable errors, and the estimation of the influence of these errors on the final result of synthesis is not possible in an obvious way.

- It is impossible to formulate criteria of approximation via the object model, since they depend on the functioning of the regulator to be designed and corresponding optimality criteria of the synthesized system. Thus, the task of identification of object is intrinsically inconsistent. Under these conditions it is necessary to address to expert estimations.

- The synthesis is carried out by methods of linear theory using linear model, while the real objects are nonlinear.

- In a real control system there are always factors that are not taken into account (for example, in the executive mechanisms of regulators with specific nonlinear characteristics), and which are difficult to handle. On the other hand, they may strongly influence the final result.

Thus we are faced with the necessity to determine the tuning parameters on working systems. Hence, any control system should be designed as adaptive. In general, service algorithms of adaptation should be included.

The process of adaptation should be accompanied by an identification of the object, and one can show that this cannot be performed merely by exploiting data under normal operating conditions. Thus, active experiments including special actions are necessary. Such actions can be [2,4,5]:

204

- Additional change of the set point or additional disturbances.
- Algorithmic actions, when the identifying actions consist in changing of algorithms of functioning of the regulator.
- Structural actions, when during the adaptation the system structure varies.

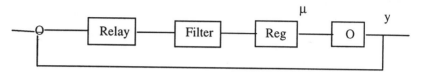

Fig. 3. The structure of adaptive system with relay in a loop of regulation

The additional change of the set point can realized by step or sinus function. In the first case the transient characteristic of the system is estimated, in the second the frequency characteristic is determined. The algorithmic action assumes the inclusion of additional nonlinear elements into the structure (usually one takes a relay with linear filter necessary for handling the parameters of auto-oscillation), see Fig. 3. The structural action can be realized by additional nonlinear feedback (Fig. 4), what, however, leads to complicated structures and auto-oscillation. All these measures decrease the quality of the system. That is why adaptation should be realized at minimal time with minimal deterioration of quality. On the other hand, this ebnforces the development of special step-by-step algorithms combining procedures of identification and optimal of tuning.

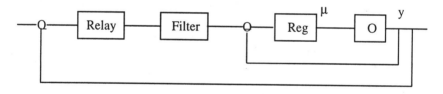

Fig. 4. The structure of adaptive system with relay in the special loop

The main difficulties are caused by random disturbances, which are always present during adaptation. To reduce their influence, methods of probability theory and random processes have to be applied. However, these methods often do not lead to the desired results. First, they are not applicable to auto-oscillatory algorithms of adaptation, since the basic process may change its character in principle. Second, probabilistic methods usually require an inadmissible duration of experiments. Last, at the final stage of synthesis, the nonlinear properties of the object should be taken into account.

Therefore, nowadays methods of adaptation have an automated character rather than an automatic. It is obvious that the participation of human operators is quite useful for formalizing expert methods that include fuzzy logic tools.

Conclusion

Control system synthesis cannot be formalized completely even in situations where the object can be described by mathematical models. The usage of expert knowledge (e.g. in heuristic form) seems to be essential at various stages of design. The demand for formalization makes the methods of fuzzy logic very attractive in a rather general environment.

In our opinion, the most promissing features for applying fuzzy logic to the synthesis of automatic systems are the following.

- Formulation of hypotheses about the structure of algorithms concerning the functioning of regulators and controllers.

- Formulation of hypotheses about the structure of information relations between regulator and object, especially for difficult control structures with many controlled variables.

- Formulation of a hypothesis about the structure of the transfer function on the base of the identification data preceeding the design of structure.

- On-line final tuning of the system with possible corrections in subsequent operations.

References

1. Horowitz, I.M.: Synthesis of Feedback Systems. Academic Press. New York and London (1963)
2. Rotach V. (ed.): Automation of Tuning of Control Systems. Energoatomizdat, Moscow (1984) (in Russian)
3. Bulgakov B.: Oscillations. GITTL., Moscow (1954) (in Russian).
4. Rotach, V. Calculation of Dynamics of Industrial Automatic Systems of Regulation. Energija, Moscow (1973), (in Russian)
5. Rotach, V.: The Theory of Automatic Control of Heat Power Processes. Energoatomizdat, Moscow. (1985) (in Russian)
6. Rotach, V.: Calculation of the Settings of Microprocessor Control Systems. Thermal Engineering. MAIK Nauka/Interperiodika Publishing 1 (1994)
7. Rotach, V.: Calculation of Cascaded Systems of Automatic Control. Thermal Engineering. MAIK Nauka/Interperiodika Publishing 10 (1997)

Comparative Analysis of Classical and Fuzzy PID Algorithms

Edik Arakeljan, Mark Panko, Vasili Usenko

Moscow Power Engineering Institute (Technical University)
Department of Automated Control Systems of Thermal Processes
pma@acsv.mpei.ac.ru

Abstract. Fuzzy PID controllers and principles of their performing are considered in a large number of publications. Some investigations deal with the comparison of PID and Fuzzy PID algorithms for application. A rather general comparative analysis of classical and Fuzzy PID algorithms will be given in the present paper.

1 Introduction

Industrial automatic control systems of complex non-linear, non-stationary objects in energetic, metallurgy, and chemical industries are performed as a rule by typical P- and PI-algorithms of regulation. These algorithms are wide-spread because of their

– tuning simplicity,
– sufficient dynamic accuracy,
– robustness.

The robust algorithms have proven their working capacity and ensure sufficient quality of control in case of system parameters variation.

However, one is faced with the situation that with increasing control system dynamic accuracy the algorithms become more complicated. The most complicated practical algorithm (among the typical ones) is the PID algorithm, which actually is not so widespread as P- and PI-algorithms. On the other hand, PID algorithms can enhance dynamic accuracy in comparison to PI regulator by one or two orders of magnitude (depending on the controlled system's characteristics).

The PID algorithm are to optimal Wiener algorithms for the controlled system without lag or with small lag and low frequency disturbances. As being suboptimal, a PID algorithm is nonrobust. Robustness of the system with PID algorithm can be achieved by proper tuning. However, PID algorithm is not effective in the case of considerable lag of controlled system.

The main reasons hampering the application of PID algorithms are the following:
– tuning complicity,
– high sensitivity to system parameters variation.

This is why there is a strong interest to develop methods its self-tuning, methods of adaptation and fuzzy PID controllers, in particular.

2 Classical (conventional) PID algorithm

The convenient form of classical PID algorithm presentation is

$$u(t) = k_P e(t) + \frac{k_P}{T_I} \int_0^t e(t) dt + k_P T_D \frac{de(t)}{dt}, \tag{1}$$

where k_P, T_I, T_D - setting (tuning) parameters.

Maximal ratio k_P / T_I gives minimal dispersion value $e(t)$ in case of low frequencies disturbances, and at the same time we get the minimal value of the linear integral quality criterion J_1. Practically, the maximal value of k_P/T_I is restricted not only by stability margins, but by robustness conditions either.

The great amount of PID regulators tuning methods (analytical, graphical, searching) [1-3] is caused by the complicity of the optimal parametric synthesis of control system with PID algorithm. The tuning parameters depends on characteristics of controlled system, explicit or implicit restrictions, which guarantee system stability margin, and accepted quality criterion. Additionally, it is necessary to ensure robustness of the system.

In the following we will present analytical method of tuning parameters determination. The analytical method gives opportunity to perform investigation of PID algorithm singularities.

One restriction is given by the stability margin leading to the consideration of closed loop characteristic equation roots. To get stability margin, it is necessary that among the roots $s_k = -\alpha_k + \omega_k$ there is none with oscillation index α_k/ω_k greater than a given value m. In accordance, stability margin criterion can be represented in the form

$$W_{OL}(m,j\omega) + 1 = W_P(m,j\omega)W_R(m,j\omega) + 1 = 0, \tag{2}$$

where $W_P(m,j\omega)$, $W_R(m,j\omega)$, $W_{OL}(m,j\omega)$ - expanded frequency responses of controlled system, regulator, and open loop system, respectively.

Expanded frequency response is defined as follows:

$$W(m,j\omega) = W(s)\big|_{s = -m\omega + j\omega} .$$

On the base of criterion (2), the following equations are obtained (for $\alpha = T_D/T_I$):

$$T_I(m,\omega,\alpha) = \frac{-A_1(m,\omega) - \sqrt{A_1(m,\omega)^2 - 4A_2(m,\omega,\alpha)A_0(m,\omega)}}{2A_2(m,\omega,\alpha)}, \qquad (3)$$

$$k_I(m,\omega,\alpha) = \frac{1}{T_I(m,\omega,\alpha)^2 B_2(m,\omega,\alpha) + T_I(m,\omega,\alpha)B_1(m,\omega) + B_0(m,\omega)}. \qquad (4)$$

The values T_I, $k_I = k_P/T_I$ are suited to describe the stability margin restriction. The right-hand sides of (3), (4) are functions of the controlled system expanded frequency responses:

$$B_0(m,\omega) = \frac{1}{\omega(1+m^2)}(m\,\mathrm{Re}[W_p(m,\omega)] - \mathrm{Im}[W_p(m,\omega)]),$$

$$B_1(m,\omega) = \mathrm{Re}[W_p(m,\omega)],$$

$$B_2(m,\omega,\alpha) = \alpha\omega(m\,\mathrm{Re}[W_p(m,\omega)] + \mathrm{Im}[W_p(m,\omega)]).$$

The gain and differential time constant are evaluated as follows:

$$k_P(m,\omega,\alpha) = k_I(m,\omega,\alpha)T_I(m,\omega,\alpha), \qquad (5)$$

$$T_D(m,\omega,\alpha) = \alpha T_I(m,\omega,\alpha). \qquad (6)$$

Equations (4-6) define the stability margin domain in the space of tuning parameters. Similar methods are used for digital PID controller

$$u[kT] = k_P\{(\frac{T}{T_I}+1+\frac{T_D}{T})e[kT] - (1+\frac{2T_D}{T})e[(k-1)T] +$$

$$+\frac{T_D}{T}e[(k-2)T]\} + u[(k-1)T]. \qquad (7)$$

Corresponding to (7), the analog PID algorithm may be written as

$$W_R(s) = \frac{k_P}{Ts}[(\frac{T}{T_I}+1+\frac{T_D}{T}) - (1+\frac{2T_D}{T})\exp(-Ts) + \frac{T_D}{T}\exp(-2Ts)]. \qquad (8)$$

where T - sampling time.
The influence of the sampling time value on the system stability margin can be considerable. An example has been considered for the controlled system with transfer function [4]

$$W_R(s) = 2/[(10s+1)(s+1)^2].$$

Calculations was made for the case of root oscillation index m=0.221, m=0.366 and different α. The quality criterion was the linear integral $(J_{1,min} \Leftarrow (k_P/T_I)_{max})$.
In Fig. 1 we illustrate the relationship between the degree of control system sensitivity and parameters variations.

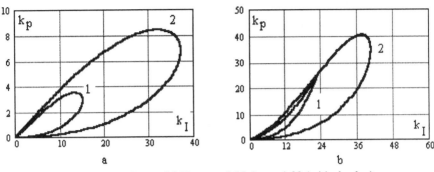

1 – m=0.366; 2 – m=0.221; a - α=0.25; b - α=1.02 (critical value)

Fig. 1. Constant stability margin lines m-const

The optimal analog PID algorithm parameters are presented in Table 1. Influence of sampling time value is shown in Table 2.

Table 1. The optimal analog PID algorithm parameters (m=0.366)

α	k_P	T_I	T_D	J_1	J_{1PI}/J_{1PID}	ω_{rez}
0	1.966	4.268	-	2.227	1	0.426
0.25	13.32	3.80	0.95	0.285	7.814	1.434
0.40	14.53	2.385	0.954	0.164	13.579	1.44
1.02*	22.69	0.933	0.952	0.041	54.317	1.434
1.20	20.70	1.015	1.218	0.049	45.449	0.9; 1.94

* - critical value

Table 2. Influence of sampling time value

			α = 0.25			
T	k_P	T_I	T_D	J_1	$J_{1,D}/J_{1,A}$	ω_{rez}
0	13.32	3.80	0.95	0.285		1.434
0.02	12.096	3.755	0.939	0.310	1.087	1.355
0.05	10.693	3.702	0.926	0.346	1.214	1.262
0.10	9.04	3.644	0.911	0.403	1.414	1.148

Our analysis shows some general properties of control systems with classical PID algorithm:
- the region of stability margin is restricted in the space of tuning parameters by complicate nonmonotonous surface;
- the quality of the system impoves, if the ratio T_D/T_I increases;
- ratio T_D / T_I is restricted and cannot be larger than the critical value (the quality of system and stability margin decreases, if this ratio exceeds the critical value);
- the system sensitivity to parameters variation increases with increasing ratio T_D/T_I;
- the ratio T_D/T_I is bounded in praxi and must be in the range 0.2 - 0.4 for sufficient system robustness;
- PID algorithm is effective in the case of low frequency external disturbances and small error;
- there is a considerable overshooting (larger than optimally adjusted PI algorithm yields) in the case of reference input step disturbance;
- adjustments reducing transient response overshooting for reference input step disturbances are not optimal with regard to other disturbances.

3 Fuzzy PID algorithms

The principles of fuzzy controllers design are investigated in a number of publications [4-6]. The main aims of fuzzy controller application are:
- improving of dynamic behaviour of the system (in comparison with standard algorithm),
- adaptation to changes of a system parameters.

Analysis shows a fuzzy PID controller is a fuzzy logic controller with three crisp input signals, which represent:
- error $e(t)$,
- integral of the error,
- derivative of the error,
and one analog crisp output signal $u(t)$ [4,5].

The Fuzzy PID is a discrete digital controller: the initial information is represented in digital form by values $e[kT]$, $e[(k - 1)T]$, $e[(k - 2)T]$; all necessary computations are digital.

Obviously, sampling time must be greater than the computation time. The quantization of signals in time reduces the stability margin of the system.
There are different variants of fuzzy controllers. Fuzzy PID controllers can differ by:

- basis of control rules,
- number of terms for each input variable,
- shapes of fuzzy sets,
-fuzzification methods,

– defuzzificatiom methods,
– structure of controller.

Fuzzy PID can be represented to respect of crisp input signal e(t) and output signal u(t) by equation (1) or in digital form as

$$u[k] = k_P e[k] + k_I e[k] + k_D [e[k] - e[k-1]] + u[k-1] .$$

In the case of fuzzy algorithm realisation, coefficients k_P, k_I, k_D are in general non-linear functions of controller input signals. Coefficients are defined by $F \Rightarrow RB \Rightarrow F^{-1}$ transformation and variation range of each input variable. The module $F \Rightarrow RB \Rightarrow F^{-1}$ realises static transformation of standardised input signal $x \in (-1,1)$ into output actuating signal $u \in (0,1)$ or its component u_i. The static characteristic of the module depends on form and number of terms in transformations F, F^{-1}, and the defuzzification method. The centre of gravity (COG) method is most frequently used.

This method is simple and can be employed for different kinds of membership functions. Defuzzification by COG method gives in general a non-linear static dependence $u = f(x)$. The character of nonlinearity depends on shape and number of terms in the membership function. Analysis shows that in the case of triangular membership function and linear rule base the dependence "input - output" is symmetrical and close to linear in the vicinity of $x = 0$ (the centre of input variable range). In the case of membership function with five terms, for example, the linear part of the characteristic practically covers the range of the input variable $-0.5 < x < 0.5$. Increasing the number of terms leads to expanding the "linear" part of the characteristic and does not changes its average slope. Module $F \Rightarrow RB \Rightarrow F^{-1}$ realises static transformation with the gain $k_M = 0.5$ in the "linear" region. The deviation from the linear dependence is increasing on the borders of the range. The length of "non-linear" parts on the borders is $2/(n - 1)$ where n is the number of terms. So in case of three terms membership function, the length of "non-linear" parts must be equal to 1, and dependence $u = f(x)$ considerably deviates from linear over the whole range. In this case , when x is small, the module gain is high, whereas increase of x leads to decrease of gain.

By correcting the borders terms, practically linear characteristics can be obtained with the same slope on the whole range.

Analysis of characteristics "crisp input signal x - crisp output u" shows that systems with classical digital and fuzzy PID controllers are equivalent in dynamic behaviour and actually differ by tuning procedures.

It is possible to obtain any given non-linear transformation "crisp input signal x - crisp output u" by selection of corresponding membership functions. But the main problem is how to find the optimal dependencies of tuning parameters upon e(t).

An interesting approach is the fuzzy gain adaptation of PID controller [5], which combines the advantages of classical PID algorithm and fuzzy adaptation.

Conclusions

The optimal adjustments of classical PID regulator can be determined analytically or by experienced specialist through "searching" methods on the base of analysis of system response on some definite input signal.

The adjustments of fuzzy PID may not be represented in explicit form. They are determined by the choice of the input and output signals variation range for given fuzification-defuzification module. It maybe interesting to illustrate the dynamic properties of fuzzy PID controller by set of its step and/or frequency responses.

Tuning of fuzzy PID controllers requires experienced specialists and tools for full scale modelling of the system.

References

1. Rotach V.J.: Theory of Automatic Control of Thermoenergetics Processes. Energoatomizdat, Moscow (1985) (in Russian)
2. Davidov N.I., Idzon O.M., Simonova O.V.: Definition of Tuning Parameters of PID Regulator by Controlled System Transient Response. Thermal Engineering no. 10 (1995) 17-22 (in Russian)
3. Rotach V.J.: Calculation of Setting of Real PID- Controllers. Thermal Engineering. Vol. 40, No.10. (1993) 775-779 (in Russian)
4. Pivonka P., Brejl M.: Use of Fuzzy PID Controllers in Fuzzy Control of Coal Power Plants. In: Proc. FUZZY 96, Zittau, Germany (1996) 441-448
5. Stegemann H., Worlitz F., Hampel R.: Fuzzy-Control Applications for Active Magnetic Supported Drives. In: Proc. 5-th Zittau Fuzzy Colloquimum Zittau, Germany (1997) 139-148
6. Lucas, V.A.: Introduction to Fuzzy Control. University of Mines, Ekaterinburg (1997) (in Russian)

Non-Fuzzy Knowledge-Rule-Based Controllers and their Optimisation by Means of Genetic Algorithms

Ivan Sekaj

Department of Automatic Control Systems, Faculty of Electrical Engineering and
Information Technology, Slovak University of Technology
Ilkovicova 3, 812 19 Bratislava, Slovak Republic
sekaj@kasr.elf.stuba.sk

Abstract. A non-fuzzy rule-based system for process control and general applications is described. The system uses a similar rule-base as fuzzy systems, but it
does not use fuzzy variables. The items of the knowledge-base are only real
numbers and the evaluation mechanism is based on a n-dimensional interpolation. It is very simple and computationally fast. Next a genetic algorithm-based
optimisation of the rule-base is shown. The proposed approach is demonstrated
on process control simulations and real-time examples as well.

1 Introduction

In various types of decision processes or in control applications, which use information and knowledge expressed in form of rules different approaches for interpretation and processing of such information can be applied. In this case very often fuzzy
systems are used. In this paper a new type of a rule-based system is described, which
makes use of a numeric-type knowledge evaluation. The rule-base is very similar to
rule-bases known from fuzzy systems. However, the proposed system does not use
fuzzy sets, but only crisp, real-number variables and the complete fuzzy mechanism is
replaced by a n-dimensional interpolation. It preserves a similar character and transparency of the interpreted information. The benefit of this method consists in its simplicity and robustness. The knowledge-data can be obtained by an empirical way, but
in this paper also an other optimisation method is shown, which is based on genetic
algorithms. The mentioned approach is explained on applications oriented to process
control.

2 The Numerical Rule-Based System and Rule-Based Controller

Consider a rule-based system, which inputs x_i are in general real variables from
bounded input intervals

$$I_i = \left\langle I_{i,\min}, I_{i,\max} \right\rangle \in R,$$

where $i=1,2,...,n$ is the number of input variables.

The system is defined on the bounded space $I_1 \times I_2 \times ... \times I_n$, which is the subspace of R^n. If each interval I_i will be defined by a discrete number of m_i points $\{\xi_{i,1}, \xi_{i,2}, ..., \xi_{i,m_i}\}$ we can create an orthogonal grid, which consists of $m_1 \times m_2 \times ... \times m_n$ points. These produce an n-dimensional rule-base or table containing the knowledge. To each point of the grid $[x_1, x_2, ... , x_n]$ an output value $b_{x_1, x_2, ..., x_n}$ is assigned, which represents a rule of the type

if x_1 is A_1 and x_2 is A_2 and ... and x_n is A_n then u is B

where A_i are values of the inputs x_i, u is the output variable, B is the output value, where A_i, B are real numbers .

However, usually, when using the system, the input vector $[x_1, x_2, ... , x_n]$ adresses not directly the nodal points of the grid, but the points among them. The basic mechanism for the system output evaluation is a simple interpolation among 2^n surrounding nodal points. For simplicity, let us assume a 2-dimensional case ($n=2$). The output value of a point located among the defined nodal values can be approximated by means of a simple linear expression

$$u_a = \frac{1}{d^2}[b_{p,q}(\xi_{1,p+1} - x_1)(\xi_{2,q+1} - x_2) + b_{p+1,q}(x_1 - \xi_{1,p})(\xi_{2,q+1} - x_2)$$
$$+ b_{p+1,q+1}(x_1 - \xi_{1,p})(x_2 - \xi_{2,q}) + b_{p,q+1}(\xi_{1,p+1} - x_1)(x_2 - \xi_{2,q})] \tag{1}$$

where d is the mutual distance of the nodal points on both axes (raster of the grid), b_{ij} are the appropriate values in the nodal points or elements of the rule-base, p is the index in the direction of the variable x_1 and q is the index in the direction of the variable x_2. For an n-dimensional problem the computation is analogical.

Such an interpretation of the knowledge-rule-based system is very simple, computationally very fast and, if the division of the input universes (the size of m_i) is properly chosen, it is also exact enough.

Remarks:
1. For the output value evaluation, any other n-dimensional interpolation method can be used.
2. To make the design of the rule-base easier, it is possible to name the areas around the values ξ like in fuzzy systems by lingvistic terms ("Positive Big", "Cold","Too fast" , etc.)

For illustration let us consider a simple example. Let a two-dimensional system have the inputs x_1 and x_2, which are defined on intervals $I_1 = I_2 = \langle 0;4 \rangle$ and let the rule-base which is defined in 5 nodal points $\{0, 1, 2, 3, 4\}$ with a unity raster ($d=1$) is displayed in Table 1.

Table 1. Example of a rule-base

x1/x2	0	1	2	3	4
0	7	5.2	4	1	2
1	7.5	8	7	3	4
2	8	9.8	9	4	5
3	5	4	1	7	5
4	2.1	3	2	5	6

Let the actual inputs have the values $x1=2.3$, $x2=2.2$. In the table, the co-ordinates of this input point, marked A, fall among the points $\xi_{p,q}$, $\xi_{p,q+1}$, $\xi_{p+1,q+1}$ and $\xi_{p+1,q}$, where $p=2$, $q=2$, with the corresponding output values $b_{p,q}=9$, $b_{p,q+1}=4$, $b_{p+1,q+1}=7$ and $b_{p+1,q}=1$ (Fig.1).

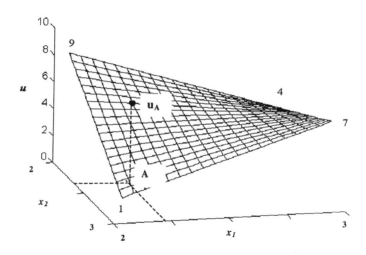

Fig.1. Graphic representation of an elementary field of the characteristic surface

The depicted elementary field is a part of the characteristic surface (input/output sur-face) of the system. Using (1) we can calculate an approximation of the output value (value of the appropriate surface point)

$$u_A = 9(3-2.3)(3-2.2)+4(2.3-2)(3-2.2)+7(2.3-2)(2.2-2)+1(3-2.3)(2.2-2) = 6.56.$$

The behaviour of the above described rule-based system is similar to a fuzzy system with symmetrical and triangular input membership functions, singleton output mem-bership functions and the COG defuzzification method.

Let us now apply the proposed algorithm for the replacement of the PID controller. Assume that the PID controller in the continuous-time domain is in the form

$$u(t) = Pe(t) + I\int e(t)dt + D\frac{de(t)}{dt} \tag{2}$$

where e is the control error and P, I and D are proportional, integrating and derivative gain, respectively. Let the inputs to the rule-based system $x_1(k)$, $x_2(k)$, $x_3(k)$ correspond to the discretized variables $e(k)$, $ie(k)$ and $de(k)$ of the signals $e(t)$, $\int e(t)dt$ and $de(t)/dt$, respectively, in the step k and let the control output is $u(k)$. Assume that the space of all input variables is $\langle min_i, max_i \rangle$, $i=1,2,3$, and, for simplicity, let the rule-base be of the size 3x3x3, which means each input space being defined in 3 points. In our case the rule-base is a 3-D grid with 27 points, which can be also expressed by means of three 2-D sections (see Tab.2). Each point of the grid represents a rule of the type

if x_1 is H_e and x_2 is H_{ie} and x_3 is H_{de} then u is U

where H_e, H_{ie}, H_{de} are actual values of the 3 inputs, and U is the corresponding output value.

Next let us show how to obtain a rule-base of a controller, which should have a behaviour equivalent with the conventional PID controller (2). If to each point of the grid is assigned a value

$$u = Px_1 + Ix_2 + Dx_3 \tag{3}$$

where x_1, x_2, x_3 gets the values $\{min_i, (min_i + max_i)/2, +max_i\}$, $i=1,2,3$, the controller step response will correspond to the PID one. In this manner we can obtain an equivalent rule-base also to other controller types.

However, the purpose of the rule-based controllers is not only to replace the conventional types of controllers with more complicated algorithms. The presented procedure can be used as a data initialisation, to which we can apply expert's knowledge with the aim to improve the control performance (similarly as in fuzzy systems). For illustration consider a single rule. If a maximal set-point jump appears, then the control error is maximal, but its derivative is still zero and its integral is big. The absolute value of the value at the position in the table corresponding to this situation, where originally has been the control value (or the control value increment) of an equivalent PID controller, will be increased. Thereby we can achieve an acceleration of the transient process. However, this can cause some overshoot. This overshoot can be eliminated by application of a different rule, which will slow down the process in a proper moment. After some similar modifications based on empirical knowledge it is possible to achieve the required, often very specific, closed loop behaviour in various phases of the transient process or in various operating points. Moreover it is also possible to increase the resistance towards noise and disturbances, etc.

3 Generation or Optimization of the Rule-Base by Means of Genetic Algorithms

The main disadvantage of the above mentioned approach is firstly a generally time consuming and often not a simple procedure of the rule-base tuning. Secondly, the obtained system is practically never optimal. If we have an appropriate model of the controlled object, it is possible to automate the rule-base data design procedure, or to optimise it according to the chosen requirements. For this purpose it is very advantageous to employ an optimisation procedure based on genetic algorithms (GA) which is sufficiently described in literature [2,3], and others. A more detailed description of the concrete genetic algorithm, which has been applied, practically without changes, also in this case is in [4]. A typical chromosome for solution to this problem has the form of a real-number string $ch=[U_{11}, U_{12},...,U_{33}]$, where $U_{i,j}$, $i,j=1,2,3$ contains all elements of the rule-base coded into the string. The objective function which is to be minimised consists from a computer simulation of the controlled process and from a quantification of a chosen performance index (e.g. integral control performance criterions). By choosing an appropriate criterion function it is possible to achieve e.g. time-optimised control processes, different dynamics at the set-point jump up and down, the reference response tracking, input energy minimisation, to design robust controllers, etc. A more detailed analysis of similar problems is given in [5,6].

4 Case Study

Now let us demonstrate all the mentioned phases of the controller design on an example. Consider a simple feedback configuration with the controlled object described by a linear model with the transfer function

$$G(s) = \frac{0.5s+1}{10s^3 + 8s^2 + 5s + 1} \quad .$$

For its control a PID controller (2) with parameters $P=97$, $I=1.7$, $D=56$ has been applied. In this case the controller parameters have been obtained via GA [5]. The input signal universes are $e \in \langle -1;1 \rangle; ie \in \langle -0.5;0.5 \rangle; de \in \langle -4;4 \rangle$. The closed-loop response under this controller after set-point jumps from 0 to 0.6 and then to -0.6 and after a disturbance affecting the input of the system is depicted in Fig.2, marked "a". The equivalent rule-based controller obtained using (3) and yielding the data in Tab.2 is marked "b" and is identical with the response "a". The corresponding control signals are in Fig.3.

Table 2. Controller rule-base equivalent with the PID controller

		e= -max	e = 0	e = max
de = -max	ie = -max	-321.85	-321.0	-320.15
	ie = 0	-224.85	-224.0	-223.15
	ie = max	-127.85	-127.0	-126.15
de = 0	ie = -max	-97.85	-97.0	-96.15
	ie = 0	-0.85	0.0	0.85
	ie = max	96.15	97.0	97.85
de = max	ie = -max	126.15	127.0	127.85
	ie = 0	223.15	224.0	224.85
	ie = max	320.15	321.0	321.85

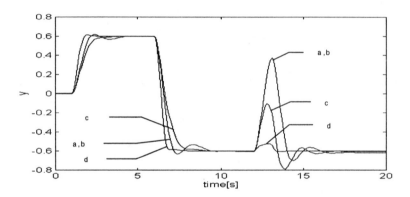

Fig.2. Closed-loop response with a PID controller (a), an equivalent rule-based controller (b), with manually optimised rule-base (c) and with a GA-optimised rule-base (d)

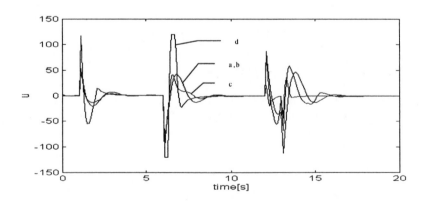

Fig.3. Control signals with a PID controller (a), an equivalent rule-based controller (b), with manually optimised rule-base (c) and with a GA-optimised rule-base (d)

Table 3. Manually optimised controller rule-base

		e= -max	e = 0	e = max
	ie = -max	-321.85	-321.0	-320.15
de = -max	ie = 0	-224.85	**-1000.0**	-223.15
	ie = max	-127.85	-127.0	-126.15
	ie = -max	**-300.0**	-97.0	-96.15
de = 0	ie = 0	-0.85	0.0	0.85
	ie = max	96.15	97.0	**300.0**
	ie = -max	126.15	127.0	127.85
de = max	ie = 0	223.15	**1000.0**	224.85
	ie = max	320.15	321.0	321.85

During the application of the empirical process knowledge some initial rules have been changed (Tab.3, the changed items are marked with a heavy line). The closed-loop response of this "manually" optimised controller is in Fig.2 marked "c".

In the last step the genetic algorithm optimisation has been employed. For this case the following performance index has been used

$$J = \int_0^T \left(\alpha |e(t)| + \beta |e'(t)| + \gamma |e''(t)| \right) dt \rightarrow \min \; ; \quad \alpha = \beta = \gamma = 1$$

where e' and e'' are the first and the second derivative of the control error. The solution obtained after 300 generation is listed in Tab.4. The system response with the corresponding controller is in Fig.2 marked "d". The control signals are in Fig.3.

Table 4. GA-optimised controller rule-base

		e = -max	e = 0	e = max
	ie = -max	-548.24	-2687.39	194.60
de = -max	ie = 0	-2571.13	-493.63	-214.85
	ie = max	-2021.92	21.23	149.12
	ie = -max	-1603.28	-546.98	-140.13
de = 0	ie = 0	-2.65	1.54	-0.43
	ie = max	1872.45	249.92	97.52
	ie = -max	495.40	43.69	1562.04
de = max	ie = 0	753.14	928.47	2259.02
	ie = max	-611.62	-417.20	334.19

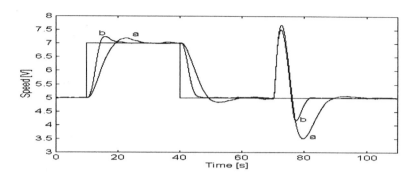

Fig.4. DC motor speed response under the PID controller (a) and the GA-optimised rule-based controller (b)

The proposed approach of the rule-based controller with the GA-optimisation procedure has been applied in real time for the DC motor speed control. In order to ensure sufficient resistance of the control loop towards the noise, the measured noise has to be added into the simulation model during the parameter evolution. Fig.4 depicts the response of the motor speed after two set-point step changes and after a pulse disturbance at the system input under the PID controller (a) and under the GA-optimised rule-based controller (b).

In the presented examples a rule-base of the size 3x3x3 has been considered. If the controlled object is non-linear or if a more complex control strategy is to be used, the size of the rule-base (refinement of the input range division) or also its dimension (in dependence on the used number of inputs) can be enlarged. It is also possible to use a non-linear transformation of the input signals as some analogy of the non-symmetric membership functions in fuzzy systems.

Besides the examples presented in this paper, the proposed rule-based contoller has been succesfully applied to control of different linear systems (oscillatory, non-mini-mum-phase etc.), for MIMO systems (2-inputs 2-outputs with interactions) as well as for nonlinear systems. A nonlinear system has been controlled, which gain and dynamics depend on the working point (output y). In this case a 4-dimensional rule-base with the inputs e, de, ie and y with sucess has been used.

All these examples show the application of the rule-based system for the direct generation of the control action. It is possible to use it in different control structures for the realisation of various functions. An application of a supervisory knowledge-based adaptive system for tuning classical PID controllers is described in [7]. Here the rules of the type

$$\text{if } x_1 \text{ is } H_e \text{ and } x_2 \text{ is } H_{ie} \text{ and } x_3 \text{ is } H_{de} \text{ then } P \text{ is } H_P, I \text{ is } H_I \text{ and } D \text{ is } H_D$$

are used, where the inputs x_1, x_2, x_3 relate to variables $e(k)$, $ie(k)$, $de(k)$, H_e, H_{ie} and H_{de} are their corresponding actual values and the values H_P, H_I, H_D are the new values of the controller parameters P,I and D (or their increments). Using another type of adaptation the inputs can represent e.g. maximal overshoot, settling time, system damping etc., where the rule-based system is not evaluated in each step of the control algorithm realisation but only in the time after the decay of the transient period [7,8]. If the system dynamics changes, the knowledge-based system evaluates new controller parameters.

5 Conclusion

This paper describes a new knowledge-based system which uses only real variables. The system employs a similar rule-base like a fuzzy system. It can be used for applications where the knowledge can be expressed in form of numeric information. It is very simple, computationally fast, well parametrisable and robust. There is also outlined a procedure of the rule-base data optimisation by means of genetic algorithms. It provides very good results in control applications as knowledge-based PID controller, or when used as a knowledge-based adaptive system in connection with conventional type controllers. Using this approach very good result has been achieved in simulation as well as in real time.

References

1. Sekaj, I.: Fuzzy Logic Controllers and their Replacement by a Non-Fuzzy Approximation Algorithm. In: Proc. of the IFAC Workshop – NTDCS'97, September 7-10, Smolenice, Slovak Rep. (1997) 354-360
2. Man, K.F., Tang, K.S., Kwong, S., Halang, W.A.: Genetic algorithms for control and Signal Processing (Advances in Industrial Control). Springer Verlag (1997)
3. Goldberg, D.: Genetic Algorithms in Search, Optimization and Machine Learning. Addison-Wessley Publ. (1989)
4. Sekaj, I.: Usage of Genetic Algorithms for Knowledge-Based Adaptive Controller Data Generating and Optimisation. In: Proc. of the PC'99 Conferrence, May 31-June 3, Tatranske Matliare, Slovak Rep. (1999) 323-327
5. Sekaj, I.: Genetic Algorithm-Based Control System Design and System Identification. In: Proc. of the conference Mendel'99, June 9-12, Brno, Czech Rep., (1999) 139-144
6. Husák, J.: Application of Genetic Algorithms for Control System Design. Diploma Thesis, FEI STU, Bratislava, Slovak Rep. (1999) (in slovak)
7. Sekaj, I.: Knowledge-Based Selftuning Controllers. In: Proc. of the Conference Mendel'98, June 24-26, Brno, Czech Rep. (1998) 202-207
8. Pfeiffer, B.M., Isermann, R.: Selftuning of Classical Controllers with Fuzzy Logic. In: Proc. of IMACS Symp. "Mathematical and Intelligent Models in System Simulation", Brussels (1993)

A New Method of Fuzzy Petri Net Synthesis and its Application for Control Systems Design

Jacek Kluska and Lesław Gniewek

Department of Electrical Engineering and Computer Science,
Rzeszow University of Technology,
W. Pola 2, PL-35959 Rzeszow, Poland
{jacklu, lgniewek}@prz-rzeszow.pl

Abstract. A new method of fuzzy Petri net synthesis and its application for control systems design is presented. The work of Misiurewicz [5] that describes directions for use some class of Petri net for binary control systems synthesis, became the inspiration for our work. Our approach is based on fuzzy logic [4] and provides significant benefits in comparison with the classical concept mentioned above. Both an architecture and dynamics of fuzzy Petri net are described. An example of the control systems design and its software implementation on PLC is presented.

1 Introduction

Complex industrial processes can be often decomposed into many parallel subprocesses, which can, in turn, be modelled using Petri nets. Such models belong to very useful in the engineering practice, if we have the possibility to use a net model to the control system design.

Misiurewicz's method described in [5] is a very attractive method for engineers from industry, because it provides a relatively simple transformation procedure of Petri net to a logic circuit. Many examples of systems may be given, for which after drawing the Petri net as a control algorithm, it is possible to obtain the logic scheme of the control system. The resulting control system may be assembled using conventional flip-flops and binary gates. However, this method is based only on boolean logic in Petri net description and therefore the control system as a computer program or hardware device, is capable only for binary information processing.

2 Fuzzy Petri Net

There is essential difference between classical binary and fuzzy Petri net (see [1]). In the fuzzy Petri net, both with the places, and the transitions, we associate some fuzzy events. This means that each transition corresponds to degree of fulfillment of some condition, which is the number from the interval [0,1].

The place may be fuzzy as well, i.e. the value of a marker belongs to the interval $[0,1]$. Logical interpretation of truth values associated with transitions and places should follow from the modelled situation. The examples taken from engineering practice will be given in the sequel, where the logical interpretation should not raise doubts.

As an algebraic system, the considered fuzzy Petri net (FPN for short) is 9-tuple

$$FPN = \langle P, T, D, G, R, \Delta, \Gamma, \Theta, M_0 \rangle \qquad (1)$$

where $P = \{p_1, \ldots, p_r\}$ is a finite set of places, $T = \{t_1, \ldots, t_s\}$ - a finite set of transitions, $D = \{d_1, \ldots, d_r\}$ - a finite set of statements, $G = \{g_1, \ldots, g_s\}$ - a finite set of conditions, $R \subseteq (P \times T) \cup (T \times P)$ - incidence relation, $\Delta : P \longrightarrow D$ - the function assigning statements for the places, $\Gamma : T \longrightarrow G$ - the function assigning conditions for transitions, $\Theta : T \longrightarrow [0, 1]$ - the function determining the degrees to which conditions related to transitions are satisfied, $M_0 : P \longrightarrow [0, 1]$ - the initial marking function.

We restrict ourselves to clean Petri nets for which the following conditions

$$\{(t_1, p) \in R, (t_2, p) \in R \Rightarrow t_1 = t_2\}, \{(p, t_1) \in R, (p, t_2) \in R \Rightarrow t_1 = t_2\} \qquad (2)$$

are satisfied. This eliminates conflicts and simplifies the method of the synthesis, which will be characterized in the sequel. Furthermore, the arcs have unity weights and places have unity capacity.

3 Firing a Transition

Let us assume that the marking $M : P \longrightarrow [0, 1]$ is given. The transition $t \in T$ is enabled from the moment at which the following conditions

$$\forall p \in {}^{\bullet}t, \quad M(p) = 1 \quad and \quad \forall p \in t^{\bullet}, \quad M(p) = 0 \qquad (3)$$

are satisfied, to the moment, when the following conditions

$$\forall p \in {}^{\bullet}t, \quad M(p) = 0 \quad and \quad \forall p \in t^{\bullet}, \quad M(p) = 1 \qquad (4)$$

hold, where ${}^{\bullet}t = \{p \in P \mid (p, t) \in R\}$ and $t^{\bullet} = \{p \in P \mid (t, p) \in R\}$.

If the transition $t \in T$ is enabled by the marking M and $\vartheta \in [0, 1]$ is the degree to which the condition associated with this transition is satisfied, then the new marking $M'(p)$ of the place $p \in P$ is calculated by the following rule

$$M'(p) = \begin{cases} M(p) \wedge (1 - \vartheta) & \Longleftrightarrow & p \in {}^{\bullet}t \setminus t^{\bullet} \\ M(p) \vee \vartheta & \Longleftrightarrow & p \in t^{\bullet} \setminus {}^{\bullet}t , \quad (\wedge = \min, \vee = \max) \\ M(p) & & \text{otherwise} \end{cases} \qquad (5)$$

The dynamics of such FPN were investigated by authors in the other work, and theorems which enable one to calculate the next marking for all places of the net (using some matrix operations) were proved.

4 Synthesis of FPN

We propose a new synthesis method that enables the transformation of FPN to logical scheme in which a fuzzy JK flip-flop belongs to the most important elements. Unfortunately, the fuzzy JK flip-flops described as yet in the literature can not be used for realization of considered FPN. Therefore a new family of fuzzy JK flip-flops were developed and described in [2], which can be applied for the FPN realization. Fuzzy JK flip-flops use three operations: bounded sum, bounded product and complementation. The following four fuzzy JK flip-flops were worked out: $JK_{SA}, JK_{AA}, JK_{AB}, JK_{SB}$, for which the outputs are given correspondingly by:

$$Q_{SA}(t+1) = \{[J \otimes n(Q)\} \oplus Q] \otimes [n(K) \oplus n(Q)] \tag{6}$$

$$Q_{AA}(t+1) = [(J \oplus Q) \otimes n(Q)] \oplus [n(K) \otimes Q] \tag{7}$$

$$Q_{AB}(t+1) = \{[Z_1 \oplus n(Z_2)] \otimes Z_2\} \oplus n(Z_3) \otimes Z_3 \tag{8}$$

$$Q_{SB}(t+1) = \{[V_1 \oplus n(V_2)] \otimes V_2\} \oplus n(V_3) \otimes V_3 \tag{9}$$

where

$$Z_1 = (J \otimes K) \oplus n(K), Z_2 = [n(K) \otimes Q] \oplus J, Z_3 = [J \otimes n(Q)] \oplus n(K) \tag{10}$$

$$V_1 = (J \otimes K) \oplus n(K), V_2 = [J \otimes n(Q)] \oplus Q, V_3 = [n(K) \otimes Q] \oplus n(Q) \tag{11}$$

and the operations are defined by

$$n(X) = 1-X, \quad X \oplus Y = 1 \wedge (X+Y), \quad X \otimes Y = 0 \vee (X+Y-1), \quad \forall X, Y \in [0,1]$$

with $\wedge = \min$ and $\vee = \max$.
Properties of the flip-flops are described in details in [2].

5 Logical Scheme of the Control System

In order to obtain a logical scheme of the control system based on FPN, we should assign for each place in the net - the fuzzy JK flip-flop JK_{AA} or JK_{AB}, one for one place. Each transitions corresponds to the fuzzy gate "$\otimes$" together with the flip-flop of T-type. This flip-flop responds to only 2 input signals: "exactly 0" and "exactly 1".

An example of transformation the FPN into the logical scheme is shown in Fig. 1.

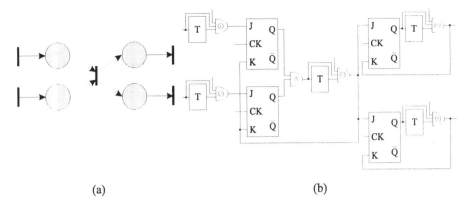

Fig. 1. (a) - The fragment of the FPN, (b) - The corresponding logic scheme.

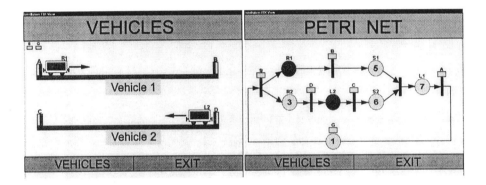

Fig. 2. Two vehicles controlled by binary Petri net.

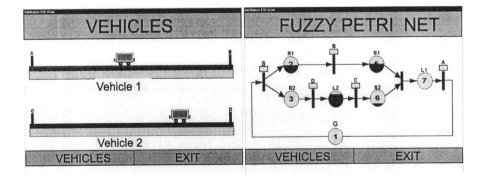

Fig. 3. Two vehicles controlled by fuzzy Petri net.

6 Examples of Control Systems Design

Consider two vehicles as in Fig. 2 and Fig. 3. They are starting simultaneously from the points A and C, correspondingly. Each of them should travel there and back; the first vehicle between points A and B, and the second one between C and D. However, the second vehicle can not return before the first one will return.

The control algorithms have been modeled and experimentally proved using both binary (Fig. 2) and fuzzy Petri net (Fig. 3). Next, they have been realized using Programmable Logic Controller S7-314 (Siemens, Step-7 language).

7 Summary

As a result of using the proposed synthesis method one obtains a multi-input-multi-output control system, that is rather difficult to obtain if we base only on the methods offered by classical control theory or classical digital circuits design techniques. The resulting control system may be realized as software for Programmable Logic Controller (or universal computer) or hardware device assembled on the Programmable Logic Device. In spite of the systems are designed using methods rooted in the logic (fuzzy logic), we get the control device capable of receiving, processing and generating both binary, and analog signals. As yet, similar methods are not well worked out; their development seems to be very important from engineering practice point of view.

We address the method of synthesis rather to large processes that can be decomposed for many parallel operating subprocesses. In the case when analog sensors or actuators are applied instead of binary ones, the FPN as the control systems operates equally well as the net obtained from classical Misiurewicz's method. So, if in the control system there are binary sensors instead of analog ones, no changes in the hardware or software are needed to ensure correct work of the control systems. Furthermore, in the presence of some disturbances in measurement paths, e.g. vibration of input liquid level or temporarily absence of information from the distance sensors, the FPN turned out to be more robust than the classical binary Petri net. The FPN was realized both as computer program for PLC and hardware (on ispLSI chips of Lattice [3]) in the Department of Control and Computer Science at Rzeszow University of Technology. In the future, the design process on PLD will be probably fully automated.

References

1. Cardoso J., Camargo H. (eds): Fuzziness in Petri nets. Kacprzyk J. (ed.-in-chief): Studies in Fuzziness and Soft Computing, Vol. 22. Springer-Verlag, Physica-Verlag Heidelberg New York (1999)
2. Gniewek L., Kluska J.: Family of fuzzy J-K flip-flops based on bounded product, bounded sum and complementation. IEEE Trans. on Systems, Man a. Cybernetics, Part B: Cybernetics 28, 6 (1998) 861-868

3. Gniewek L., Hołota B.: Implementation of fuzzy logic circuits using ispLSI structures of Lattice (in Polish). Measurement-Automation-Control. Part I 3 (1997) 72-75; Part II 4 (1997) 108-110
4. Kluska J.: Fuzzy logic control (in Polish). Rzeszow University of Technology. Rzeszow (1992).
5. Misiurewicz P.: Problems of digital control systems design (in Polish). Proc 8-th National Automation Conference in Szczecin 1 (1980) 664-670

Fuzzy Control Applications

Improving Control Behaviour with Set-Point Pre-Processors

Peter Vogrin and Wolfgang A. Halang

Faculty of Electrical Engineering, FernUniversität
D-58084 Hagen, Germany

wolfgang.halang@fernuni-hagen.de

Abstract. The design principle of control algorithms working with set-point pre-processors, called SPP controllers, is described. Set-point pre-processors calculate internal set-point graphs of controlled variables in such a way that very high controller gains are attainable and, thus, stability is increased. The behaviour of SPP controllers is normally much closer to the "best physically possible" controller performance, and much more predictable than the one of conventional control structures. Furthermore, otherwise conflicting design objectives can nearly all be achieved, e.g., stability, safety, high speed, small energy consumption, or steady and harmonic temporal controller output values. Since SPP control algorithms allow for extremely fast reactions, it was obvious to approximate their mathematical models by rule base tables allowing their implementation by safety licensable fuzzy controllers.

1   Introduction

With regard to the following two main reasons, the performance of state-of-the-art controllers employed in safety critical systems is often very unsatisfying:

1. Owing to the (limited) possibilities for safety licensing and the guidelines of the licensing authorities, the design principles of control equipment have to be quite elementary. In particular, according to the *"List of Type Approved Programmable Electronic Systems"* [1] published by TÜV Rheinland, *it is prohibited to use PID or other more complex control algorithms in safety related applications.*

2. Another detriment to present controllers is their inherent lack of two aspects of speed:
   - Owing to their mathematical models, control systems react slowly after modifications of their set-points, and after unforeseen events, such as disturbances, defects and alterations of parameters of the technical processes being controlled.
   - Loop execution time.

Present controllers often exhibit a big gap between real and desired control system performance, which has its reason in their mathematical models. The set-point pre-processor (SPP) controllers introduced in this paper promise a clear improvement of controller behaviour in many application cases. A set-point pre-processor computes the internal set-point graphs of a controller in such a way, that its real behaviour is as close as possible to its desired behaviour. The effort to design such a controller is relatively small if the mathematical model of the technical process does not contain any considerable dead time or lag elements, as is the case for robots. The attainable speed and stability of SPP robot controllers are several times higher than for conventional controllers, even if the mathematical model of the technical process to be controlled could only approximately be taken into consideration in designing their set-point pre-processors.

In this paper, the algorithm of a SPP robot controller is designed as a fuzzy controller with rectangular input membership functions, called IF controller [2]. The rectangular classification scheme is implemented by linear or non-linear analogue-to-digital converters. The controller operation is based on an inference scheme which requires just look-ups in a table containing a fuzzy rule set. Software only takes the form of rules in a table, lending itself to rigorous verification by diverse back translation [3]. Generally, IF controllers have two major benefits as compared to controllers with conventional structures. Even though IF controllers are relatively simple to be safety licensable, they can, in principle, approximate any control algorithm with sufficient precision. The second great advantage of IF controllers is their extremely short loop execution time.

2 Controllers with Set-Point Pre-Processors

The set-point pre-processor of a SPP controller calculates the internal set-point graph in such a way that the maximum difference between the set-point graph and the real position is much smaller than the maximum control error of a conventional controller. Furthermore, the temporal graph of this difference is steady and harmonic. Thus, higher controller gain is possible and, therefore, feedback control systems become more stable. Another great advantage of SPP controllers is the continuous temporal behaviour of their actuations. Moreover, compared with conventional controllers such as PD or PID, the influence of disturbances on controller performance is very small.

The only differences between the design principle of a conventional controller of a single loop feedback system as shown in Figure 1

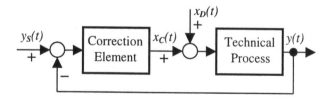

Fig. 1. Principle of a common conventional control system

and the one of a SPP controller as depicted in Figure 2 are that a SPP controller contains a set-point pre-processor and that its correction element gains are higher.

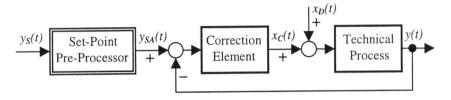

Fig. 2. Principle of a controller working with a set-point pre-processor (SPP controller)

The set-point pre-processor input $y_S(t)$ is the desired temporal graph of the controlled value. Its output is the internal set-point graph $y_{SA}(t)$. Using the difference $y_{SA}(t) - y(t)$ instead of the control error $y_S(t) - y(t)$ as static controller relation inputs leads to the great advantage that the behaviour of the controlled system is much more predictable as the one of a conventional controller. Moreover, the real performance of a properly designed SPP controller is usually much closer to the "best physically possible" controller behaviour. Thus, in controller design priorities and objectives are almost freely selectable.

2.1 Example: Control of a Sleigh on a Horizontal Line

As displayed in Figure 3, we consider the very elementary example of position control of a sleigh on a horizontal line. We shall only show the basic design principle of controllers working with set-point pre-processors. The design and the behaviour of a conventional PD controller (PD controller 1) is compared with the ones of two PD controllers working with set-point pre-processors (SPP controllers 2 and 3).

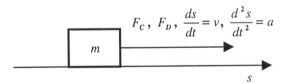

Fig. 3. Sleigh on a horizontal line

The mathematical model is characterised by the following differential equation:

$$F_C(t) + F_D(t) = m \bullet \frac{d^2 s(t)}{dt^2} = m \bullet a(t) \tag{1}$$

The total force $F_C + F_D$ and the speed of the sleigh v have to be zero in the stationary state. Discrete degrees of freedom of robots often have similar properties. The control problem is to guide the sleigh's position s by adjusting the force F_C. The controlled system has an integrative behaviour. Thus, the performance of the best PD controller is better than the performance of the best PID controller, if the disturbing force F_D is zero.

Design of the Conventional PD Controller 1. In this example, the force controlling the sleigh $F_C(t)$ is limited between -25 N and 25 N. It was tried to design controllers with the best performance assuming that the maximum amount of the set-point jump is 100 m. Moreover, the control errors due to disturbing forces $F_D(t)$ should be as small as possible. The control deviation $s_S(t) - s(t)$ of the conventional PD controller 1 may be very large. The force $F_C(t)$ is about $(s_S(t) - s(t))$ multiplied by the proportional amplification P. Thus, P has to be rather small, otherwise the controlled system will be unstable. The chosen mathematical algorithm of the correction element is described by the following equation:

$$F_D(t) = P \bullet (s_S(t) - s(t)) + PD \bullet v(t) = 3.5 \text{ Nm}^{-1} \bullet (s_S(t) - s(t)) - 3 \text{ Nsm}^{-1} \bullet v(t) \qquad (2)$$

These assumptions lead to some great disadvantages, viz., inherent tardiness after a set-point jump and after a sudden change of $F_D(t)$. Another detriment is the large per-manent control error if a permanent force disturbs the sleigh.

Construction of the SPP Controllers. As described above, controller design objec-tives are nearly freely selectable. In this example, we consider the desired temporal graph of the force controlling the sleigh $F_A(t)$. The task of the set-point pre-processor is, therefore, the computation of the internal set-point graph $s_{SA}(t)$ by assuming the de-sired temporal graphs $s_S(t)$ and $F_A(t)$, whereas s_S is the set-point of the controlled value.

There is a static relationship between the position of the sleigh $s(t)$ and the force accelerating it. So, lag or dead time elements are not contained in the mathematical model of the controlled system. We do not consider occasionally occurring disturbing forces $(F_D = 0)$. The sleigh mass is $m = 1$ kg. These considerations lead to a simple equation of the set-point pre-processor's calculator:

$$s_{SA}(t_1) = s_{SA}(t_0) + \frac{1}{m} \bullet \int_{t0}^{t1} \left\{ \int_{t0}^{t1} F_A(t) \bullet dt \right\} \bullet dt \qquad (3)$$

Set-Point Pre-Processor of the SPP controller 2: The set-point pre-processor calculates the fastest internal set-point graph physically possible by considering that the controlling force is within the controller's actuation range $(-25 \text{ N} \leq F_C(t) \leq 25 \text{ N})$. Therefore, $F_A(t)$ is either the maximum or the minimum of the controlling force $F_C(t)$ or zero. This property has the great disadvantage that the controller has small force reserves if the real technical process is different from the one considered in the design of the controller, e.g., due to disturbances or aging processes. An example of an internal set-point graph $s_{SA}(t)$ is shown in Figure 4.

Set-Point Pre-Processor of the SPP controller 3: The design principles of the SPP controllers 2 and 3 resemble each other. The only difference is that $F_A(t)$ is limited between −6.25 N and 6.25 N. Thus, the theoretical reserve force of the SPP controller 3 is 25 N − 6.25 N = 18.75 N. On the other hand, as displayed in Figure 4, the time after which the internal set-point graph $s_{SA}(t)$ reaches the set-point $s_S(t)$ is twice as long.

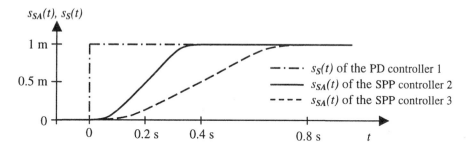

Fig. 4. Correction element inputs leading to a set-point jump from $s(t) = 0$ to 1 m

Correction Element of the SPP controllers: Generally, the design principle of controllers working with set-point pre-processors is to distinguish between the desired temporal graph of the controlled value (in this example, $s_S(t)$) and its internal set-point graph ($s_{SA}(t)$). The mathematical algorithm of a set-point pre-processor is designed in such a way that the difference between the internal set-point graph $s_{SA}(t)$ and the actual controlled value $s(t)$ is rather small. Due to this property, it is both possible and essential that the amplifications of the controller are high. Owing to the small difference $s_{SA}(t) - s(t)$, it is possible to design a controller with high amplification, even though its actuations are limited. Additionally, also in order to decrease this difference ($s_{SA}(t) - s(t)$), the controller's amplification has to be large. The temporal graph $s_{SA}(t) - s(t)$ and, thus, the performance of the controller primarily depends on the controller's proportional amplification P_P. Therefore, SPP controllers should be designed according to the following procedure:

1. Design of the set-point pre-processor. The controlled values should be able to follow the internal set-point graphs even if the controlled technical process is disturbed by any anticipated fault.

2. Coarse determination of the controller's proportional amplification P_P. Afterwards, its differential and integral amplifications P_D and P_I are set. Meaningful values of P_P, P_D and P_I mutually affect each other.

These considerations lead to the following actuation function of the SPP controllers:

$$F_D(t) = P_P \bullet (s_{SA}(t) - s(t)) + P_D \bullet v(t) = 250 \text{ Nm}^{-1} \bullet (s_S(t) - s(t)) - 26 \text{ Nsm}^{-1} \bullet v(t) \quad (4)$$

Their proportional amplification P_P is 250 Nm^{-1} instead of $P = 3.5$ Nm^{-1} by the conventional PD controller 1.

Performance of the Controllers: As shown in Figures 4 and 5, at the moment t_1 (in this example, $t_1 = 0$) a set-point jump immediately leads to a sudden change of the conventional controller's control error. At the time t_1 the control deviation jump $s_S(t_1) - s(t_1)$ corresponds to the amount of the set-point jump (in this example, 1 m). Moreover, also the actuation $F_C(t_1)$ jumps at this time as displayed in Figure 6. With regard to the stability of the controlled system, the amplification of the controller has to be small, particularly if the maximum amount of the set-point jump is high. Therefore, conventional controllers are usually slow and not very stable. The aperiodic border case is assumed in the design of the conventional PD controller 1. Thus, the sleigh reaches its desired values $s_S(t)$ without overshoots if it is not greatly disturbed. However, only small enlargements of the controller's amplifications result in big overshoots, and still, the time after which the set-point is reached the first time is much longer than the one for an SPP controller which exhibits a minute overshoot, at most.

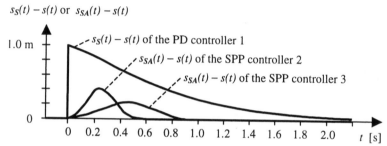

Fig. 5. Control errors after a set-point jump from zero to $s = 1$ m

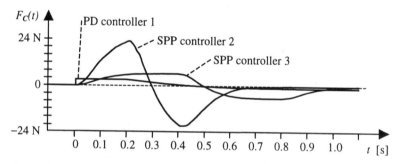

Fig. 6. Controlling force $F_C(t)$ after a set-point jump from zero to $s = 1$ m

In contrast to the control deviation of a conventional controller, the SPP controller's temporal graph of its difference $s_{SA}(t) - s(t)$ is continuous. Therefore, the actuations $F_C(t)$ of an SPP controller are very continuous. In this example, the actuation graph $F_C(t)$ roughly has the shape of one sine wave. The positive force halfwave speeds up the sleigh towards the set-point. Afterwards, the negative force halfwave decreases the sleigh's speed and guides it to the desired position. The temporal graph of the controlling force $F_C(t)$ and, therefore, its maximum and minimum amounts can be well determined. Due to these properties, the maximum amount of $F_C(t)$ and the reaction speed of SPP controllers can be much larger than the ones of conventional controllers as depicted in Figures 6 and 7 and Table 1. Nevertheless, the sleigh reaches its desired position without overshoots.

Table 1. Comparison of the performance of the PD controller 1 **(1)**, the behaviour of the SPP controller 2 **(2)**, and the performance of the SPP controller 3 **(3)** after a set-point jump from zero to $s_S = 1$ m or to $s_S = 100$ m, respectively. s_B is the amount of the first overshoot. t_D is the time after which the output error is smaller than 0.1 m (t_E: $|\Delta s| < 0.01$ m; t_F: $|\Delta s| < 0.001$ m)

type	s_S [m]	overshoot s_B [m]	$|\Delta s| < 0.1$ t_D [s]	$|\Delta s| < 0.01$ t_E [s]	$|\Delta s| < 0.001$ t_F [s]	range of F [N] F_{C-}	F_{C+}
(1)	1	-	1.4	2.6	3.6	-0.76	3.5
(2)	1	-	0.43	0.56	0.67	-20.92	22.92
(3)	1	0.0004	0.73	0.88	0.96	-6.28	6.27
(1)	100	13.3	7.0	8.0	9.0	-25	25
(2)	100	-	4.03	4.16	4.27	-25	25
(3)	100	-	7.9	8.14	8.35	-6.25	6.25

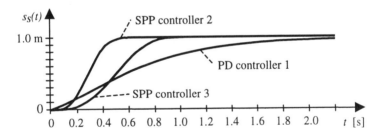

Fig. 7. Controlled value $s(t)$ after a set-point jump from zero to $s = 1$ m

3 Example

We consider **position control of a robot with two degrees of freedom**. The robot arm rotates on a horizontal plane around a pivot with the angular velocity ω. The distance r between the burden and the pivot of the robot is changeable with the velocity v. Thereby, the axis of the translation movement runs through the pivot. The task is to control the position of the burden with the mass m_L, which is located at the end of the robot arm, by adjusting the force F_C and the torque M_C.

3.1 Design of a Conventional Robot Controller 1

The principle of non-linear decoupling of systems as presented in [4] is used as design principle for this controller. Its controlling force $F_C(t)$ and its controlling torque $M_C(t)$ are not limited. It was tried to find the parameters of the fastest controller that leads to the assumption of the desired values without considerable overshoots by considering that $m_L = 20$ kg. Thus, the conventional controller 1 is described by the following equations for the distance r and the angle of rotation φ as functions of time:

$$F_C(t) = 70\text{Nm}^{-1} \bullet (r_S(t) - r(t)) - 119\text{Nsm}^{-1} \bullet v(t) - \{70\text{Nm}^{-1}\text{s}^2 \bullet r(t) - 25\text{Ns}^2\} \bullet \omega^2(t) \quad (5)$$

$$M_C(t) = \{17.67\text{kgm}^2 - 50\text{kgm} \bullet r(t) + 70\text{kg} \bullet r^2(t)\} \bullet [1\text{s}^{-2} \bullet (\varphi_S(t) - \varphi(t)) - 1.75\text{s}^{-1} \bullet \omega(t)] +$$

$$+ \{140\text{kg} \bullet r(t) - 50\text{kgm}\} \bullet v(t) \bullet \omega(t) \quad (6)$$

3.2 Design of an SPP Robot Controller 2

The conventional robot controller 1 and the SPP robot controller 2 compensate for the centrifugal force and the Coriolis torque of the robot. Thus, the translation dynamic controller only affects the dynamics of the robot arm translation, and the dynamics of the robot rotation is only influenced by the rotation dynamic controller. As shown in Figure 8, apart from the facts that the dynamic controllers work with a set-point pre-processor, and that the control algorithm is approximated by an IF controller, the SPP robot controller 2 has the same design as the robot controller 1. The set-point pre-processor of the SPP robot controller 2 calculates the internal set-point graphs of the controlled values $r_{SA}(t)$ and $\varphi_{SA}(t)$ by assuming that the necessary actuations controlling the robot are within the ranges of possible actuations of the control system. Thus, the position of the burden $r(t)$ and $\varphi(t)$ is, in principle, able to follow the internal set-point graphs. With this example, we shall show that SPP controllers exhibit excellent performance under any type of stress, even if the technical process to be controlled is insufficiently understood when constructing the set-point pre-processor. Thereby, the following simplifications are made: The non-linearities of the mathematical process model, the disturbing force $F_D(t)$ and the disturbing torque $M_D(t)$, and the influences of the controllers' differential parts on the actuations are all not considered. Thus, $r_{SA}(t)$ and $\varphi_{SA}(t)$ have similar shapes. An example of $r_{SA}(t)$ is shown in Figure 9. As described above, the real position of the burden is very close to $r_{SA}(t)$ and $\varphi_{SA}(t)$. Thus, the burden motion is of utmost continuity.

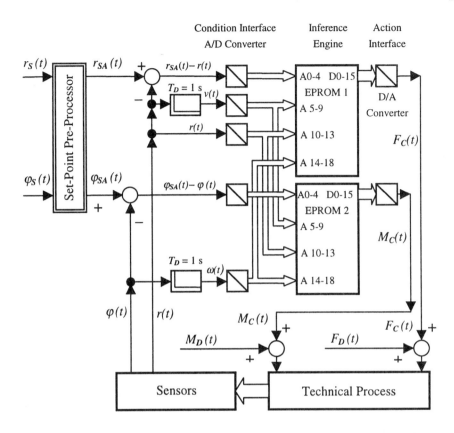

Fig. 8. Principle of the SPP robot controller 2

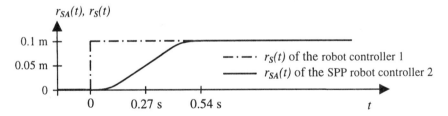

Fig. 9. Correction element inputs leading to a set-point jump from $r(t) = 0$ to 0.1 m

The mathematical model of the SPP robot controller 2 is characterised by the following equations:

$$F_C(t) = P_{Pr} \bullet (r_{SA}(t) - r(t)) + P_{Dr} \bullet v(t) - \{70 \text{ Nm}^{-1}\text{s}^2 \bullet r(t) - 25 \text{ Ns}^2\} \bullet \omega^2(t) \qquad (7)$$

$$\underbrace{\hspace{4.5cm}}_{\text{Dynamic Controller}} \qquad \underbrace{\hspace{5.5cm}}_{\text{Decoupling of the Centrifugal Force}}$$

$$M_C(t) = P_{P\varphi} \bullet (\varphi_{SA}(t) - \varphi(t)) + P_{D\varphi} \bullet \omega(t) + \{140 \text{ kg} \bullet r(t) - 50 \text{ kgm}\} \bullet v(t) \bullet \omega(t) \qquad (8)$$

$$\underbrace{\hspace{4.5cm}}_{\text{Dynamic Controller}} \qquad \underbrace{\hspace{5.5cm}}_{\text{Decoupling of the Coriolis Torque}}$$

It is both possible and essential that the constants P_{Pr} , $P_{P\varphi}$, P_{Dr} and $P_{D\varphi}$ of the dynamic controllers are high. Furthermore, meaningful values of the proportional amplification, e.g., P_{Pr} , and of the differential amplification, e.g., P_{Dr} , of a single dynamic controller mutually affect each other. However, otherwise these amplifications are nearly freely selectable. Since the controller compensates for the centrifugal force and the Coriolis torque of the robot, it is possible to design a single dynamic controller without consideration of the robot's other degree of freedom. The compared controllers are optimised for simultaneous set-point jumps from zero to $r = 0.1$ m and $\varphi = 1$ rad. The rotation and the translation set-point jumps need about the same build-up periods and start at the same time to provide for the most difficult case to control. It was tried to find the parameters of the best controller with the properties described above. They are:

$$P_{Pr} = 5800 \text{ Nm}^{-1}, \quad P_{P\varphi} = 4000 \text{ Nm}, \quad P_{Dr} = -1260 \text{ Nsm}^{-1} \quad \text{and} \quad P_{D\varphi} = -450 \text{ Nms}$$

In this example, the proportional amplifications of the SPP controller 2 are more than 80 times larger than the proportional amplifications of the conventional controller 1.

As displayed in Figure 8, the dynamic controllers and the decoupling of the controller (centrifugal force and Coriolis torque) are designed as safety licensable IF controllers. The rule base table controlling the translation is stored on EPROM 1, and the rule base table controlling the rotation is stored on EPROM 2. A section of the rule base table of EPROM 1 is shown in Table 2.

Table 2. Part of the rule base table (cause effect table) stored in EPROM 1

Cause				Effect
if	*and*	*and*	*and*	*then*
interval of the input ω	interval of the input r	interval of the input v	interval of the input $r_{SA} - r$	actuation F_C (N)
30	11	5	28	122.69
30	11	5	29	234.05
30	11	5	30	591.33
30	11	6	0	−570.18
30	11	6	1	−212.90
30	11	6	2	−101.54

For the signals $r_{SA}(t) - r(t)$, $v(t)$, $\varphi_{SA}(t) - \varphi(t)$ and $\omega(t)$ analogue-to-digital converters with unequally spaced input intervals featuring higher precision around zero as shown in Figure 10 are selected. Due to the unequally spaced input intervals of the A/D converters, under the given conditions the EPROM storage space required is by orders of magnitude less than the storage space required for a controller with equally spaced input intervals. The SPP robot controller 2 contains two 16 bits wide 0.5 Mwords EPROMs. A further effective way to reduce the storage space required is cascading IF controllers (rule base tables). For instance, it is also possible to design the SPP robot controller 2 with one 16 Kwords EPROM, two 1 Kwords EPROMs and one 0.5 Kwords EPROM, only.

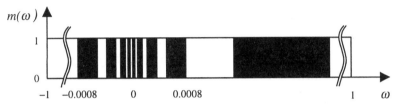

Fig. 10. Typical unequally spaced input intervals (e.g., angular velocity ω in rads^{-1})

It was tried to design the best controllers by considering that $m_L = 20$ kg. In Table 3, we compare the performance of the conventional robot controller 1 **(1)** with the performance of the SPP robot controller 2 **(2)** under a number of circumstances: mass of the burden $m_L = 20$ kg **(a)**; $m_L = 0$ kg **(b)**; $m_L = 50$ kg **(c)**; $m_L = 20$ kg by considering damping due to springs **(d)**; $m_L = 20$ kg by assuming coincident disturbing impulses $F_D(t)$ and $M_D(t)$ **(e)**.

Table 3. Behaviour after simultaneous set-point jumps from zero to $r_S = 0.1$ m and $\varphi_S = 1$ rad **(a-d)**, and after disturbing impulses (100 N and 100 Nm) with the impulse width 1 s **(e)**

simulation	translation movement						rotation movement					
	r_O (10^{-3} m)		t_r (s)		$\Delta r_{(t=\infty)}$ (10^{-3} m)		φ_O (10^{-3} rad)		t_φ (s)		$\Delta\varphi_{(t=\infty)}$ (10^{-3} rad)	
	(1)	(2)	(1)	(2)	(1)	(2)	(1)	(2)	(1)	(2)	(1)	(2)
(a)	0.09	0	5.8	1.21	0	0	0.16	0	9.7	1.21	0	0
(b)	0	0	9.3	1.49	0	0	0.08	0	7.8	1.27	0	0
(c)	3.7	3.8	11.6	1.40	0	0	0.32	0	10.7	1.34	0	0
(d)	0	0	never	never	59	1.7	0	0	never	never	113	0.4
(e)	570	13.2	10.48	0.96	0	0	3340	26.3	12.08	0.41	0	0

Performance is expressed in terms of the amount of the first overshoot r_O (φ_O), the time t_r (t_φ) after which the output error is smaller than 0.0001 m (rad), and the permanent control error $\Delta r_{(t=\infty)}$ ($\Delta\varphi_{(t=\infty)}$).

The desired actuation functions of the SPP robot controller 2 are approximated by a safety licensable IF controller. It works very well and shows excellent performance. This example reveals that the SPP robot controller 2 has, in addition to the ones mentioned in the Conclusion and the Introduction, many advantages as compared to the conventional robot controller 1:

- The time after which the control error is smaller than 0.0001 m after a translation set-point jump is between 4 and 8 times shorter, and the time after which the control error is smaller than 0.0001 rad after a rotation set-point jump is between 6 and 8 times shorter.

- A permanent force disturbing the robot arm leads to an about 80 times smaller control error, and a permanent disturbing torque leads to a 200 times smaller control error.

- The permanent control error due to damping because of springs $\Delta r_{(t=\infty)}$ is about 35 times smaller, and $\Delta \varphi_{(t=\infty)}$ is more than 200 times smaller.

- The overshoot after a force impulse disturbing the robot arm is 43 times smaller, and a disturbing torque impulse leads to an about 120 times smaller overshoot. Furthermore, the length of the period of time between the time after which the disturbing force F_D (torque M_D) vanishes after an F_D (M_D) impulse and the time after which the control error is smaller than 0.0001 m (rad) is about 10 times (30 times) shorter.

Conclusion

SPP controllers make use of an advanced mathematical control algorithm. As described above and in Table 4, the use of SPP controllers normally leads to essential performance improvements, even if the mathematical model of a technical process assumed in the design of a controller is very coarse and inaccurate. In contrast to the latter, IF controllers are hardware structures. The expenditure to design IF controllers is reasonably small. Nevertheless, IF controllers approximate any control algorithms with sufficient precision. Especially if the structure of an IF controller is customised to the mathematical model of a given controller, the memory requirements for the corresponding rule base table often become surprisingly small. An IF controller consists of a few relatively simple, industry standard hardware modules. Hence, safety licensing of its hardware can follow well-understood and long-established procedures. The main task of a safety proof is to verify a rule set's correct implementation. For this, inspection or diverse back translation is used. This method is essentially informal and immediately applicable without any training. So, people with heterogeneous educational backgrounds are able to use diverse back translation very effectively due to its ease of understanding. A prototype of an IF controller with a 64 Kwords EPROM has already been built. Table 5 summarises the benefits achieved with IF controllers. The SPP robot controller 2 described in this paper has the advantages of both SPP controllers and IF controllers.

Table 4. Comparison of the properties of SPP controllers and of conventional controllers

Properties of **SPP controllers**	Properties of **conventional controllers**
The control behaviour can be determined almost freely within the physical limits.	The control behaviour can be determined only coarsely within small ranges.
The objectives of controller design can be selected almost freely, e.g., stability, speed, reduction of expenses and ecological damages due to a technical process.	Often the main design problem is the stability of control systems, and it is not possible to meet other design objectives as well.
The controller gain is very high. Thus, the stability of control systems against disturbances and changes of technical processes is extremely high. Furthermore, the process speed attainable is high.	The controller gain is small. Control system are, therefore, very sensitive to disturbances and parameter changes of the processes to be controlled, and control systems react slowly.
The temporal graphs of actuations are steady and continuous.	The actuation shapes are unsteady. This leads to clear performance degradation.

Table 5. Comparison of the properties of IF controllers and of conventional controllers

Properties of **IF controllers**	Properties of **conventional controllers**
They are licensable for the highest safety requirements, viz., Safety Integrity Level 4 of IEC 61508, although they can approximate any arbitrarily complex control algorithm.	According to the authorities, it is not admitted to use PID or other complex control algorithms in safety related applications.
The use of appropriate, complex control algorithms enhances the inherent safety of technical processes to be controlled.	Very simple safety techniques lead to reductions of the inherent safety of controlled systems.
Operating in a strictly cyclic fashion, IF controllers exhibit fully predictable real-time behaviour.	The time needed for operation is usually not predictable. Thus, proper real-time behaviour cannot be guaranteed for these controllers.
Their loop execution times are extremely short, e.g., the one of the prototype is 0.2 microseconds.	The loop execution times normally exceed 1000 microseconds.

The use of SPP controllers promises a clear improvement of controller performance in many application cases. Furthermore, the use of IF controllers for safety related functions eliminates the disadvantages due to the restrictions imposed by the licensing authorities, e.g., TÜV [1].

References

1. Melchers, W.: List of Type Approved Programmable Electronic Systems (PES), Version 3.2. TÜV Rheinland Sicherheit und Umweltschutz GmbH, Institute for Software, Electronics, Railroad Technology (ISEB), Cologne (1998)
2. Halang, W.A., Colnari, M., Vogrin, P.: Safety Licensable Inference Controller. In: Proc. 6th Zittau Fuzzy Colloquium. Wissenschaftliche Berichte Heft 54, Nr. 1643-1670, pp. 18-23, Hochschule für Technik, Wirtschaft und Sozialwesen Zittau/Görlitz, 1998
3. Krebs, H., Haspel, U.: Ein Verfahren zur Software-Verifikation. Regelungstechnische Praxis rtp 26 (1984) 73-78
4. Hoyer, H., Freund, E.: The Principle of Non-linear Decoupling of Systems with Application to Industrial Robots. Regelungstechnische Praxis rtp 22 (1980) 80-87 and 116-126

Automation of Process Pretuning by a Fuzzy-Similar Method

Edik Arakeljan, Vasili Usenko

Moscow Power Engineering Institute (Technical University)

eka@acsv.mpei.ac.ru

Abstract. The combined base of rules consisting from crisp and fuzzy components is developed. With its help the procedure of adjustment smoothing the filter ensuring condition for optimisation of systems of automatic control (crisp and fuzzy) is automated. The opportunity of use for these purposes a priory of the information contained in the special points of random process is shown.

1 Introduction

Today it is widely accepted that the optimisation of systems of regulation by PID controllers consists in finding the optimal values of parameters k_p, k_i, k_D

$$u(t) = k_p\, e(t) + k_i \int_0^t e(t)dt + k_D \frac{de(t)}{dt}. \qquad (1)$$

Though this is an important stage of work, in practice, however, pre-tuning of the period should preceed this stage, since otherwise optimisation of the system (crisp or fuzzy) will be impossible.

At the pre-tuning stage it is necessary to estimate real random noise and to find measures to decrease it. Usually this is carried out by smoothing pertubations, for what in conventional regulators a filter with adjusting parameter β is applied. Moreover, the tolerance region Δ and duration of signal Δt is given beforehand. Parameter Δt is necessary for PID-regulators, which are realised as a relay impulse device, what is widespread (at least, in Russia). Adjustment of β, Δ and Δt is not easy to formalise. Usually, their concrete values are determined by the operator via practice-oriented considerations.

The adjustment of the system starts with process observation, which is subject to automation. First of all, the properties of noise and its level (more precisely the ratio "signal - noise") are estimated. As a result of observations the operator gets information about the character of random noise and about the reaction of the object to operator's actions. Then he forms up some ideas on models of object and operating noises. In the simplest case the model of process is given by an additive mix of two components without correlation

$$y(t) = x(t) + r(t), \qquad (2)$$

where $x(t)$ – desired signal, $r(t)$ – noise.

Experience shows that in the process of adjusting preferbly qualitative characteristics of signal and noise like "the level of a noise high", "low", "commensurable with a signal" (frequency ranges of noise and signal are decribed by "close", "varying", and "essentially varying") are used rather crisp numerical values. On the basis of such kind of information the operator adjusts the parameter of the filter etc., applying linguistic rules like:

if "noise is essential" *then* "strongly increase β".

Though building up such rule bases is possible in principle, one is faced with several difficulties. First, there is no simple automatic procedure to transform the operator's visual estimations into usable information [1]. Second, the generated rules are valid only to for the particular object under consideration, whereas other objects may require a considerable updating. Thus we aim at building rule bases reflecting adjustment strategies of the filter in a rather general environment. In particular, we will replace the operator's visual estimations by special characteristics of the [2]. Therefore, we use a ratio characterizing the intersections between the real (measured) process and some smoothed value (here we take the mathematical expectation). Another important indicator is the average number of escapes from the tolerance region.

2 Smoothing Random Noise

Since as a rule, high-frequency noise makes the system uncontrollable, some smoothing (filtering) procedure should be applied to reduce the influence of disturbances. Theoretically, the best way of filtering a signal with noise is to use an optimal Wiener filter. However, for its use it is necessary to have the exact information about frequency properties of the signals and disturbances, and variances, what unrealistic in practice. A proven substitute is given by the so-called pseudo-optimal filters [3]. Their filtering properties only negigibly differ from the optimal. However, for discrete realisation, exponential smoothing (from now on denoted as filter) is most simple [4]. The transformation of the discrete input signal $y(n)$ to the target signal Y (n) is given by

$$Y(n) = \beta \cdot Y(n-1) + (1-\beta) \cdot Y(n) \qquad (3)$$

where $\beta \in (0,1)$ is the adjusting parameter.

The optimal adjusting parameters could be found if the knowing the ratio of signal - noise variances and frequency ranges of $x(t)$, $r(t)$ were known. Since this information is usually absent, during the preparation of the automatic system the operator visually estimates the specified parameters. Thus, the level of noise can be decreased to an acceptable value. In this case, the accepted parameter value of the filter is a compromise between optimisation and restrictions [4]. It is well known that with increasing parameter values the noise level is reduced. On the other hand, the probability of distorting the desired signal is increased (Fig.1). Changing β to 85% does not influence the character of the signal essentially (Fig.1a). For other objects with other frequency properties (Fig.1b) the filter may deform the signal already for β= 0.7. Further increase β will change the signal qualitatively, what is infeasible. We can conclude that complete noise suppression very often leads to an essential falsification of the signal (obviously, that for β = 1 the filter output always will be zero). Therefore the acceptable value

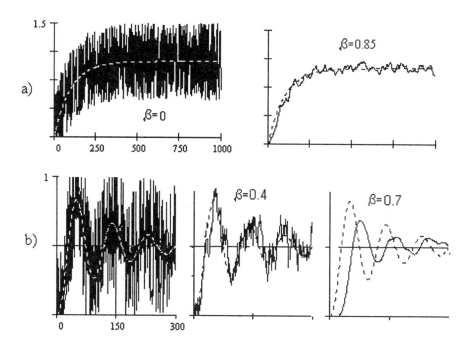

Fig.1. Influence of β

of β (close to the optimum meaning) supposes a small level of a disturbance. Further reduction can be achieved by introducing dead zones.

3 Estimating the Noise

As an estimation of the frequency properties of a process we use information of the number of its intersections with the mathematical expectation over a given time horizon (Fig.1), namely

$$n_0 = \frac{1}{\Delta t} \cdot card\, \Omega \qquad (4)$$

where $\Omega = \{j : r_j \cdot r_{j-1} \le 0.01\}$ and *card* denotes cardinality; Δt - interval of observation; r_j - noise.

248

Over a long time period, n_0 can be estimated by mathematical statistics. This, however, limits its practical use, because the time horizon need not be sufficiently large in reality. On the other hand, in a fuzzy environment great accuracy of n_0 is not necessary. It is enough to have qualitative estimations like "much" or "little" that allows to use rather "small" intervals of observation (5-10 min). As it is seen from Fig.2, n_0 of course depends on β.

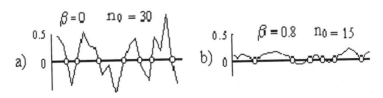

Fig.2. Relationship between β and n_0

For $\beta = 0.8$ the quantity of intersections has decreased twice with simultaneous reduction of the variance whose value can be estimated through

$$D = -\frac{C^2}{2 \ln (n_c/n_0)} \qquad (5)$$

Here, n_c is the number of escaping the tolerance region by the process and C is a suited contant. It is described in more detail below by (7). The estimate of D can be used for first estimation of β. If $0.5 \le n_c/n_o \le 1$, the standard deviation of the process can be approximately found by

$$\sigma \approx \frac{0.5 \cdot C}{1 - n_c/n_o} \qquad (6)$$

Fig. 3 shows the process behavior with repsect to the tolerance region.

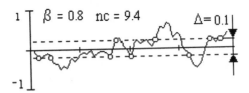

Fig.3. Process escaping the tolerance region

The number of crossing the tolerance border is computed as follows:

$$nc = \frac{1}{\Delta t} \cdot \sum_j nc1(j) + \frac{1}{\Delta t} \cdot \sum_j nc2(j) \qquad (7)$$

where $nc1(j))$ - condition of an output(exit) from a zone $\Delta/2$
$nc2(j)$ - condition of an output(exit) from a zone $-\Delta/2$.

where $nc1(j))$ - condition of an output(exit) from a zone $\Delta/2$
$\qquad nc2(j)$ - condition of an output(exit) from a zone $-\Delta/2$.
This important characteristic of regulation quality of a process shows the number of times a regulator is switched on.

4 Rule Bases for Filter Adjustment

For adjustment of the filter the following rule based system has been developed.

$$
\begin{aligned}
\text{CB: if } |\Delta\beta| &\geq 0.02 \text{ and region} = 0 \text{ then RB1} \\
\text{if } |\Delta\beta| &< 0.02 \text{ and region} = 0 \text{ then region} = \Delta \text{ and RB2} \\
\text{if } |\Delta\beta| &\geq 0.02 \text{ and region} = \Delta \text{ then region} = 0 \text{ and RB1} \qquad (8) \\
\text{if } 0.01 &\leq |\Delta\beta| < 0.02 \text{ and region} = \Delta \text{ then RB2} \\
\text{if } |\Delta\beta| &\leq 0.01 \text{ and region} = \Delta \text{ then STOP.}
\end{aligned}
$$

$$
\begin{array}{ll}
\text{RB1: 1. if } n_o = S \text{ then } \Delta\beta = S & \text{RB2: 1. if } n_c = s \text{ then } \Delta\beta = s \\
\quad\ 2. \text{ if } n_o = Z \text{ then } \Delta\beta = Z & \quad\ \ 2. \text{ if } n_c = z \text{ then } \Delta\beta = z \qquad (9) \\
\quad\ 3. \text{ if } n_o = B \text{ then } \Delta\beta = B & \quad\ \ 3. \text{ if } n_c = b \text{ then } \Delta\beta = b
\end{array}
$$

The memberships for the granulations in RB1 and RB2 are given as follows.

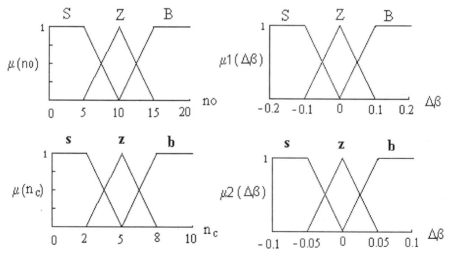

Fig.4. Granulations in RB1 and RB2

The crisp values of $\Delta\beta$ are calculated with the help of standard center-of-gravity methods.

$$CG1 := \frac{\int\limits_{-0.2}^{0.2} \Delta\beta \cdot \mu1(\Delta\beta) \cdot d\Delta\beta}{\int\limits_{-0.2}^{0.2} \mu1(\Delta\beta) \cdot d\Delta\beta} \quad ; \quad CG2 := \frac{\int\limits_{-0.1}^{0.1} \Delta\beta \cdot \mu2(\Delta\beta) \cdot d\Delta\beta}{\int\limits_{-0.1}^{0.1} \mu2(\Delta\beta) \cdot d\Delta\beta} . \tag{10}$$

The algorithm works as follows.

1. Put initial values for βo and , compute

$$\beta = \beta o + \Delta\beta \tag{11}$$

2. Depending on new the values $\Delta\beta$, regulation is performed by RB1 or RB2. In the former case, the tolerance region is set to zero, otherwise to Δ.
3. New correction of β will be carried out after period Δt with updating n_o and n_c.
4. If the $\Delta\beta \leq 0.01$, the adjustment of the filter is interupted or is transfered into adaptation mode.

In Fig.5 an example of automatic adjustment of the filter with $\Delta = 0.1$ is given.

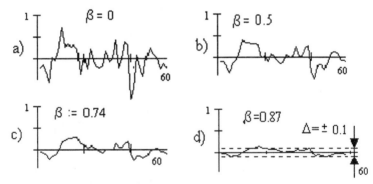

Fig.5. Automatic adjustment of the filter

5 Conclusion

For the optimization of fuzzy or crisp controllers the pre-tuning stage is investigated in detail. Particularly, we introduce tolerance regions and apply random noise smoothing. This part of work can be automated by using intersection points of the mathematical expectation. The use of combined rule bases enables an adaptation system for the adjustment of filter smoothing under realistic conditions. Obviously, automation of any process or object requires adequate information about the system under consideration whereas the question "crisp or fuzzy" is of less importance if the available amount of information is suffient This is due to the simple fact that in these cases both approaches can be applied. However, if we are faced with an essential lack of information or measuring is incomplete (or impossible), fuzzy tools may be of limited applicability, as well [6].

References

1. Meiritz, A., Schildt, G.H.: Model of an Adaptive Fuzzy Controller with Explicit Transfer Function. In: Proc. FUZZY'96 (1996) 58-66. Zittau, Germany
2. Bendat, J.: Principles and Applications of Random Noise Theory. John Wiley (1962)
3. Volgin, V., Karimov, R., Usenko, V.: Optimisation of Parameters of Smoothing of the Filter Current Average. In: Proc. MPEI (1969) (in Russian)
4. Volgin, V., Usenko, V.: Application of the Filter of Exponential Smoothing in Tasks of Definition of the Statistical Characteristics of Industrial Processes. In: Proc. MPEI 136 (1972) (in Russian)
5. Bellman, R.E., Zadeh, L.A.: Decision-Making in a Fuzzy Environment. Management Sci. 17 (1970) B141-B164
6. Demenkov, N, I.Motcalov, I.: Fuzzy LogicRegulator in Tasks of Management. Industrial Controllers of ACU 2 (1999) 30-35 (in Russian)

Selection of Mathematics Model for Daily Diagram Prediction According to Box-Jenkins Method

Jaroslav Baláte[1] and Bronislav Chramcov[1]

[1]Technical University Brno
Faculty of Technology in Zlín
Nám. TGM 275, 762 72 Zlín, Czech Republik
{balate,chramcov}@zlin.vutbr.cz

Abstract: The paper concerns selection of mathematical statistical model for prediction of time series of Heat Supply Daily Diagram (HSDD). This forecast is utilized for control of technological process in real time. The solved prediction is specially determined for qualitative-quantitative method of hot-water piping heat output control – Baláte system which enables to eliminate transport delay by the qualitative control only. Importance of it is increased namely for controlled systems having great time constants and great transport delay. The model according to BOX-JENKINS is used for calculation of time series prediction. The credibility of the tested model is estimated from three points of view.

1   Introduction

An improvement of technological process control level can be achieved by creating mathematical model of investigated process (in our case of heat supply diagram – HSD) and verifying the results of the model with the real course and using this mathematical model for prediction. Calculation of future HSD course prediction enables then to get new information of better quality for control of the process itself. Importance of it increases namely for controlled systems having great time constants and great transport delay. It is suitable to use the knowledge from the probability theory and mathematical statistics for calculation because the HSDD course has a form of time series with random behavior. The methodology according to Box-Jenkins (BJ) can be used for calculation of forecast of HSDD course. This methodology is based on the correlation analysis of time series and it works with stochastic models, which enable to give a true picture of trend component and also of periodic components.

2   The Methods of Prediction

The course of time series of HSDD (see Fig.1) contains two periodic components (daily and weekly period), but general model according to BJ enables to describe only one periodic component. We can propose two possible accessions to calculation of forecast to describe both periodic components.

- The method, that uses the model with double filtration
- The method – superposition of models

First we introduce simplified form (2) of general model according to BJ for the next using, when there is used substitution in the form (1). We can find more detailed analysis of general model in [1].

$$F = \Phi_P^{-1}\left(B^s\right) \cdot \varphi_p^{-1}(B) \cdot \Theta_Q\left(B^s\right) \cdot \theta_q(B) \cdot \nabla_s^{-D} \cdot \nabla^{-d} \tag{1}$$

$$z_t = F \cdot a_t \tag{2}$$

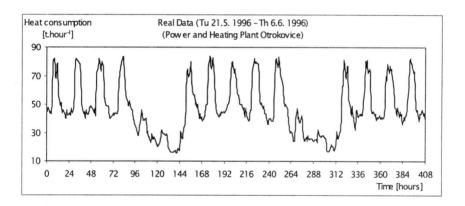

Fig. 1. Heat supply daily diagram

2.1 The Method that Uses Model with Double Filtration

We can describe model with double filtration through the substitution (1). The model in the form (3) is the result of it.

$$z_t = F \cdot \nabla_{s^*}^{D^*} \cdot a_t \tag{3}$$

where:

D   - degree of seasonal difference – daily (in equation 1)
D*   - degree of seasonal difference – weekly
s   - daily period (in equation 1)
s*   - weekly period

It is important to adhere to this general plan for using the method that uses model with double filtration for calculation of HSDD prediction.

- The filtration of time series is executed for the reason of elimination of weekly periodic component.
- This filtered time series can be described by means of general model according BJ and then calculation of forecast by means of course can be executed; that is provided in [1].
- It is important to do back transformation, that is inverse to the point a), because we have executed elimination of weekly periodic component.

The model in the form (3) enables to describe the HSDD course (i.e. it describes daily periodic component and also weekly one). It can be used for analysis and prediction of following regular influence of calendar (Saturday, Sunday).

2.2 The Method – Superposition of Models

We can use second method i.e. superposition of models for elimination of regular influence of calendar. This method was published in the work [2]. This method works with two time series, whence it follows, that this method uses on two models in the form (2). These time series are discerned by means of symbols * and **. The time series inscribed with symbol * is series of values of HSDD outputs in every hour (the sampling period is 1 hour) and the time series inscribed by means of symbol ** is series of values of heat consumption per day (the sampling period is 1 day). The plan of calculating prediction by means of the method of superposition of models is shown on the Fig. 2. We can find more detailed analysis in [2].

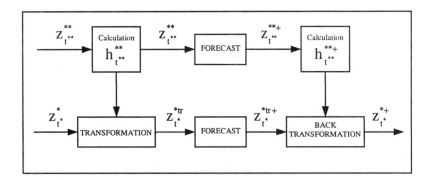

Fig. 2. Superposition of models – plan of calculating prediction

3 Evaluation of Prediction

Many real data measured in various localities are used for the purpose of testing mathematical statistical model (e.g. Heating Plant Olomouc). These obtained real data enable to execute comparison of the predicted values of HSDD with the real measured data. This comparison can be executed for various lead time (L). The credibility is estimated from three points of view. It is evaluated:

- average relative error $v(L)$ expressed in the form (4)

$$v(L) = \frac{100}{L} \sum_{i=1}^{L} \frac{\left| z_{N+i} - z_{N+i}^{+} \right|}{z_{N+i}} \quad [\%] \tag{4}$$

where:

z   - a real value
z^{+} - a predicted value

- square root of mean square error $\lambda(L)$ expressed in the form (5)

$$\lambda(L) = \sqrt{\frac{1}{L} \sum_{i=1}^{L} (z_{N+i} - z_{N+i}^{+})^2} \quad \left[t \cdot hour^{-1} \right] \tag{5}$$

- ratio of areas under the curve of real data and predicted data.
- for information it is found also the maximum variance of real data and predicted data.

4   Selection of Parameters of Mathematical Model - Program in Matlab

Pursuant to the mentioned theory and literature a program was created in Matlab, which enables to choose available mathematical statistical model for calculation of prediction of HSDD course. All testing is based on lot of real data. These data were obtained in specific locality and they are processed for next using in text file form. The program is drawn in user's menu and by help of that it is possible to choose many parameters of this program. After start of this program, first it is necessary to load pack of real data, which are saved as text file. Then we can set up parameters of calculation through menu "Option". It belongs to setting:

Prediction Length – number of predicted values
Sampling Period – selection of sampling period (10 minutes, 15 minutes, 30 minutes and one hour; selection is dependent upon period of real data, which is available)
Historical Area – selection of historical area beginning
Probability Limit – zone of future values occurrence with selected probability (50%, 75%, 90%, 95%, and 99%)

After opening file we get number of data, which are at disposal and also at what sampling period they were obtained. This information is displayed in input window (see in Fig. 3).

256

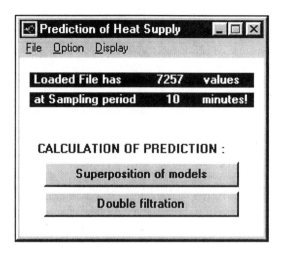

Fig. 3. Input window

Selection of calculation method of prediction of HSDD course is a other possibility of submitted program. We can realize the calculation of prediction by means of the method that uses model with double filtration and the method – superposition of models. Difference in solution process of both methods was indicated in Part 2.

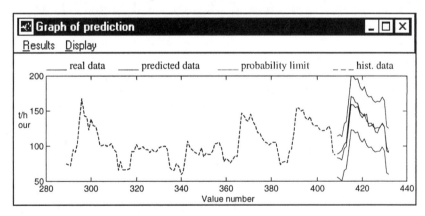

Fig. 4. Resulting graphic window

After choosing one of the methods the calculation of prediction is started. At first in the course of calculation it is searched for the most suitable model, it is for optimum number of autoregression parameters and optimum number of parameters of moving average process. After following calculation of prediction, resulting graphic window is displayed (see Fig. 4.). In this window there is drawn course of HSDD, course of predicted data and probability limit, which can be displayed through the menu "Display".

The result can be represented in concrete value form. These values are followed by calculation and they can be displayed in resulting window, which is shown on Fig. 4. This window can be activated through the menu "Results" in graphic window.

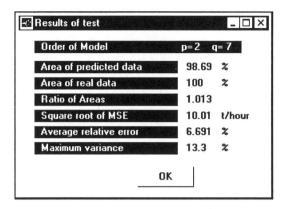

Fig. 5. Window of results

In this window it is possible to find also optimum number of autoregression parameters and optimum number of parameters of moving average process.

5 Conclusion

The calculation of time series prediction of HSDD:
- is suitable and utilizated for operation preparation of load distribution between cooperative production sources and production units inside these sources
- is necessary for qualitative-quantitative control method of hot-water piping heat output – Baláte system in real time.

The determination of corresponding mathematical statistical model is the main presumption of successful prediction. This determinate model has to represent investigated process. Because nowadays computer techniques are available, which enable to process a lot of data, the searching of this model is simpler.

References

1. Box, G.E.P., Jenkins, G.M.: Time Series Analysis, Forecasting and Control. Holden Day, San Francisco (1976)

2. Dostál, P.: Machine Processing of Daily Diagram Course Prediction of Loading the Centralized Heat Supply System. Ph.D Thesis, Faculty of Mechanical Engineering of TU Brno, Brno (1986)

3. Dostál, P.: Mathematical stochastic linear model. Students' Scientific Work (20 pp). Faculty of Mechanical Engineering of TU Brno, Brno (1978)

4. Chramcov, B.: Control of Hot-Water Piping Heat Output – Program for Calculation of Time Series Prediction. Diploma Thesis. Institute of Automation and Control Technology, Faculty of Technology in Zlín, TU Brno, Zlín (1998)

5. Baláte, J.: Design of Automated Control System of Centralized Heat Supply. PhSc Thesis. Faculty of Mechanical Engineering of TU Brno, Brno (1982)

6. Cipra, T.: Time Series Analysis with Applications in Economy SNTL, Prague (1986)

7. Andel, J.: Statistical Analysis of Time Series. SNTL, Prague (1976)

The Application of Fuzzy Set to Analysis of Cultural Meaning

Ikuhiro Koyauchi[1], Norihide Sano[1], Hiroyuki Kadotani[1], Ryoichi Takahashi[2]

[1] Shizuoka Sangyo University, Dept. of Communications and Informatics, Surugadai 4-1-1,
Fujieda, Shizuoka 426-8668, Japan
{koyauchi,sano,kadotani}@fujieda-ssu.ac.jp
[2] Tokyo Institute of Technology (Emeritus Professor),
Shizuoka Sangyo University, Dept. of Communications and Informatics, Surugadai 4-1-1,
Fujieda, Shizuoka 426-8668, Japan
ryo3@fujieda-ssu.ac.jp

Abstract. The meaning of any word contains two main elements which might be overlapped in some part, that is, literal meaning and pragmatic meaning. The latter is deeply associated with human activities and surroundings. Therefore, by examining that, we could realize the personal mentality of the reader and social tendency which surrounds him. There have been ever been many studies that tried to realize functions and semantic meanings of hedges. It is assumed that they usually intensify or weaken the overall meaning of a sentence and affect each attribute of words which constitutes a sentence in different ways. In this paper, we take up the word *safe* as primitive term and combine it with various hedges, chiefly hedges of Type II, to determine the degree of comprehensibility of pragmatic or cultural meaning and calculate it by using Fuzzy measure Learning Identification Algorithm (FLIA) and analyzed it.

1 Introduction

In the beginning, we will consider the meaning of the word *safe*. According to a dictionary, the first definition is that *secure from liability to harm, injury, danger, or ri*sk. It is possible to say that this meaning is very literal meaning (denotation), while in case of some persons saying 'A nuclear plant is safe.' we might surmise that each one says so from various thoughts. For example, she works for a nuclear plant, he majors in nuclear physics, and they may need a large quantity of nuclear energy to lead a comfortable life in their homeland. Hence, the meaning of a word is twofold: literal meaning (knowledge about language), but also psychological meaning. We would like to put them in other words: pragmatic meaning or cultural meaning (world knowledge about language).

Next, the boundary between *Safe* and *Unsafe* is not discrete but rather is fuzzy. As a basis for semantic intersection in cultural meaning, we would like to begin with establishing the sets of members defined by semantic components as fuzzy sets.

In order to make each component remarkable, we will take hedged sentences as examples.

2.1 Cognitive Process for Determining the Degree of the Comprehensibility of Cultural Meaning

The realization of the human cognitive process is an ill-defined area, that is to say, it is difficult to describe the human cognitive process completely with commonly used differential equations or algebraic equations. Nevertheless, we will try to think the human cognitive process for determining the overall degree of comprehensibility of cultural meaning.

When deciding the priority of alternatives, the summation of area for each component can be used it. It is hypothesized that the reader of the sentences determines the comprehensibility of meaning on the basis of the comprehensive value of each of the semantic components. The process for determining the degree of overall comprehensibility can be described as follows:

(1) evaluation of the comprehensive value of each component of a sentence

(2) integration of the comprehensive value of each component with the weights for a reader as the measures

(3) taking the integral values as the degree of overall comprehensibility

The combination of weights affects the degree of overall comprehensibility superadditively or subadditively.

3 Application

To determine the degree of the comprehensibility of the cultural meaning of the word *safe* for the examinee, we proposed the three groups, that is, twelve sentences, which included the hedges of Type II. The examinee reads them and the semantic components mentioned below, and the importance of each component which he held according to each hedge was digitized.

(1) a. A nuclear plant **typically** *safe*.
    b. A nuclear plant **technically** *safe*.
    c. **Strictly speaking**, a nuclear plant is *safe*.
    d. **Loosely speaking**, plant is *safe*.

(2) a.   An automobile is **typically** *safe.*
      b.   An automobile is **technically** *safe.*
      c.   **Strictly speaking,** an automobile is *safe.*
      d.   **Loosely speaking,** an automobile is *safe.*

(3) a.   An airplane is **typically** *safe.*
      b.   An airplane is **technically** *safe.*
      c.   **Strictly speaking,** an airplane is *safe.*
      d.   **Loosely speaking,** an airplane is *safe.*

4.   Problem Setting

The word *safe* concerned about a state of emergency, such as 'accident' or 'disaster' or 'happening', was used, because the extent to which we hold the conception and emotion for 'safety' or 'security' is reflected by various semantic components of the sentences including the primitive term *safe* plus some hedges.

We will assume five of semantic components which consist of a semantic meaning of a sentence in terms of linguistics. Moreover, they could affect the overall degree of comprehensibility of meaning when it is analyzed linguistically. The five semantic components are as follows:

1. **Scale of casualty:**
    means how great influence the accident causes on the people concerned, chiefly indicates economic influence on the society.

2. **Number of the killed and casualties:**
    means how many people were killed and injured in the accident.

3. **Probability of death:**
    means of what percents the people involved in the accident would be killed if the accident happened.

4. **Frequency of occurrence:**
    means how often the accidents by each vehicle would occur statistically according to the past data.

5. **Continuation of influence:**
    means how long the influence of the accident will have lasted socially or personally or mentally etc. since the accident occurs.

4 Results and Discussion

Step 1. The fuzzy integral that determines the degree of overall comprehensibility of the cultural meaning of the word *safe* as mathematical model can be defined as follows:

$$Z_c = \int_K h \cdot g(\cdot)$$
$$= \max\left[\min h_i \wedge g(E)\right] \tag{1}$$

Z_c : Fuzzy Integral K: $\{X_1, X_2, ..., X_5\}$

h_i: $h(X_i)$ g: Fuzzy Measure

$E : E \subset K$

Step 2. The degree of overlap among the components was determined by the following steps. At first the coupling coefficient between each attribute was obtained from the weights $g(K_i)$ as follows,

$$\mu \in [-1, \infty]$$

$$\mu_{ij} = \frac{g(X_i \cup X_j) - [g(X_i) - g(X_j)]}{g(X_i) \wedge g(X_j)} \quad (i \neq j) \tag{2}$$

$$\mu_{ij} = 0 \quad (i = j)$$

Step 3. Next the overlap coefficient between each component was determined from coupling coefficient μ_{ij} as follows.

$$m \in [-1, 1]$$

$$m_{ij} = \mu_{ij} \quad (\mu_{ij} \leq 0)$$

$$m_{ij} = \frac{\mu_{ij}}{\mu_{ij} + 1} \quad (\mu_{ij} \geq 0) \tag{3}$$

Step 4. Finally the degree of overlap among the components was obtained from the overlap coefficient m_{ij} as follows,

$$\eta \in [-1, 1]$$

$$\eta_i = \sum_{j=1}^{n} m_{ij}^{3} / (n-1) \qquad (i = 1, 2, \cdots\cdots, n) \qquad (4)$$

When the degree of overlap η_i has a minus value, the component i has a subadditive property among the other components. And if η_i has a plus value then the component i has a superadditive property. Table 1 shows the degree of overlap among the components which were identified from fuzzy measures weights $g(K_i)$. In Table 1, all attributes have a subadditive property and the component of η_4 has the largest overlap area of all the components. The necessity coefficient was determined by using the degree of overlap η_i as follows.

$$\xi_i \in [-1, 0]$$

$$\xi_i = \eta' [1 - g(X)] \qquad \begin{aligned} \eta' &= \eta_i \ (\eta_i \le -0.4) \\ \eta' &= 0 \ (\eta_i > -0.4) \end{aligned} \qquad (5)$$

If the necessity coefficient ξ_i is zero then the component i is necessary for the cognition process. And if ξ_i is equal to -1 then the component i is unnecessary. Table 2 shows the necessity coefficients which were obtained from the degree of overlap η_i and weights $g(K_i)$. In Table 2, the component of ξ_4 is seen to be indispensable for the cognitive process of comprehensibility of the word.

Table 1. Degree of Overlap η_i

Overlap	Degree
η_1	-0.65
η_2	-0.65
η_3	-0.50
η_4	-0.26
η_5	-0.40

Table 2. Necessity of Coefficient ξ_i

Necessity of Coefficient	Value
ξ_1	-0.53
ξ_2	-0.53
ξ_3	-0.45
ξ_4	0.0
ξ_5	-0.46

5 Conclusion

A method for identifying the fuzzy measures in deciding the degree of comprehensibility of cultural meaning with FLIA is proposed using calculations of degree of overlap between the components. The word 'safe' was taken an example of a word which everyone could not be indifferent to and which could reflect his/her personal or local mentality.

(1) The cognition process for the degree of overall comprehensibility of cultural meaning could be described by FLIA(Fuzzy measure Learning Identification Algorithm). The degree of overlap between the components was obtained.

(2) When the reader of a sentence decides the degree of overall comprehensibility of cultural meaning, he/she attaches importance to the forth component (frequency of occurrence).

References

1. Zadeh, L. A.: A Fuzzy-Set-Theoretic Interpretation of Linguistic Hedges. In: Fuzzy Sets and Applications. John Wiley & Sons (1987) 467-498
2. Lakoff, G.: Hedges: A Study in Meaning Criteria and the Logic of Fuzzy Concepts. In: Papers from the 8th Regional Meeting, Chicago Linguistic Society, Chikago (1972) 183-228.
3. Mac Cormac, E. R.: A Cognitive Theory of Metaphor. MIT Press (1987)
4. Kay, P.: Linguistic Competence and Folk Theories of Language: Two English Hedges, from Cultural Model in Language and Thought. Cambridge University Press, Cambridge (1987)
5. Koyauchi, I., Sano, N., Kadotani, H., Takahashi, R.: Fuzzy Set Operators for Type II Linguistic Hedges. In: Proc. 6th Zittau Fuzzy-Colloquium, Zittau, Germany (1998) 39-43

Natural Language Input for Fuzzy Diagnosis

Norihide Sano[1], Ryoichi Takahashi[2]

[1] Shizuoka Sangyo University, Dept. of Communications and Informatics,
Surugadai 4-1-1, Fujieda, Shizuoka 426-8668, Japan
nsano@shizuokanet.ne.jp
[2] Shizuoka Sangyo University, Dept. of Communications and Informatics,
Surugadai 4-1-1, Fujieda, Shizuoka 426-8668, Japan
bzq06072@nifty.ne.jp

Abstract. Several kinds of diagnosis techniques based on the kinetic or statistical model of a nuclear plant have been already utilized in operation. Rule-based approaches have been proposed as well to realize a flexible diagnostic method with use of fuzzy sets, which pay attention at the rules describing the relationship between the cause and symptom of the failures. These might be able to infer the failure by modus tollens using implications to represent the relation between the cause and symptom. In the first step of a diagnosis, it needs a broad outline of the failure from slightly symptoms. The present paper attempts to express the natural language input for fuzzy diagnosis with a dialog box on Visual C++.

1 Introduction

Recently, it has been reported that safe and effective operation of nuclear power plants is achieved by means of automatic control during normal operation. It may be heavy burden to an engineer to deal with a highly-automated system in case of emergency until the cause and consequence are clarified by examining the signal flow. A simplified technique for diagnosis should be incorporated in order to support the human operators, which is capable of identifying only the dominant failure with a small number of symptoms defined by the fully-experienced engineer.

In past we tried to express the solutions on fuzzy diagnosis by inverse process for a broad outline of the failure and to use it in an example which describes the leakage of the cooling system in a Boiling Water Reactor plant [6]. First, the fuzzy diagnosis by inverse process is explained. Second, it is applied to identify leaking locations in the coolant pipes of a Boiling Water Reactor plant. The present paper attempts to improve the input method for diagnosis system with natural language.

2 Basic concepts and its formulations

Human thinking is characterized by the argument that the symbol ($\rightarrow$) always appears in the universal proposition. Modus ponent and modus tollens in the binary logic are written as follows

$$P, P \rightarrow Q , \tag{1}$$

$$Q .$$

$$\neg Q, P \rightarrow Q , \tag{2}$$

$$\neg P$$

where the symbol ($\neg$) stands for negation. Eq.(1) is then read practically as "P is true" and "If P is true then Q must be true" then "Q is true". Eq.(2) infers "not P is true(P is false)" from "Q is false" and "If P is true then Q must be true".

The linguistic truth values [1] of fuzzy sets are defined on the truth values space. The truth values of T is

$$T_\alpha = (1 - P_\alpha + Q_\alpha) \wedge 1, \tag{3}$$

where the symbol (α) stands for an alpha-cuts defined by the range

It is sufficient for diagnosis to consider only the case where the linguistic truth values of T, P and $\neg$Q are close to "true". Therefore, the alpha-cuts of T, P, and Q are as follows

$$\begin{aligned} T_\alpha &= (t_l, 1] , \\ P_\alpha &= (p_l, 1] , \end{aligned} \tag{4}$$

$$Q_\alpha = [0, q_u) ,$$

with the suffixes l and u standing for the lower and upper bound respectively.
Then, the Eq.(3) gives a solution Q_α as follow,

$$Q_\alpha = ((t_l + p_l -1) \vee 0, 1], \tag{5}$$

and a solution P_α as follow,

$$P_\alpha = [0, (1 + q_u - t_l) \wedge 1), \tag{6}$$

Eqs(5) and (6) are the fuzzified modus ponens and modus tollens.

Conventional fuzzy diagnosis[1-3] consists of two implications what means "For all j, there exists a failure x_i that correlates with a symptom y_j if y_j is recognized" and "For all i, there is a symptom y_j such that should be observed if x_i appears".

Here the diagnosis system is given by the propositions

$$P(y_j) : B(y_j) \rightarrow \exists x_i (A(x_i) \wedge R(x_i, y_j)) \qquad (j=1,2,...) \tag{7}$$

$$P(x_i, y_j) : A(x_i) \rightarrow \exists\, y_j\,(B(y_j)) \quad (i=1,..., m, j=1,...,) \tag{8}$$

where the function $A(x_i)$ means that the failure x_i appears, and $B(y_j)$ gives that the symptom y_j is being recognized. $R(x_i, y_j)$ indicates the correlation between x_i and y_j.

For simplicity, Eqs.(7), (8) have been derived from the binary logical point of view. Here fuzzification of these equations shall be performed in accordance with fuzzy set theory, since recognition depends on the subjective tasks of human. Defining the fuzzy sets on the spaces X, Y and $X \times Y$.

$$f = \text{(set of appearing failures)},$$

$$s = \text{(set of recognized symptoms)},$$

$$r = \text{(set of relations } (x_i, y_j)). \tag{9}$$

We can obtain fuzzy propositions A_i, B_j and R_{ij} from $A(x_i)$, $B(y_j)$ and $R(x_i, y_j)$ in the form

$$A_i = (x_i \text{ is } f),$$

$$B_j = (y_j \text{ is } s),$$

$$R_{ij} = ((x_i, y_j) \text{ is } r), \tag{10}$$

where the truth values of these propositions are represented by linguistic truth values of A_i, B_j and R_{ij}. Substituting Eq.(10) into (7) and (8), the fuzzification of Eqs. (7) and (8) are written in the logical form

$$P_j : B_j \rightarrow \exists_i (A_i \wedge R_{ij}) \qquad (j = 1,...,n), \tag{11}$$

$$P_{ij} : A_i \rightarrow \exists_j B_j \qquad (i = 1,...,m, j = 1,...,n). \tag{12}$$

3 Utilization of implications to diagnosis

Assuming that the implications of Eq.(11) holds the truth value of 'completely true'

$$P_{j\alpha} = [1, 1], \tag{13}$$

and Eq. (12) has the truth value of 'very true'

$$P_{ij\alpha} = (p_{ijl}, 1], \tag{14}$$

the failure, the symptom, and the fuzzy relation are written

$$A_{i\alpha} = (a_{il}, a_{iu}),$$

$$B_{j\alpha} = (b_{jl}, b_{ju}),$$

$$R_{ij\alpha} = (r_{ijl}, r_{iju}), \tag{15}$$

Substituting of Eqs.(5) and (6) into Eqs.(11) and (12) generated the solution

$$B_{j\alpha} = (\bigvee_{i=1}^{m} (a_{il} \wedge r_{ijl}), 1], \tag{16}$$

$$A_{i\alpha} = [0, (1 + b_{ju} - p_{ijl}) \wedge 1). \tag{17}$$

The inferred upper bound of $A_{i\alpha}$ is obtained from Eq.(17) to be form of

$$a_{iu} = (1 + b_{ju} - p_{ijl}) \wedge 1. \tag{18}$$

Eq. (16) allows us to obtained the lower bound of $A_{i\alpha}$ by solving the fuzzy inverse problem [1] of the form

$$b_{jl} = \bigvee_{i=1}^{m} (a_{il} \wedge r_{ijl}). \tag{19}$$

4 Step of solutions by inverse process

Exercise Assuming that $B=\{b_j\}$ and $R=\{r_{ij}\}$, $(i=1,\ldots,m, j=1,\ldots,n)$ are given, and calculate $A=\{a_i\}$ which satisfies with the following conditions

$$\vee (a_i \wedge r_{ij}) = b_j \qquad (i=1,\ldots m). \tag{20}$$

Solutions: **Step1.** Produce the matrices u_{ij} and v_{ij} with r_{ij}, b_j by ω-composition [1] and ϖ-composition [1] as follows

$$u_{ij} = \{r_{ij} \ \omega \ b_j\}, \ v_{ij} = \{r_{ij} \ \varpi \ b_j\}. \tag{21}$$

Step 2. Produce the matrices w_{ij} with u_{ij} and v_{ij}, where w_{ij} is made from the following procedures. First, select the *i-th* element, which is not zero, from each column in u_{ij}. Second, put the element u_{ij} in that *i-th* raw and *j-th* column of w_{ij}. Third, exchange the rest element of w_{ij} for v_{ij}. There exist some kind of matrices w_{ij}, if there are some none-zero elements in the same column in the u_{ij}.
Step 3. Obtain the a_i as follows,

$$a_i = \cap_j \{w_{ij}\}. \tag{22}$$

There exist some kind of a_i, if there are some kind of w_{ij}.

5 Identification of leakage location in a boiling water reactor plant

The present fuzzy diagnosis was applied to identify the leakage location in the cooling system of a Boiling Water Reactor Plant. Figure 1 shows a sketch of the cooling system of this power plant. Here typical examples were assumed for the leaking locations and the induced symptoms in conformity with the text[4]. In this exercise, it was assumed that the leakage inside the dry-well and buildings were found generally by Eqs.(16), (17) derived from the implications of Eqs.(7), (8). Table 1 and Table 2 define the failure- and symptom-vector in the cooling system. Table 3 shows the matrices r_{ijl} and p_{ijl} appearing in Eqs. (16), (17).

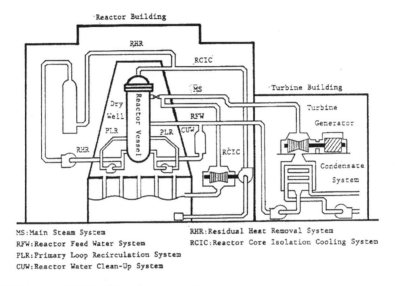

MS:Main Steam System
RFW:Reactor Feed Water System
PLR:Primary Loop Recirculation System
CUW:Reactor Water Clean-Up System

RHR:Residual Heat Removal System
RCIC:Reactor Core Isolation Cooling System

Fig. 1. Diagram of Boiling Water Reactor cooling system

Table 1. Definition of failure vector

(a) *Element of failure vector*
$x_1 = \{$ the upper reaches of main steam line $\}$
$x_2 = \{$ the lower reaches of main steam line $\}$
$x_3 = \{$ residual heat removal system $\}$
$x_4 = \{$ reactor core injection cooling system $\}$
$x_5 = \{$ reactor water clean-up system $\}$
$x_6 = \{$ feed water $\}$

Table 2. Definition of symptom vector

(b) *Element of symptom vector*
y_1 = { flow rate increase in dry-well sump }
y_2 = { flow rate increase in air condenser drain }
y_3 = { pressure increase in dry-well }
y_4 = { pressure decrease in steam line }
y_5 = { flow rate increase in main steam line }
y_6 = { flow rate increase in residual heat removal system }
y_7 = { high differential flow rate of reactor water clean-up system }

Table 3. Matrices of fuzzy relation of failures with symptoms

(a) *Lower bound of* $R_{ij\alpha}$ *appearing in Eq.* (16)

$$r_{ijl} = \begin{bmatrix} 0.0 & 0.0 & 1.0 & 1.0 & 1.0 & 0.0 & 0.0 \\ 0.0 & 0.0 & 1.0 & 1.0 & 0.0 & 0.0 & 0.0 \\ 1.0 & 1.0 & 0.0 & 0.0 & 0.0 & 1.0 & 0.0 \\ 0.0 & 1.0 & 1.0 & 1.0 & 0.0 & 0.0 & 0.0 \\ 1.0 & 0.0 & 0.0 & 0.0 & 0.0 & 0.0 & 1.0 \\ 1.0 & 1.0 & 0.0 & 0.0 & 0.0 & 0.0 & 0.0 \end{bmatrix}$$

(b) *Lower bound of* $P_{ij\alpha}$ *appearing in Eq.* (17)

$$p_{ijl} = \begin{bmatrix} 0.0 & 0.0 & 0.6 & 0.5 & 0.5 & 0.0 & 0.0 \\ 0.0 & 0.0 & 0.6 & 0.5 & 0.0 & 0.0 & 0.0 \\ 0.8 & 0.5 & 0.0 & 0.0 & 0.0 & 0.5 & 0.0 \\ 0.0 & 0.8 & 0.3 & 0.3 & 0.0 & 0.0 & 0.0 \\ 0.8 & 0.0 & 0.0 & 0.0 & 0.0 & 0.0 & 0.5 \\ 0.8 & 0.0 & 0.0 & 0.0 & 0.0 & 0.0 & 0.0 \end{bmatrix}$$

6 Example of diagnosis through fuzzy inverse process

Example 1. The symptom "pressure decrease in steam line", "pressure increase in dry-well" and "flow rate increase in dry-well sump" are sharply observed. The value of $b_{1l} = 0.95$, $b_{3l} = 0.95$ and $b_{4l} = 0.95$ were substituted for the lower bounds as these symptoms. Then the natural language input is shown in Fig.2,

The solutions A_i ($i = 1,...6$) is obtained as:

$$A_{1\alpha} = [0.0, 0.50), \quad A_{2\alpha} = [0.95, 1.0], \quad A_{3\alpha} = [0.0, 0.50)$$

$$A_{4\alpha} = [0.0, 0.20), \quad A_{5\alpha} = [0.0, 0.50), \quad A_{6\alpha} = [0.0, 0.50) \tag{23}$$

The solution indicates a leakage in the lower reaches of main steam line (Fig.3) by reading the α-cuts of $A_{2\alpha}$.

Example 2. "The flow rate slightly increased in dry-well sump and the differential flow rate of the clean-up system for reactor water has risen". The truth value of

$$b_{1l} = 0.60, \quad b_{7l} = 0.60, \quad b_{jl} = 0.0 \ (j=2,...6), \tag{24}$$

are substituted into the symptoms. The solution A_i is

$$A_{1\alpha} = [0.0, 0.4), A_{2\alpha} = [0.0, 0.4), A_{3\alpha} = [0.0, 0.5)$$

$$A_{4\alpha} = [0.0, 0.2), A_{5\alpha} = (0.6, 0.8), A_{6\alpha} = [0.0, 0.5) \tag{25}$$

This indicates that "a leakage of the clean-up system for reactor water exists" is fairly true (Fig.4).

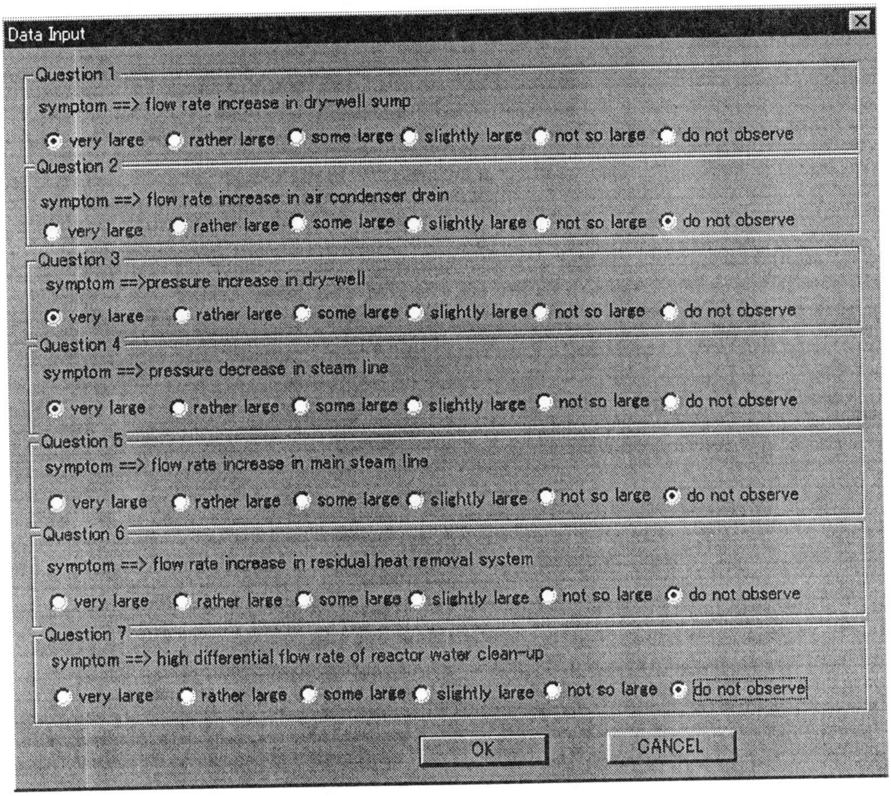

Fig. 2. The Natural Language Input with a Dialog Box

272

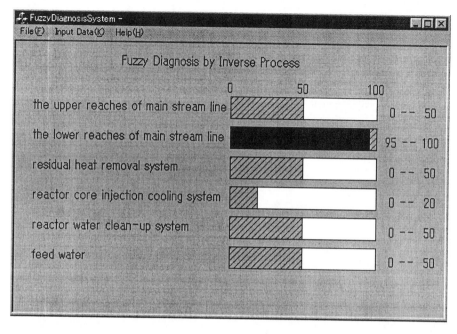

Fig. 3. Result of Diagnosis for Example 1

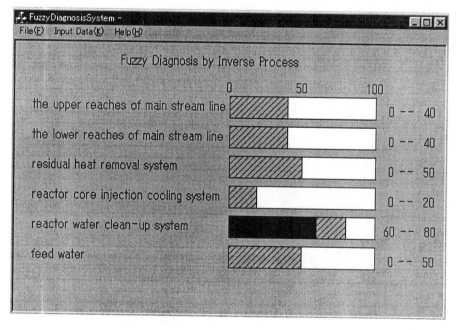

Fig. 4. Result of Diagnosis for Example 2

7 Conclusion

The present paper attempts to express input data as natural languages for fuzzy diagnosis. The results of several diagnosis examples led us to the conclusion that the fuzzy diagnosis using natural language input has potential enough to describe the broad outline failure by uncertain data.

It is subject of further research to express also output with natural languages instead of bar graphs, and to indicate the sequence and the reason why the failure was finally selected.

References

1. Tsukamoto, Y.: Fuzzy Logic Based on Lukasiewicz Logic and its Implications to Diagnosis and Control. Dissertation, Tokyo Institute of Technology (1977)
2. Maruyama, Y., Takahashi, R.: Application of Fuzzy Reasoning to Failure Diagnosis. J. Atomic Energy Soc. Japan 27 (1985) 851-860 (in Japanese)
3. Takahashi, R., Maruyama, Y.: Practical Expression of Exception to Diagnosis. Bull. Res. Lab. Nucl. Reactors Tokyo Inst. Technol. 12 (1987) 50-53
4. Suguri, S.: Reactor Safety Engineering. Nikkan Kogyo Shinbunsha, Tokyo (1975) (in Japanese)
5. Maruyama, Y., Takahashi, R.: Practical Expression for Exception and its Utilization to Leakage Identification of Power Plant. Fuzzy Sets and Systems 34(1990) 1-13
6. Sano, N., Takahashi, R.: Solutions on Fuzzy Diagnosis by Inverse Process. Proc. 6th Zittau Fuzzy-Colloquium, Zittau, Germany (1998) 7-12

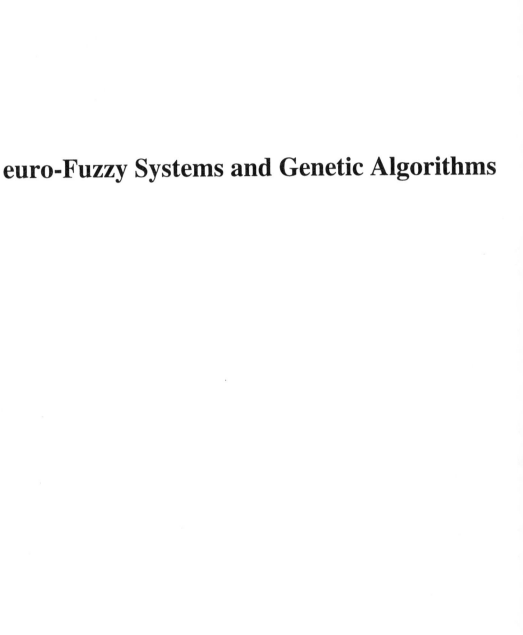

euro-Fuzzy Systems and Genetic Algorithms

Neural Network Architecture of Fuzzy Systems

Danuta Rutkowska

Department of Computer Engineering
Technical University of Czestochowa, Poland
drutko@kik.pcz.czest.pl

Abstract. The paper deals with representations of fuzzy systems by means of neural networks. Two different kinds of neural networks are employed for fuzzy systems modeling. One approach applies feed-forward neural networks which are similar to the well known architecture of a multi-layer perceptron. Another one is based on the equivalence of fuzzy systems and radial basis function neural networks. The two approaches are compared taking into account the assumptions needed to perform a role of a fuzzy system by a neural network. More general representations of fuzzy systems by neural network architectures, which are called fuzzy inference neural networks, are described at first. It is shown that a special case of these networks can be presented in the form of the radial basis function neural network as well as the multi-layer perceptron architecture. The neural network representations of fuzzy systems with singleton and non-singleton fuzzifier are depicted in this paper.

1 Introduction

There is a great deal of research about combining fuzzy systems (e.g. [7]) and neural networks (e.g. [14]). These approaches take advantage from the both techniques. Their main merits are the linguistic rule base description of the fuzzy systems and the learning ability of the neural networks. There has been a large amount of interest and practice in the synergistic combination of the both methods (see e.g. [12]). Moreover, there are some theoretical researchs which refer to an equivalence of fuzzy systems and neural networks (e.g. [4]). This paper develops the latter idea and describes two different types of neural networks (the radial basis function network and the multi-layer perceptron: RBFN and MLP, for short, respectively), which under some assumptions can be treated as equivalent to fuzzy systems. Other, more general, representations of fuzzy systems by means of neural networks, known as fuzzy inference neural networks, are depicted in the next section. Besides the fuzzy system with the singleton fuzzifier, which is commonly considered in the literature, also an interesting case of a neuro-fuzzy system with the non-singleton fuzzifier is presented in this paper.

2 Fuzzy Inference Neural Networks

Integration of fuzzy systems and neural networks can be realized by building connectionist architectures of fuzzy systems in the form similar to the feed-forward multi-layer neural networks. Figure 1 presents an example of these neuro-fuzzy archi-tectures. This kind of network has been described by Wang [13] and studied by many researchers. Various modifications of such a network with the learning algorithms has been developed in [10]. These neuro-fuzzy systems can be trained by means of the gradient procedures based on the steepest descent optimization method. The same idea is employed in the back-propagation algorithm, commonly used learning of neu-ral networks [14]. These procedures can be supported by genetic algorithm [5] in or-der to find a proper starting point, which leads to the global optimum and avoids to get trapped in local optima. In this way the hybrid learning algorithm (genetic algorithm + gradient method) tune the weights of neural networks or the parameters of the fuzzy systems.

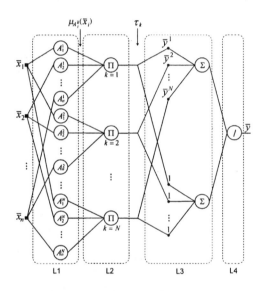

Fig.1. Connectionist network of the neuro-fuzzy system

The neuro-fuzzy system illustrated in Fig.1 is described by the following expres-sion

$$\overline{y} = \frac{\sum\limits_{k=1}^{N} \overline{y}^k \tau_k}{\sum\limits_{k=1}^{N} \tau_k} \quad , \tag{1}$$

where $\bar{y}^k$ is a center of the membership function $\mu_{B^k}(y)$ and τ_k is given as

$$\tau_k = \prod_{i=1}^{n} \mu_{A_i^k}(\bar{x}_i) , \qquad (2)$$

with μ_{A^k} and μ_{B^k} as membership functions of fuzzy sets A_i^k and B^k, for $i = 1,...,n$; $k = 1,...,N$, in the fuzzy rule base, which consists of a set of N rules in the form

$$R^{(k)}: \text{ IF } x_1 \text{ is } A_1^k \text{ AND } ... \text{ AND } x_n \text{ is } A_n^k \text{ THEN } y \text{ is } B^k, \qquad (3)$$

where x_i are linguistic variables, and $\bar{x}_i$ are crisp input variables. Fuzzy rule base (3) can be written, for short, as follows

$$R^{(k)}: \text{ IF } x \text{ is } A^k \text{ THEN } y \text{ is } B^k, \qquad (4)$$

with $x = [x_1,...,x_n]^T$ and

$$A^k = A_1^k \times \cdots \times A_n^k. \qquad (5)$$

The architecture of the neuro-fuzzy system illustrated in Fig.1 is composed of four layers: L1, L2, L3, L4. The elements of the first layer realize the membership functions of fuzzy sets A_i^k, $i = 1,...,n$; $k = 1,...,N$, of the antecedent part of fuzzy IF-THEN rules (3). The elements of the second layer perform the multiplication functions, according to formula (2). The last two layers realize the defuzzification task; in this case center average defuzzification method has been employed. Parameters $\bar{y}^k$, in the third layer, denote the centers of the membership functions of fuzzy sets B^k, which appear in the conclusion part of fuzzy IF-THEN rules. Layer L4 performs the division function. As a fuzzifier the commonly used singleton fuzzification method has been applied. Formula (1), which describes the neuro-fuzzy system presented in Fig. 1, has been derived based on the singleton fuzzifier, the product-inference rule (Larsen's rule) with the product operation as a definition of Cartesian product in formula (6), and center average defuzzifier [9,13].

If the product operation in the reasoning of the fuzzy system is replaced by the minimum operation (Mamdani's inference rule and the minimum operation as a definition of Cartesian product), the elements of the second layer in Fig. 1 should realize the minimum functions instead of the multiplication ones [13,9,10]. However, the neuro-fuzzy systems based on the product operation perform better than their counterparts which employ the minimum operation, because the minimum function is not differentiable, so this kind of system is more difficult to learn by means of the gradient methods.

The membership functions of fuzzy sets A_i^k and B^k are usually Gaussian or triangular shape functions. Both of them are employed very often but Gaussian membership functions seem to be more convenient and wider applicable, for example in RBF networks (see Section 3) and in the systems with non-singleton fuzzifier (see Section 5).

3 RBF Network Architecture of Fuzzy Systems

In [4] an equivalence between fuzzy systems and radial basis function (RBF) networks is shown. However, some conditions must be fulfilled. One of them tells that the consequent part of each of fuzzy IF-THEN rules is composed of a constant. Let us notice that fuzzy rule base (3) does not fulfill this condition. These rules contain fuzzy sets in the conclusion part of the rules.

In this section we show the equivalence between the neuro-fuzzy system described in Section 2 (illustrated in Fig.1) and RBF neural network. Let us emphasize that this system uses the rule base with non-singleton fuzzy sets in the consequent part of IF-THEN rules.

Another condition, concerning the equivalence between the fuzzy system and RBF network, which has to be fulfilled in [4], tells that the membership functions within each rule are chosen as Gaussian functions with the same variance. Let us assume that the membership functions of fuzzy sets A_i^k, which appear in the antecedent part of fuzzy IF-THEN rules (3), are Gaussian functions with the same width parameter within each rule, i.e.

$$\mu_{A_i^k}(x_i) = G_i^k(x_i) = \exp\left[-\left(\frac{x_i - \overline{x}_i^k}{\sigma_i^k}\right)^2\right], \tag{6}$$

where $\sigma_i^k = \sigma^k$ for all $i = 1,...,n$. It is easy to see that in case when the elements of the first layer of the network shown in Fig.1 perform Gaussian functions, expressed by formula (6), and the elements of the second layer realize the multiplication functions, then the architecture illustrated in Fig.1 can be presented in the form portrayed in Fig. 2. This is a normalized version of RBF neural network [6].

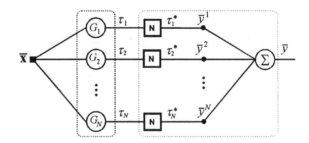

Fig. 2. Normalized RBF neural network

The network depicted in Fig. 2 consists of two main parts (layers). The first one is composed of the elements which perform Gaussian radial basis functions G_k, $k = 1,...,N$. This part corresponds to layers L1 and L2 in the architecture shown in Fig. 1. The second part is the same as layers L3 and L4 in Fig. 1. For simplicity, some connections between the elements which realize radial functions G_k and the normalization units (denoted as squares with letter **N**), are omitted in Fig. 2. The number of radial basis functions is equal to the number of fuzzy IF-THEN rules. It is very important because the same condition must be fulfilled in [4] concerning the equivalence between the fuzzy system and RBF network. The normalization units perform the following operation

$$\tau_k^* = \frac{\tau_k}{\sum\limits_{j=1}^{N} \tau_j} . \tag{7}$$

The radial basis functions are expressed by

$$G_k(\overline{x}) = \exp\left[-\left(\frac{\|\overline{x} - \overline{x}^k\|}{\sigma^k}\right)^2\right] = \prod_{i=1}^{n} \exp\left[-\left(\frac{\overline{x}_i - \overline{x}_i^k}{\sigma^k}\right)^2\right], \tag{8}$$

where $\overline{x} = [\overline{x}_1, ..., \overline{x}_n]^T$ and $\overline{x}^k = [\overline{x}_1^k, ..., \overline{x}_n^k]^T$, for $k = 1, ..., N$. Output $\overline{y}$ of the normalized RBF neural network, illustrated in Fig. 2, is given by

$$\overline{y} = \frac{\sum\limits_{k=1}^{N} \overline{y}^k G_k(\overline{x})}{\sum\limits_{k=1}^{N} G_k(\overline{x})} . \tag{9}$$

Formula (9) describes the normalized RBF neural network with N radial basis functions G_k, expressed by equation (8), in the hidden layer, and weights $\overline{y}^k$ in the output layer. The weights of this RBF network are equal to the centers of the membership functions of fuzzy sets B^k. The value of radial basis functions G_k, for given input vector $\overline{x}$, equals to τ_k defined by formula (2) and is called the degree of activation of rule $R^{(k)}$, presented in the form of (3) or (4). Expression (9) is the same as formula (1). Thus, the neuro-fuzzy system described in section 2 is equivalent to the normalized RBF neural network.

Let us emphasize that, similarly as in [4], the fuzzy inference system is equivalent to the RBF network only if some conditions are fulfilled. Besides the ones mentioned above, also T-norm operator used to compute each rule's firing strength must be the multiplication function. It means that elements of the second layer, in the architecture illustrated in Fig.1, have to realize the multiplications (not the minimum operators). Moreover, both the RBF network and the fuzzy inference system have to use the same method to derive their overall outputs. It determines the defuzzification operation which can be applied in the neuro-fuzzy system. Taking into account all these assumptions, we can conclude that only special cases of the neuro-fuzzy systems are equivalent to the RBF network. The architecture presented in Fig. 1, with Gaussian membership functions performing by the elements of the first layer , is equivalent to the normalized neural network depicted in Fig. 2. Many other types of neuro-fuzzy systems, based on different inference rule, defuzzification method, and membership functions are not equivalent to the RBF network.

It is worth reminding that the fuzzy inference system studied in this paper incorporates fuzzy rules in which the consequent part can contain the fuzzy sets with Gaussian membership functions (or others, in general), not necessary singletons. It is very important to emphasize the difference between the fuzzy inference system under consideration in [4], which uses the fuzzy rule base with constants in the conclusion parts, and the fuzzy systems discussed in this paper which employ IF-THEN rules with fuzzy sets in the antecedent as well as the consequent parts.

4 MLP Architecture of Fuzzy Systems

The connectionist network illustrated in Fig.1 can be treated as a neural network representation of the fuzzy system. This kind of neuro-fuzzy system has different name in literature, for example: neural network representation of fuzzy logic system, neural network-based fuzzy logic system, connectionist model of fuzzy logic system, adaptive fuzzy system, fuzzy inference neural network, etc. [12]. The architecture of the neuro-fuzzy system shown in Fig.1 is composed of layers, similar to the feed-forward multi-layer neural networks but the elements of the layers differ from the typical neurons in neural networks. In the standard neural networks each neuron multiplies its input signals by the corresponding weights and adds these results. Then a transfer function changes it to an output of the neuron. As we observed in Section 2, the elements of the architecture depicted in Fig.1 perform different operations.

In Section 3 we observed another representation of a fuzzy system by means of a neural network. A special kind of the neuro-fuzzy systems, described in Section 2, was represented by the normalized RBF network architecture.

Now let us try to find a standard neural network which could be equivalent to the fuzzy system (or the neuro-fuzzy system depicted in section 2). From the results described in section 3, it is easy to show that under some additional assumptions there is a standard neural network architecture which is equivalent to the networks presented in the previous sections.

Referring to Fig. 2, and taking into account formulas (2), (6), and (8), we obtain the following equations

$$\tau_k = G_k(\bar{x}) = \prod_{i=1}^{n} \exp\left[-\left(\frac{\bar{x}_i - \bar{x}_i^k}{\sigma^k}\right)^2\right] = \exp\left\{\sum_{i=1}^{n}\left[-\left(\frac{\bar{x}_i - \bar{x}_i^k}{\sigma^k}\right)^2\right]\right\}$$

$$= \exp\left\{-\frac{1}{\left(\sigma^k\right)^2}\left[\sum_{i=1}^{n}(\bar{x}_i)^2 - 2\sum_{i=1}^{n}\bar{x}_i\bar{x}_i^k + \sum_{i=1}^{n}\left(\bar{x}_i^k\right)^2\right]\right\} . \qquad (10)$$

Let us assume that vectors $\bar{x}$ and $\bar{x}^k$ are normalized, i.e.

$$\left|\overline{\mathbf{x}}\right| = \sqrt{\sum_{i=1}^{n}\left(\overline{x}_i\right)^2} = 1 \qquad (11)$$

and

$$\left|\overline{\mathbf{x}}^k\right| = \sqrt{\sum_{i=1}^{n}\left(\overline{x}_i^k\right)^2} = 1 . \qquad (12)$$

Under these assumptions formula (10) becomes

$$\tau_k = \exp\left\{-\frac{2}{\left(\sigma^k\right)^2}\left[1-\sum_{i=1}^{n}\overline{x}_i\,\overline{x}_i^k\right]\right\} . \qquad (13)$$

Thus, the network illustrated in Fig. 2 can be presented in the form depicted in Fig. 3.

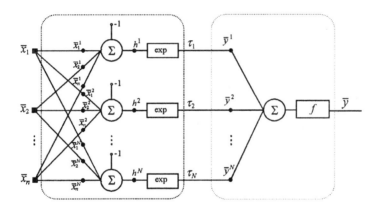

Fig. 3. Standard neural network representation of fuzzy systems

The network shown in Fig. 3 is a typical neural network with an input layer, one hidden layer, and an output layer. There are n inputs in the input layer, N neurons in the hidden layer, and one neuron in the output layer. The weights of this neural network are the centers of Gaussian membership functions. The transfer functions in the hidden layer are exponential functions multiplied by

$$h^k = 2/\left(\sigma^k\right)^2 \qquad (14)$$

and the transfer function in the output layer is expressed by following formula

$$f(s) = s \Big/ \sum_{k=1}^{N} \tau_k \, , \tag{15}$$

where s is an input of the unit which performs the transfer function f.

Let us remind that the neuro-fuzzy systems described in section 2 can be represented by the neural network depicted in Fig. 3, if the conditions mentioned in section 3 and moreover assumptions (11) and (12) are fulfilled.

A similar neural network architecture of fuzzy systems has been derived in [8] and [9]. The standard neural network can be also applied as a defuzzifier [9,10].

5 NN Architecture of Fuzzy Systems with Non-Singleton Fuzzifier

The neuro-fuzzy systems under consideration in Sections 2-4 most often employ the singleton fuzzifier, which maps crisp input $\overline{x}$ into fuzzy set A' characterized by the following membership function

$$\mu_{A'}(x) = \begin{cases} 1 & \text{if} \quad x = \overline{x} \\ 0 & \text{if} \quad x \neq \overline{x} \end{cases} . \tag{16}$$

The membership function of non-singleton fuzzifier can be defined as follows

$$\mu_{A'}(x) = \exp\left[-\left(\frac{\|x - \overline{x}\|}{\sigma}\right)^2\right] = \prod_{i=1}^{n} \exp\left[-\left(\frac{x_i - \overline{x}_i}{\sigma}\right)^2\right] \, , \tag{17}$$

where $\overline{x} = [\overline{x}_1, \ldots, \overline{x}_n]^T$ and σ denote centers and widths of the non-singleton membership functions, respectively.

The fuzzy inference neural network based on the non-singleton fuzzifier, given by formula (17), the product-inference rule (Larsen's rule) with the product operation as a definition of Cartesian product in formula (6), center average defuzzifier, and Gaussian membership functions of the fuzzy sets in the antecedent as well as the consequent parts of fuzzy IF-THEN rules, is described by expression (1) with τ_k defined as

$$\tau_k = \prod_{i=1}^{n} \exp\left[-\left(\frac{\overline{x}_i - \overline{x}_i^k}{\overline{\sigma}_i^k}\right)^2\right] \, , \tag{18}$$

where $\overline{x}_i^k$ are centers of Gaussian membership functions μ_{A^k}, given by formula (6), and $\overline{\sigma}_i^k$ depend on width parameters σ and σ_i^k of Gaussian membership functions $\mu_{A'}$ and μ_{A^k}, respectively, $i = 1, \ldots, n$; $k = 1, \ldots, N$, according to the following equation

$$\overline{\sigma}_i^k = \sqrt{\sigma^2 + \left(\sigma_i^k\right)^2} \, . \tag{19}$$

Formula (19) has been derived in [10] and the description of the neuro-fuzzy system with the non-singleton fuzzifier is presented in [11]. The connectionist network of this kind of system is similar to that depicted in Fig. 1, except the first layer, which is different in the both architectures. The elements of layer L1 in Fig. 1 perform the membership functions of fuzzy sets A_i^k of the antecedent part of fuzzy IF-THEN rules (3), while the elements of the first layer of the non-singleton fuzzifier based network realize Gaussian functions with centers $\bar{x}_i^k$, which are the same as centers of Gaussian membership functions $\mu_{A_i^k}$, but different width parameters, given by formula (19).

Since the connectionist architecture of the described above neuro-fuzzy system with the non-singleton fuzzifier is similar to the network illustrated in Fig. 1, it is easy to see that the fuzzy inference neural network based on the non-singleton fuzzifier can be presented in the form of the normalized RBF neural network shown in Fig. 2. In this case the radial basis functions are Gaussian functions with parameters $\bar{x}_i^k$ and $\bar{\sigma}_i^k$, according to expression (19). The learning procedure for this kind of network has been depicted in [11].

Under assumptions (11) and (12) the neuro-fuzzy system with the non-singleton fuzzifier can be presented in the form of the standard neural network, illustrated in Fig. 3, where

$$h^k = 2/\left(\bar{\sigma}_i^k\right)^2. \tag{20}$$

From formula (18), according to equations (10), and assumptions (11), (12), we conclude that in this case τ_k, for $k = 1,...,N$, become as follows

$$\tau_k = \exp\left\{-\frac{2}{\left(\bar{\sigma}_i^k\right)^2}\left[1 - \sum_{i=1}^{n} \bar{x}_i \bar{x}_i^k\right]\right\}. \tag{21}$$

Based on expression (21) we obtain the neural network (NN) architecture depicted in Fig. 3 with h^k and f given by formulas (20) and (13), respectively.

6 Conclusions and Remarks

It is important to know that in some practical applications the input data are normalised and scaled in order to achieve better performance of the system. In this case it is easier to fulfil assumptions (11) and (12).

If the neuro-fuzzy system takes a form of the RBF network or the standard (perceptron type) neural network, we can directly employ the learning procedures which are usually applied for these kind of networks. This is the main advantage of the neural network representation of fuzzy systems.

It should be emphasized that the weights of the neural network shown in Fig. 3 have interpretation of the centers of the membership functions. This fact is very important because it gives an explanation of the system performance. Such a network is not a "black box" which works without any explanation facilities, as most of neural networks do.

From this paper one can draw a conclusion about an equivalence of the radial basis function network and the multi-layer perceptron neural network (described in Section 3 and Section 4, respectively). However, it is worth mentioning that in [2] we can find a sentence which tells that MLP and RBF networks are functionally equivalent but there is only little insight into how the networks may be designed or learned efficiently.

There are also other neural network representations of fuzzy systems, for example the hybrid neuro-fuzzy architecture proposed by Ayoubi [1], with both the radial basis neurons and the perceptrons in one network. Let us also mention fuzzy neural networks, i.e. neural networks with fuzzy inputs, fuzzy weights, fuzzy transfer functions - another kind of neuro-fuzzy systems [3,12].

References

1. Ayoubi M.: Rule Extraction for Fault Diagnosis with a Neural-Fuzzy Structure and Application to a Turbocharger. Proc. of 2nd European Congress on Intelligent Techniques, EUFIT'94. Aachen, Germany (1994)
2. Carse B., Fogarty T. C.: Fast Evolutionary Learning of Minimal Radial Basis Function Networks Using Genetic Algorithm. In: Lecture Notes on Computer Science, Vol. 1143. Springer-Verlag . Berlin Heidelberg New York (1996) 1-22
3. Hayashi Y, Buckley J. J., Czogala E.: Fuzzy Neural Network with Fuzzy Signals and Weights. International Journal of Intelligent Systems 8 (1993) 527-537
4. Jang J. -S. R., Sun C. -T.: Functional Equivalence Between Radial Basis Function Networks and Fuzzy Inference Systems. IEEE Transactions on Neural Networks 4 (1993) 156-159
5. Michalewicz Z.: Genetic Algorithms + Data Structures = Evolution Programs. Springer-Verlag. Berlin Heidelberg New York (1992)
6. Moody J., Darken C.: Fast Learning in Networks of Locally-Tuned Processing Units. Neural Computation 1 (1989) 281-294
7. Pedrycz W.: Fuzzy Control and Fuzzy Systems. John Wiley & Sons Inc. New York Chichester Toronto (1993)
8. Rutkowska D.: Neural Structures of Fuzzy Systems. 1997 IEEE International Symposium on Circuits and Systems (ISCAS'97). Hong Kong. June (1997) 601-604.
9. Rutkowska D., Pilinski M., Rutkowski L.: Neural Networks, Genetic Algorithms, and Fuzzy Systems. PWN, Warsaw (1997) (in Polish)
10. Rutkowska D.: Intelligent Computational Systems. PLJ Academic Publishing House. Warsaw (1997) (in Polish)
11. Rutkowska D.: RBF Neuro-Fuzzy System with Non-Singleton Fuzzifier and Hybrid Learning Procedure. 7th European Congress on Intelligent Techniques and Soft Computing. Aachen, Germany. September (1999); Proceedings on CD-ROM.
12. Rutkowska D., Hayashi Y.: Neuro-Fuzzy Systems Approaches. International Journal of Advanced Computational Intelligence 3 (1999) 177-185
13. Wang L. -X.: Adaptive Fuzzy Systems and Control. PTR Prentice Hall, Englewood Cliffs, New Jersey (1994)
14. Zurada J. M.: Introduction to Artificial Neural Systems. West Publishing Company. New York (1992)

A Neuro-Fuzzy Inference System Optimized by Deterministic Annealing

J.Leski and E.Czogała

Institute of Electronics

Technical University of Silesia

Akademicka 16, 44-100 Gliwice, Poland

e-mail: jl@biomed.iele.polsl.gliwice.pl

Abstract. In this paper artificial neural network based fuzzy inference system (ANNBFIS) learned by deterministic annealing has been described. The system consists of the moving fuzzy consequent in if-then rules. The location of this fuzzy set is determined by a linear combination of system inputs. This system also automatically generates rules from numerical data. The proposed system operates with Gaussian membership functions in premise part. Parameter estimation has been made by connection of both deterministic annealing and least squares methods. For initialization of unknown parameter values of premises, a preliminary fuzzy c-means clustering method has been employed. The application to prediction of chaotic time series is considered in this paper.

1 Introductory Remarks

In approximate reasoning realized in fuzzy systems the if-then fuzzy rules or fuzzy conditional statements play an essential and up to now the most important role [2], [4], [5], [6], [8], [10], [14]. Often they are also used to capture the human ability to make a decision or control in an uncertain and imprecise environment. In this section we will use such fuzzy rules to recall the important approximate reasoning methods which are basic in our further considerations. We assume that m numbers of n-input and one-output (MISO) fuzzy implicative rules or fuzzy conditional statements are given. The i-th rule may be written in the following form:

$$R^{(i)} : \text{if } X_1 \text{ is } A_1^{(i)} \text{ and ... and } X_n \text{ is } A_n^{(i)} \text{ then } Y \text{ is } B^{(i)} \tag{1}$$

Membership functions of fuzzy sets $B^{(i)}$ can be represented by the parameterized functions in the form:

$$B^{(i)} \sim f^{(i)}[\text{Area}(B^{(i)}), y^{(i)}] \tag{2}$$

where $y^{(i)}$ is the center of gravity (COG) location of the fuzzy set $B^{(i)}$:

$$y^{(i)} = \text{COG}\left(B^{(i)}\right) = \frac{\int y B^{(i)}(y) dy}{\int B^{(i)}(y) dy} \tag{3}$$

Now let us consider a system with the conjunctive interpretation of fuzzy if-then rule (Larsen's product) which may be equivalent (under some conditions) to the logical implication interpretation (Reichenbach fuzzy implication). It should be pointed out that different defuzzifiers are used for those interpretations [3]. A general form of final output value can be put in the form:

$$y_0 = \frac{\sum\limits_{i=1}^{m} y^{(i)} \, \text{Area} \left(B^{(i)}\right)}{\sum\limits_{i=1}^{m} \text{Area} \left(B^{(i)}\right)} \tag{4}$$

For symmetric triangle (isosceles triangle) fuzzy values we can write the formula:

$$y_0 = \frac{\sum\limits_{i=1}^{m} \frac{w^{(i)} R_i(\underline{x}_0)}{2} y^{(i)}}{\sum\limits_{i=1}^{m} \frac{w^{(i)} R_i(\underline{x}_0)}{2}} \tag{5}$$

where $w^{(i)}$ is the width of the triangle base, $R_i(\underline{x}_0)$ denotes the degree of activation of the i-th rule.

Usually the value describing the location of COG's consequent fuzzy set in if-then rules is constant and equals $y^{(i)}$ for i-th rule. A natural extension of the above described situation is an assumption that the location of the consequent fuzzy set is a linear combination of all inputs for i-th rule:

$$y^{(i)}(\underline{x}_0) = p^{(i)T} \underline{x}_0' \tag{6}$$

where $\underline{x}_0' = \left[1 \; \underline{x}_0^T\right]^T$. Hence we get the final output value in the form:

$$y_0 = \frac{\sum\limits_{i=1}^{m} \frac{w^{(i)} R_i(\underline{x}_0)}{2} p^{(i)T} \underline{x}_0'}{\sum\limits_{i=1}^{m} \frac{w^{(i)} R_i(\underline{x}_0)}{2}} \tag{7}$$

Additionally, we assume that $A_1^{(i)}, ..., A_n^{(i)}$ have Gaussian membership functions:

$$A_j^{(i)}(x_{j0}) = \exp\left[-\frac{\left(x_{j0} - c_j^{(i)}\right)^2}{2(s_j^{(i)})^2}\right] \tag{8}$$

where $c_j^{(i)}$, $s_j^{(i)}$; $j = 1, 2, ..., n$; $i = 1, 2, ..., m$ are the parameters of the membership functions. On the basis of (8) and for explicit connective AND taken as product we get:

$$A^{(i)}(\underline{x}_0) = \prod_{j=1}^{n} A_j^{(i)}(x_{j0}) \tag{9}$$

Hence, we have:

$$R_i\left(\underline{x}_0\right) = \exp\left[-\sum_{j=1}^{n}\frac{\left(x_{jo} - c_j^{(i)}\right)^2}{2(s_j^{(i)})^2}\right] \qquad (10)$$

2 Learning Methods

Obviously, the number of if-then rules is unknown. Equations (7), (10) describe a radial neural network. The unknown parameters (except the number of rules m) are estimated by means of a gradient method performing the steepest descent on a surface in the parameter space. Therefore the so called learning set is necessary, i.e. a set of inputs for which the output values are known. This is the set of pair $[x_0\left(k\right), \ t_0\left(k\right)]$; $k = 1, 2, ..., N$. The measure of the error of output value may be defined for a single pair from the training set:

$$E = \frac{1}{2}\sum_{n=1}^{N}\left[t_0\left(n\right) - y_0\left(n\right)\right]^2 \qquad (11)$$

The minimization of error E is made iteratively (for parameter α):

$$\left(\alpha\right)_{new} = \left(\alpha\right)_{old} - \eta\frac{\partial E}{\partial\alpha}\bigg|_{\alpha=(\alpha)_{old}} \qquad (12)$$

where η is the learning rate.

The unknown parameters may be modified on the basis of (12) after the input of one data collection into the system or after the input of all data collections (cumulative method). Additionally, the following heuristic rules for changes of η parameter have been applied [7]. If in four sequential iterations the mean square error has diminished for the whole learning set, then the learning parameter is increased (multiplied by n_I). If in four sequential iterations the error has been increased and decreased commutatively then the learning parameter is decreased (multiplied by n_D). Another solution accelerating the convergence of the method is the estimation of parameters $p^{(i)}$; $i = 1, 2, ..., m$ by means of the least square method. The method presented above may lead to the achievement of local minimum. Therefore an application of global optimization (simulated annealing) seems to be reasonable. However, such a method is characterized by a significant cost of computation. An alternative method characterized by less computation is the deterministic annealing [9], [12], [13]. Such a method is based on minimization of energy error of a modelled system while its entropy is controled. This corresponds to the minimization of:

$$F = E - TS \qquad (13)$$

where T is a parameter controlling the entropy of the system (equivalent of temperature). Entropy of the system is defined as:

$$S = \sum_{n=1}^{N} \sum_{i=1}^{m} \mu_i \left[\underline{x}_0 (n) \right] \ln \left\{ \mu_i \left[\underline{x}_0 (n) \right] \right\} \tag{14}$$

where $\mu_i (\cdot) := \frac{w^{(i)} R_i(\cdot)}{2}$. In this case equation (12) takes the form:

$$(\alpha)_{new} = (\alpha)_{old} - \eta \frac{\partial E}{\partial \alpha} \bigg|_{\alpha = (\alpha)_{old}} + \eta T \frac{\partial S}{\partial \alpha} \bigg|_{\alpha = (\alpha)_{old}} \tag{15}$$

The optimization method consist in iterational determining of the parameters based on equation (15) for definite "temperature" T. After the execution of the definite number of iterations parameter T is diminished according to the formula: $(T)_{new} = \beta (T)_{old}$. In tests presented below $\beta = 0.95$ was applied, however the "temperature" was changed after each iteration of equation (15).

Another problem is the estimation of the initial values of membership functions for premise part. This task is solved by means of preliminary clustering of training data, for which fuzzy c-means method has been used [1], [11]. This method assigns each input vector $\underline{x}_0 (k)$; $k = 1, 2, ..., N$ to clusters represented by prototypes v_i; $i = 1, 2, ..., c$ measured by grade of membership $u_{ik} \in [0, 1]$. A $c \times n$ dimensional matrix called a partition matrix fulfils the following assumptions:

$$\begin{cases} \forall_k \sum_{i=1}^{c} u_{ik} = 1 \\ \forall_i \sum_{k=1}^{N} u_{ik} \in (0, N) \end{cases} \tag{16}$$

The c-means method minimizes the scalar index for parameter r > 1:

$$J_r = \sum_{k=1}^{N} \sum_{i=1}^{c} u_{ik}^r \left\| \underline{x}_0 (k) - v_i \right\|^2 \tag{17}$$

Defining $D_{ik} = \left\| \underline{x}_0 (k) - v_i \right\|^2$, where $\|\cdot\|$ is a vector norm (the most frequent Euclidean norm), we get an iterative method of commutative modification of partition matrix and prototypes [1]:

$$\forall_i \qquad v_i = \frac{\sum_{k=1}^{N} u_{ik}^r \, x_0 (k)}{\sum_{k=1}^{N} u_{ik}^r} \tag{18}$$

$$\forall_{i,k} \qquad u_{ik} = \left[\sum_{j=1}^{c} \left(\frac{D_{ik}}{D_{jk}} \right)^{\frac{2}{r-1}} \right]^{-1} \tag{19}$$

According to the above written equations the obtained calculations are initialized if we take into account a random partition matrix U which fulfils conditions (16). Such a method leads to the local minimum of index (17). Therefore the most frequently used solution is multiple repeated calculations in accordance with equations (18, 19) for various random realizations of partition matrix initializations. As a termination rule we have applied the execution of the set number of iterations (in our case 500) or when in sequential iterations the change of index value J_r is less than the set value (in our case 0.001) the computation has been completed. As a result of preliminary clustering the following assumption for ANNBFIS initialization can be made: $c^{(j)} = v_j; \ j = 1, 2, ..., m$ and:

$$s^{(i)} = \frac{\sum\limits_{k=1}^{N} u_{ik}^r \left[x_0\left(k\right) - v_i\right]^2}{\sum\limits_{k=1}^{N} u_{ik}^r} \tag{20}$$

For calculations presented in the next section, the following parameter values: $\eta = 0.01$, $n_I = 1.1$, $n_D = 0.9$, $r = 2$ have been applied.

3 Application to Chaotic Time Series Prediction

A chaotic time series (a discrete signal) obtained on the basis of the solution of the Mackey-Glass equation was investigated:

$$\frac{\partial x(t)}{\partial t} = \frac{0.2x(t - \tau)}{1 + x(t - \tau)^{10}} - 0.1x(t) \tag{21}$$

Prediction of the time series generated by means of equation (21) was realized by many authors [2], [5]. To make a precise comparison we applied data generated by Jang and obtained via anonymous ftp (ftp:// ftp.cs.cmu.edu/ users/ ai/ areas/ fuzzy/ systems/ anfis). For obtaining such a time series Jang applied fourth-order Runge-Kutta method with the following parameters: time step 0.1, $x(0) = 0.1$, $\tau = 17$ [5], [7]. Such generated data are combined in the embedded vector $[x(n) \quad x(n-6) \quad x(n-12) \quad x(n-18)]^T$. The goal is the prediction of value $x(n+6)$ for the embedded vector as input. The data consist of 500 pairs of input-output data of the learning set and 500 pairs of the testing set. By means of the system described in section 2, 500 iterations were carried out, the number of rules changing from 2 to 17. Prediction quality has been evaluated with mean square error. Table 1 shows the results multiplied by 10^6 (GM - gradient method, DA - deterministic annealing). From the Table it can be inferred that the learning method based on deterministic annealing leads to smaller prediction error in comparison with gradient learning method. The presented method is useful for more than three rules and becomes even more useful for more if-the rules.

Number of Rules	2	3	4	5	6	7	8	9
Training - GM	91.42	18.44	14.32	8.62	7.32	6.91	7.55	3.02
Training - DA	95.61	21.64	12.05	8.14	7.32	5.46	5.25	2.23
Testing - GM	90.92	18.53	15.16	9.02	7.49	7.29	7.89	4.43
Testing - DA	95.93	21.67	12.53	8.91	7.48	6.27	7.06	3.69
Number of Rules	10	11	12	13	14	15	16	17
Training - GM	3.21	1.94	1.57	1.35	1.38	1.20	1.01	0.95
Training - DA	1.79	1.43	1.12	1.08	1.42	0.93	0.83	0.69
Testing - GM	4.28	3.34	3.15	2.60	2.48	2.28	2.01	1.98
Testing - DA	3.06	2.94	2.90	2.12	2.47	1.76	1.81	1.77

Table 1. Simulation results for Mackey-Glass chaotic time series prediction.

4 Conclusions

In this paper a new Artificial Neural Network Based Fuzzy Inference System (ANNBFIS) has been described. Such a presented system can be used for the automative if-then rule generation. The novelty of that system in comparison with the one well known from literature is a whole moving fuzzy consequent. A particular case of our system is Jang's ANFIS (moving consequent considered as singleton) or Cho and Wang AFS with a constant fuzzy consequent. A connection of the deterministic annealing and least squares methods of parameter optimization for ANNBFIS has been used. For initialization of calculations preliminary fuzzy c-means clustering has been used. Promising applications of the presented system to chaotic time series prediction have been shown. The results show the advantage of the learning method based on deterministic annealing in comparison with classic gradient learning methods.

References

1. J.C. Bezdek: Pattern recognition with fuzzy objective function algorythms, New York, Plenum (1981)
2. K.B. Cho, B.H. Wang: Radial basis function based adaptive fuzzy systems and their applications to system identification and prediction, Fuzzy Sets & Systems 83 (1996) 325-339
3. E.Czogała, J.Leski: Fuzzy implications in approximate reasoning, in: L.Zadeh, A.Kacprzyk Eds., Computing with words in information/intelligent systems, Springer-Verlag (1999) 342-357
4. S. Horikawa, et al: On fuzzy modeling using fuzzy neural networks with the backpropagation algorithm, IEEE Trans. NN 4 (1992) 801-806
5. R.J. Jang, C. Sun: Neuro-fuzzy modeling and control, Proc. IEEE, 83 (1995) 378-406
6. R.J. Jang, C. Sun: Functional equivalence between radial basis function and fuzzy inference systems, IEEE Trans. NN 4 (1993) 156-159

7. R.J. Jang, C. Sun, E. Mizutani: Neuro-fuzzy and soft computing:, London, Prentice-Hall (1997)
8. B. Kosko: Fuzzy associative memories, in: A.Kandel, Ed., Fuzzy Expert Systems (1987)
9. D.Miller, et al: A Global optimization technique for statistical classifier design, IEEE Trans. SP 12 (1996) 3108-3121
10. S. Mitra, S. Pal, K.: Fuzzy multi-layer perceptron, inferencing and rule generation, IEEE Trans. Neur. Net., 6 (1995) 51-63
11. N.R. Pal, J.C. Bezdek: On cluster validity for the fuzzy c-means model, IEEE Trans. Fuzzy Systems, 3 (1995) 370-379
12. A.V.Rao, et al: Mixture of experts regression modeling by deterministic annealing, IEEE Trans. SP, 11 (1997) 2811-2819
13. K.Rose, et al: Vector quantization by deterministic annealing, IEEE Trans. Inf. Thoery, 4 (1992) 1249-1257
14. L. Wang, J.M. Mendel: Generating fuzzy rules by learning from examples, IEEE Trans. SMC, 22 (1992) 1414-1427

Neuro-Fuzzy Detector for Industrial Process

Marek Kowal and Józef Korbicz

Department of Robotics and Software Engineering
Technical University of Zielona Gora
ul. Podgorna 50, 65-246 Zielona Gora, Poland
{M.Kowal,J.Korbicz}@irio.pz.zgora.pl

Abstract. This paper deals with the use of neuro-fuzzy approach to fault detection in industrial processes. The general information regarding the model based fault detection and neuro-fuzzy techniques is presented. The neuro-fuzzy simulator is employed to fault detection in the boiler drum. To illustrate our approach, the simulation results are presented in the final part of the paper.

1 Introduction

A fault can be described as an unexpected change of the system functionality that is manifested as a deviation of at least one characteristic property or a variable of a technical process. It may not, however represent the failure of physical components. Such malfunctions may occur either in the sensors and actuators or in the components of the process itself. A system that includes the capability of detecting and diagnosing faults is called the fault diagnosis system. Tasks of such systems can be viewed as a sequential data processing, which contains two steps: detection and isolation. Detection is required for symptom generation (data reduction, contrast amplification). Symptom signals include information about the state of the process. However a detector can determine only that a fault occurred in the system. The second step of data processing is needed to obtain more information about a fault (type, location, and reasons). This process is described by a discrete mapping from continuous symptom space to discrete fault space. Such a task is carried out by a classifier, which determines what kind of failure is present.

The focus of this paper is the actual fault detection algorithm. For several years the idea of using artificial intelligence methods has been widely exploited in fault detection applications. Mainly, research has been done for neural networks, expert systems, genetic algorithms, fuzzy and neuro-fuzzy systems. These methods seem to be a promising solution for high complex systems where other methods cannot be used. Quantitative or qualitative knowledge is used to build up artificial intelligence applications. Such an approach does not require the mathematical model of the process. Its features make it possible to use such a solution for a wide range of industrial hardware components. They can be used both in the case of static and dynamic systems diagnosis. Especially neuro-fuzzy methods have received much attention because of their possibilities to use simultaneously qualitative and quantitative knowledge about the process. It means that numerical observations and expert observations can be employed in the supervised learning process. The neuro-fuzzy system has a layered structure like ordinary neural nets. It means that well known back-propagation learning algorithms can be applied to these structures. Tuned parameters store knowledge that is coded as fuzzy rules, which can be extracted from applications or inserted into them. This situation allows one to prepare better start values of parameters than by using random selection. After training fuzzy rules can be extracted and analysed. The learning task is critical for simulation results and depends particularly on the quality of the training data. Therefore the neuro-fuzzy application can reach good simulation results due to using both type of data.

2 Model Based Fault Detection

A prompt detection of faults is essential for reliable, safe and efficient operations of the plant for maintaining quality of the products. There are different methods of fault detection available. Generally, they can be divided into two groups: a simple process variable monitoring, and, more complex model based methods. For a simple fault that can be detected by a single measurement a conventional alarm circuit may be proper. However, since it is usually very difficult in complex industrial systems to directly measure the state of the process, more sophisticated solutions are needed. The well-known approach is to use a model based system for fault detection [4]. This requires modelling of the process and may be called an analytical redundancy because the model and the process work simultaneously. The idea of such fault detection is to compare output signals of the model and the process, thereby generating a residual or an output error, which is used to make a decision about the state of the process [3]. Such a fault detector scheme is shown in Fig. 1.

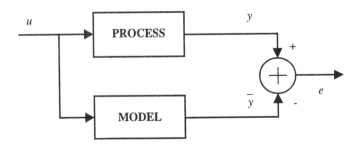

Fig. 1. Model based fault detection scheme

The approach makes it possible to detect small scale faults quickly and reliably. These are the reason of using such a system in life-critical applications where even small defects can cause big damages.

Many different methods of model designing are avaliable. The most popular are analytical methods and artificial intelligence methods. Analytical methods (the Kalman filter, the Luenberger observer, etc.) can be applied to the process that contains a mathematical model. On the other hand, artificial intelligence methods that do not require the process model have been investigated intensively in the recent years. Neural nets and neuro-fuzzy applications have received much attention due to their fast and robust implementation, their performance in learning arbitrary non-linear mappings, and their abilities of generalization.

3 Neuro-Fuzzy Systems

Recently, the neuro-fuzzy approach has been actively employed for different applications. It integrates fuzzy modeling techniques and neural nets learning procedures. Such hybrid systems are topologically designed to copy the fuzzy system. Thus, they have a possibility of emulating the fuzzy inference mechanism [5], which is shown in Fig. 2. First, the input data are fuzzificated, then the inference operation is made using the defined rule base. Finally, results are defuzzificated.

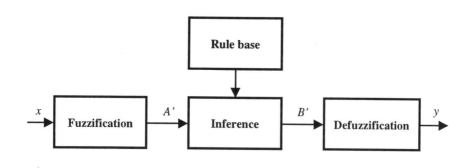

Fig. 2. Fuzzy inference mechanism scheme

The main aim of such an approach is to join the strengths of both theories in order to achieve a learning system with transparent knowledge. The relationship between the weights of the neuro-fuzzy system and the rule base renders it possible to utilize expert knowledge for initialisation of neuro-fuzzy network parameters to achieve a better and quickest training process. After this process rules may be extracted and a human expert can verify their correctness.

Generally, a neuro-fuzzy structure consists of three layers. An antecedent part includes fuzzy sets. Its outputs express the degree of membership to the implemented fuzzy sets. Parameters of the neurons can be estimated before learning using clustering process (Fuzzy C-Means, Fuzzy Bounded Classification, ISO-DATA, Mountain Method)[1]. The second layer contains elements that realize the logical conjunction of the antecedents from the first layer. Here, a membership degree of composition of antecedents to each rule is evaluated. The third layer includes conclusions and subsystem for defuzzification. Its structure depends on conclusion and defuzzification type. Conclusions may be defined as singletons, linear dependence, non-linear dependence or fuzzy sets. A system with singleton conclusions and center of average defuzzification is shown in Fig. 5.

4 Simulation Results

The neuro-fuzzy model is proposed for fault detection in the boiler drum. The boiler is an element of the steam power plant and it is responsible for steam generation [2]. The water level in the boiler is a life-critical value for the system's functionality. A simplified boiler scheme is shown in Fig. 3. The water level in the boiler depends on the rate of water inflow and the rate of discharge of steam.

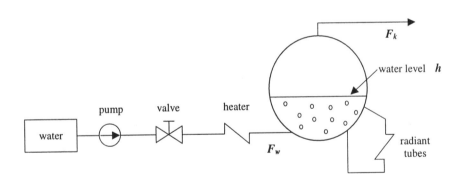

Fig. 3. Boiler drum scheme

A neuro-fuzzy simulator of the boiler is shown in Fig. 4. Delay lines and neuro-fuzzy network with singleton conclusion, center of average defuzzification, Gaussian functions of model fuzzy subsets, logical conjunction defined as a product were used to build the boiler's model. The applied neuro-fuzzy network is shown in Fig. 5.The backpropagation algorithm to train the network was used.

An identification of the structure of the model was the most important task during model designing. The boiler is a dynamic system and its output depends on the present and past values of inputs. We had to decide which delay inputs should be used.

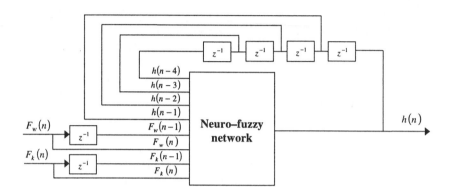

Fig. 4. Boiler simulator, F_w - rate of water inflow, F_k - rate of discharge of steam, h - water level

Different configurations of inputs were tested and the best results were obtained for the model shown in Fig. 4.

The neuro-fuzzy network (Fig. 5.) generates the output signal that can be described by the equation (1).Three types of weights are tuned during the learning process $c_{i,h_i}, w_{i,h_i}, d^{h_1,...,h_8}$. Weights c_{i,h_i} and w_{i,h_i} are parameters of membership functions of fuzzy sets, c_{i,h_i} are defined as centers of Gaussian functions, w_{i,h_i} are defined as the widths of Gaussian functions (Fig. 6). Weights $d^{h_1,...,h_8}$ are defined as singletons and express conclusions.

$$\bar{y} = \frac{\sum_{h_1}^{N_1} \cdots \sum_{h_n}^{N_n} \left(d^{h_1...h_n} \left(\prod_{i=1}^{n} \exp\left(-\left[\frac{\bar{x}_i - c_{i,h_i}}{w_{i,h_i}} \right]^2 \right) \right) \right)}{\sum_{h_1}^{N_1} \cdots \sum_{h_n}^{N_n} \left(\prod_{i=1}^{n} \exp\left(-\left[\frac{\bar{x}_i - c_{i,h_i}}{w_{i,h_i}} \right]^2 \right) \right)} \tag{1}$$

After the learning process some simulations were made. First, the quality of the implemented model was tested. Some results are presented in Fig. 7. After these tests some faults were simulated in the boiler drum and the effectiveness of the fault detection using the implemented model was tested. Some results are shown in Figs. 8 and 9.

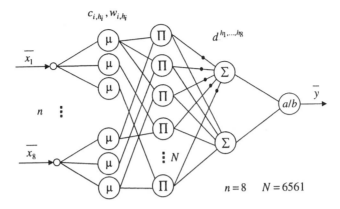

Fig. 5. Neuro-fuzzy network where $\bar{x}_1, \ldots, \bar{x}_8$ - inputs, $\bar{y}$ - output, n - number of inputs, N - number of rules, $c_{i,h_i}, w_{i,h_i}, d^{h_1,\ldots,h_8}$ - weights, i, h_i - indexes

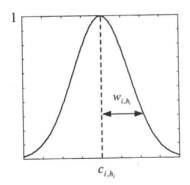

Fig. 6. Gaussian function and parameters c_{i,h_i} and w_{i,h_i}

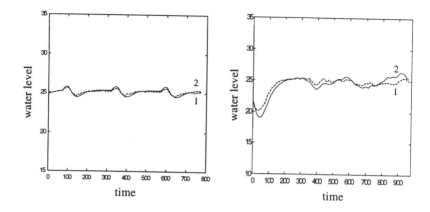

Fig. 7. Simulation results 1- neuro-fuzzy simulator response, 2- boiler response

5 Conclusions

This paper deals with neuro-fuzzy systems used as a simulators in model-based fault detection. A detection tool for non-linear dynamic systems has been proposed. The neuro-fuzzy model of the plant was developed and implemented. Simulations for the boiler drum showed the effectiveness of the implementation. All simulated faults were detected successfully and abilities to detect small defect were also observed.

Another advantage of the neuro-fuzzy approach in fault detection applications is that no mathematical model is required. In the training process, numerical observations and human expert knowledge can be used as well. These features allow building accurate models for highly reliable fault detection applications.

References

1. Bauman, E., Dorofeyuk, A., Filev, D.: Fuzzy identification of non-linear dynamical systems. Proceedings International Conference on Fuzzy Logic and Neural Nets, Iizuka, Japan (1990) 1013–1015
2. Korbicz, J., Pieczynski, A.: Dynamic model of steam power plant. Report of Department of Robotics and Software Engineering, Zielona Gora, Poland (1993) (in Polish)
3. Maki, Y., Loparo, K.A.: A neural network approach to fault detection and diagnosis in industrial processes. IEEE Trans. on Control Systems Technology, Vol. 5, No. 6 (1997) 529–541
4. Marcu, T., Mirea, L.: Robust detection and isolation of process faults using neural networks. IEEE Trans. on Control Systems Technology, Vol. 10 (1997) 72–79
5. Yager, R.R., Filev, D.P.: Essentials of fuzzy modelling and control. John Wiley and Sons, Inc. (1994)

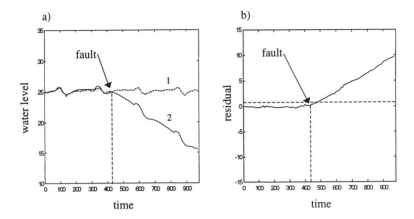

Fig. 8. Fault simulation 1- neuro-fuzzy simulator response, 2- boiler response (a) water level (b) residual

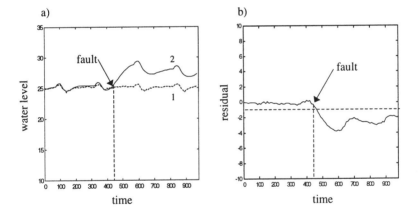

Fig. 9. Fault simulation 1- neuro-fuzzy simulator response, 2- boiler response (a) water level (b) residual

Fuzzy Setting of GA Parameters

Radek Matoušek, Pavel Ošmera

Technical University of Brno, Faculty of Mechanical Engineering
Technická 2, 616 69 Brno, Czech Republic
Department of Automation and Information Technology
{matousek,osmera}@kinf.fme.vutbr.cz

Abstract. Applications of Genetic Algorithms – GAs for optimization problems are widely known as well for their advantages and disadvantages compared with classical numerical methods. In practical tests, a GA appears as robust method with a broad range of applications. The determination of GA parameters could be complicated. Therefore, for some real-life applications, several empirical observations of an experienced expert are needed to define these parameters. This fact degrades the applicability of GA for most of the real-world problems and users. Therefore, this article discusses some possibilities with setting a GA. The setting method of GA parameters is based on the fuzzy control of values of GA parameters. The feedback for the fuzzy control of GA parameters is realized by virtue of the behavior of some GA characteristics.

1 Introduction

The using of genetic algorithms for optimization the fuzzy systems is known. We will follows the opposite way. The setting of GA parameters is often a seriously complicated procedure because it must meet two contradictory requirements:

- to search up the whole space;
- to search some parts of the space in detail.

A balance between the utilisation of the whole space and the detailed searching of some parts can be adapted to pressure [1] of selection and recombination operators. This balance is critical for a GA behavior. For this reason, it is very important to understand the influence of selection and recombination operators on the GA. The operators have a direct influence on the GA convergence.

For "suitable" behavior of a GA with respect to the problem to be solved the follow requirements can be formulated:

1. to find an acceptable solution (a global one is the best);
2. to keep number of iterations (generations) and time of computation acceptable.

These requirements will be fulfilled if "suitable" methods (selection and recombination) and "suitable" setting of a GA will be used. Only just in verbal term and mean "suitable" is a problem. When setting GA parameters, the following solutions are available for the standard or less experienced GA user:

- to try set up of GA parameters for a specific problem with the advice of experts [1,2] – can be acceptable if taken in all aspects
- to set up parameters by expert or to become an expert yourself!
- to try to set up parameters randomly by means of common antecedents [1], and make pre-sets in case of a failure – the work after a longer time having some magic and the user getting to be a wizard!

It is necessary to say that the human experts in a given branch could have a problem with "suitable" behaviour of a GA as well. The "suitability" of changes of GA parameters comes during the run of the algorithm and sometimes comes from the experiments and experiences with a GA behaviour [3].

The following criteria and fundamentals can be used to solve the problem. We try to suggest a solution on the basis of given reasons. The basis of the solution is a fuzzy control of GA parameters.

2 Genetics Algorithms

Used GA represents generation model of GA. The short formal description of GA is follows:

$$P_{n+1} = f(P_n, \xi_n), \quad \xi \in (S, C, M) \tag{1}$$

Where: P...population

ξ ...a set of GA operators: S ...selection

C ...crossover

M...mutation

n ...number of generation

3 Characteristics of GA

Fuzzy Inference System – FIS is used for the fuzzy control of GA parameters. The following must be determined for applications of FIS to control the GA:

- input values
- model of FIS

3.1 Quantitative Characteristics

The input values are extracted from quantitative characteristics of GA. Various types of these characteristics are contained, for instance, in [1-3].

The following characteristics can be used as the input:

$\rightarrow$ Variability of population - *varH*
Also called survival probabilities. This variability is represented by a ratio of identical individuals to all individuals per population (2). Since a binary string represents the individuals, Hamming distances are used to compare these individuals. The Hamming distance is given by the metric system ρ_H, (3):

$$\mathrm{var}\,H = \frac{\#kind}{\#individual}\,,\;varH \in (0,1] \tag{2}$$

$$\rho_H(\vec{a},\vec{b}) = \sum_{i=1}^{n}|a_i - b_i| \tag{3}$$

Where $\vec{a} = (a_1,\dots,a_n) \in B^n, \vec{b} = (b_1,\dots b_2) \in B^n$

$\rightarrow$ Coefficient of partial convergence - *cpc*
Value *cpc* (5) determines the average or the weight-average change of a standard value of fitness function (4) per sample of generations.
Let the population be sorted:

z_n^k ... *n*-th fitness value from *k*-th population

$\quad n = 1$ ... the best solution
$\quad k = 1$ ... the first population

$$\|\vec{z}_n\| = \frac{z_n^{(k)}}{\sum_{k \in sample} z_n}; z_n \geq 0, \|\vec{z}_n\| \in [0,1] \tag{4}$$

Where a *sample* is a discrete interval [*k*, *k+sample*] of observations of values z_n related to the generation.

$$cpc = \sum_{i=k}^{k+sample-1} f_w(z_n^{i+1} - z_n^i) \tag{5}$$

Where f_w is a weight function related to the *sample*. If $f_w=1$, then $pc \in [-1,1]$. For a small number of samples (about 5÷10 generation) $f_w=1$ is used. The value of a sample can be constant. We used ten samples in our research (it is sample$\in$ [k,k+9]). For a time behavior of the *sample* the so-called H-characteristics (see below) was used.

→ Furthermore, some characteristic based on <u>cluster analysis</u>, <u>monitoring fitness values</u> (function values z_{min}, z_{mean}, and so on) and another statistical method based on <u>fitness distribution</u> can be used.

→ So-called <u>H-characteristics</u> [3] for determination of a suitable moment to change GA parameters, which is a fuzzy control action. If a function value of H-characteristics is under stagnation, the change is required. Stagnation is tested by *cpc*.

The general formulas of **H**-characteristics are represented by the following expressions (6),(7):

$$\bullet \quad H_n^1 = \sum_{\substack{i_1 < i_2 \\ i_1, i_2 \in K}} \sum_{j=0}^{m} \left(x_{n(i_1), j} - x_{n(i_2), j} \right)^2 \tag{6}$$

$$\bullet \quad \Delta H_n^1 = H_n^1 - H_{n-1}^1, \quad H_0^1 = 0$$

$$\bullet \quad \left| \Delta H_n^1 \right|$$

$$\bullet \quad H_n^2 = \sum_{\substack{i_1 < i_2 \\ i_1, i_2 \in K}} \left(\sum_{j=0}^{m} \left(x_{n(i_1), j} - x_{n(i_2), j} \right)^2 \right) \cdots \tag{7}$$

$$\cdots \cdot \left(x_{n(i_1)0} - x_{n(i_2)0} \right)^2$$

$$\bullet \quad \Delta H_n^2 = H_n^2 - H_{n-1}^2, \quad H_0^1 = 0$$

$$\bullet \quad \left| \Delta H_n^2 \right|$$

Where K is a characterized subset of population; $x_{n(1)0} \cdots x_{n(p)0}$ represent ordered sequence of fitness values of *n*-th generation (until now denoted as z_n); $x_{n(i)j}$ denotes *i*-th co-ordinate of domain element from *n*-th generation with fitness value $x_{n(i)0}$.

Several figures are presented for the case of test and set $K=\{1,2,3\}$, (see Fig. 1,2).

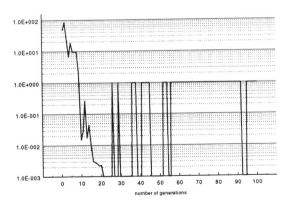

Fig. 1. The $\mathbf{H}^1$ characteristic

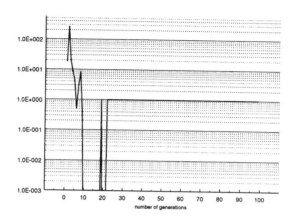

Fig. 2. The $|\Delta \mathbf{H}^2|$ characteristic

We can see that **H**-characteristics follow the behavior of z_{min}. Therefore, we observe the **H**-characteristics and adapt the GA parameters when these characteristics are stabilized (look at $\Delta \mathrm{H}$ and $|\Delta \mathrm{H}|$).

3.2 Qualitative-Quantitative Characteristic

This is a verbal characteristic with respect to the common aspect of a GA. It serves for assembling and setting of FIS by an expert, with respect to the rule of fuzzy modelling.

Verbal characteristics represent the instruction for building the fuzzy rules, as e.g. (with terms mentioned above):

- if *varH* is "hi" then *mutation* is/set "low" & *selection* is/set "hi"
- if *varH* is "low" & *pc* is "stagnation" then *mutation* is/set "hi"
- and so on …

Next a design for a fuzzy membership function:

- selection intensity = {low, mild, hi}

  Where [2]: "low"   tournament $\approx 1 \div 3$

  "mid'   tournament $\approx 3 \div 5$

  "hi"   tournament $\approx > 4$
- and so on …

4 Fuzzy Setting of GA

The design of the algorithm for fuzzy setting of GA parameters is based on the classical model GA given by Figure 3.

Our idea is the adaptation of the GA operators value $\xi\in$ (selection; crossover; mutation) during the run of GA. The fuzzy control is applied if the condition of fuzzy adaptation is true. The arguments why not to use the fuzzy control during the complete run of GA are as follows:

1. The time cost of calculating of FIS must be respected.
2. After changing the value of GA parameters it is reasonable to let some time-generations for stabilization of GA process.

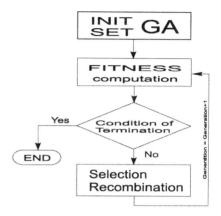

Fig. 3. The Genetic Algorithm

3. If convergence of GA is acceptable it is not possible to use fuzzy control action.

These three arguments represent *"Condition of Fuzzy Adaptation"* in our algorithm given by Figure 4.

Fig. 4. The GA with fuzzy control of parameters

A basis of fuzzy control mechanism is Fuzzy Inference System – FIS [4] assembled by an expert. Fuzzy inference is the process of formulating the mapping from a given input to an output using fuzzy logic. For our work we used Mamdani's inference method. The process of fuzzy inference involves these pieces of fuzzy logic: membership function, fuzzy logic operators, if-then rules, and defuzzification.

It follows a short description of the pieces fuzzy control mechanism and FIS:

Real inputs: Represent numerical values of observed characteristic of GA, see above *varH*, *cpc*, and so on. The inputs are crips (non-fuzzy) numbers.

Fuzzify inputs: Resolve all fuzzy statements in the antecedent to a degree of membership between 0 and 1. Suitable parameterisation is used for fuzzy sets in antecedent and consequent.

Fuzzy rules: Fuzzy sets and fuzzy operators are subjects and verbs of fuzzy logic. These IF-THEN rule statements (8) are used to formulate the conditional statements that comprise fuzzy logic

$$R_i: \quad \text{IF } (in_1 \text{ is } A_{1i}) \&\dots \&(in_m \text{ is } A_{mi}) \tag{8}$$
$$\text{THEN } (out_1 \text{ is } B_{1i}) \&\dots \&(out_n \text{ is } B_{ni}).$$

Apply fuzzy operator to multiple part antecedents: If there are multiple parts to the antecedent, apply fuzzy logic operators and resolve the antecedent to a single number between 0 and 1.

Apply implication method: Use the degree of support for the entire rule to shape the output fuzzy set. The consequence of a fuzzy rule assigns an entire fuzzy set to the output.

Aggregation: The output of each rule is a fuzzy set (sets if necessary). The output fuzzy sets B_r for each rule, these are then aggregated into a single output fuzzy set B_o.

Defuzzifcation: The input for the defuzzification process is a fuzzy set μ (the aggregate output fuzzy set) and output is a single number y^*. For defuzzification we used the method of centroid (center-of-sums) by equation (9).

$$y^* = defuzz \ (B_o) = \frac{\int_y y_n \cdot \sum_{r=1}^{n} \mu B_r (y) dy}{\int_y \sum_{r=1}^{n} \mu B_r (y) dy} \tag{9}$$

Defuzzified value is directly acceptable value of GA parameters, like for example:

Output #1 $\equiv y_1^* \equiv$ "power of tournament selection",
Output #2 $\equiv y_2^* \equiv$ "probability of bit-mutation",
$y_1^* = 2$ ... represent binary tournament selection,
$y_2^* = 0.03...$ represent 3% probability of mutation.

5 Test Problem

For the testing of the performance of evolutionary heuristic algorithms, such as GA, some set of artificial designed optimization problems was used. Usually it is a set of five famous DeJong's [5,6] functions: F1-F5 and few other testing functions named after their authors or again as F6-F12.

These functions are designed to represent a specific part or set of problems for optimization algorithms. The test functions are categorized using a taxonomy incorporating:

- Continuous v. discontinuous
- Convex v. non-convex
- Unimodal v. multimodal

- Quadratic v. non-quadratic
- Low dimensionality v. high dimensionality
- Deterministic v. stochastic

We used a test function denoted Ackley's function. Originally, it was formulated by Ackley (see [5,6]). To facilitate its use for minimization and to achieve a standardization of the global minimum to an objective function value of zero, the function is formulated as follows:

$$f_{Ackley}(\vec{x}) = -c_1 \cdot \exp\left(-c_2\sqrt{\frac{1}{n}\sum_{i=1}^{n} x_i^2}\right) - \exp\left(\frac{1}{n}\sum_{i=1}^{n}\cos(c_3 \cdot x_i)\right) + c_1 + e \quad (10)$$

$$n...dimensionality, \ c_1 = 20, \ c_2 = 0.2, \ c_3 = 2\pi$$
$$x_i \in [-20, 30]; \quad \min f_{Ackley}(\vec{x}) = 0,$$

In order to facilitate an empirical visualization analysis of this function, a 3D representation, i.e. with two parameters, $n=2$ will be used. Its graph is shown in Fig. 5.

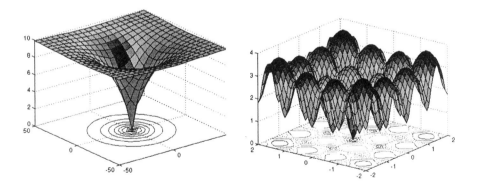

Fig. 5. Ackley's function

6 Experiment

In our tests we used following setting of GA:

method, operator, parameter	setting per test #1,#2,#3		
Model GA, # of individuals	generation, non-elite, 40		
Parameter encoding, length	gray, 20bit/parameter		
Test problem	$\min\{F_{Ac}(x_1,x_2)\}$, $x_i \in (-25,25)$		
All without elitism!	#1	#2	#3
Selection - [#] tournament	2	3	FIS
Crossover - [%], # of cut-point	60; 2	50; 3	FIS
Mutation per bit - [p_m]	0.02	0.05	FIS

An environment of MatLab®, with our own genetic toolbox was used. All results represent 100 runs of our GA per test #1, #2 and #3.

Setting of GA (S, C, M) per test representatives:

Test #1: Approximate values according to some GA literature. This setting is constant per all runs of GA. The setting corresponds to the workaday user.

Test #2: The setting corresponds to: test problem, size of population, applied operators and making use of tips from experts [1-3]. This setting is constant for all runs of GA.

Test #3: The base for the setting is FIS, which is given by an expert. For this test the GA with fuzzy control was the value of GA parameters described in Figure 4 (in detail see above). Initial settings are as for the test #2. A value of the *sample* is ten.

The short description of designed FIS:

Inputs: varH∈ (0, 1],
cpc∈ [-1,1]

Fuzzy inputs			
varH – in1	low	mid	hi
μ_{BELL} – c,σ,β	[0.3 3.278 0]	[0.238 3.28 0.5]	[0.3 3.278 1]
cpc – in2	descent	stagnation	grow
μ_{BELL} – c,σ,β	[0.862 2.5 -1]	[0.132 1.69 0]	[0.857 2.5 1]

Outputs: S∈ [1, 10],
M∈ [0, 0.1]

Fuzzy outputs			
Selection - S	Low	mid	hi
μ_{BELL} – c,σ,β	[2.64 7.12 -1.03]	[1.33 3.57 2.58]	[2.86 2.43 9.05]
Mutation – M	low	mid	hi
μ_{BELL} – c,σ,β	[0.02 3.28 -0.01]	[0.02 3.28 0.04]	[0.04 3.28 0.10]

Fuzzy rules:

output #1 – Selection - S				output #2 – Mutation - M			
in1\in2	desc.	stagn.	grow	in1\in2	desc.	stagn.	grow
low		LOW		low		MID	
mid	MID	MID	HI	mid	LOW	HI	LOW
hi			MID	hi			LOW

Membership function: by equation (11):

$$\mu_{BELL,i} = \frac{1}{1 + \left|\dfrac{x_i - c}{\sigma}\right|^{2\beta}}$$

(11)

Of course, this setting of FIS does not stand for the best solution. It presents a subjective view of an expert with respect to the objective characteristics of GA, see above. The setting is viewable as control surfaces of GA, see Figure 6.

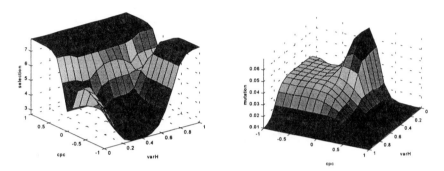

Fig. 6. Control surfaces of GA – selection, mutation

The change of behaviour (convergence) of GA in accordance with fuzzy control action can be seen from a characteristic in Figure 7.

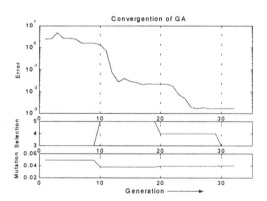

Fig. 7. A behavior of GA, test #3

7 Conclusion and Further Research

A designed GA with Fuzzy Control of GA parameters – GA-FC, let us say FIS, can be further modified with better designed FIS, and for example, different shapes and number of membership functions with better respect of GA, along with further criteria and characteristics of GA.

312

This work represents a new doorway by an algorithm given in Figure 4, and must be understood for the possibilities of Fuzzy Setting of GA Parameters.

Practical results of a GA-FC showed (see Figure 8) that the GA-FC algorithm (test #3) for our specific problem demonstrated essentially better behavior than when used with common setting – test #1. Further, GA-FC has about 10 % better convergence regarding the number of generations than GA constantly set by an expert – test #2. GA-FC is on average slower per generation with respect to the time cost.

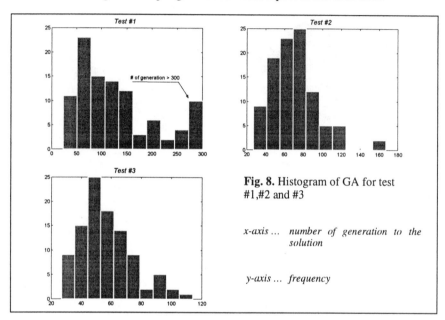

Fig. 8. Histogram of GA for test #1,#2 and #3

x-axis ... number of generation to the solution

y-axis ... frequency

The GA-FC presents a chance for every user who is only able to formulate the fitness and takes already built-up FIS. Other areas of application are the dynamic variable systems where FIS can provide a good description of the system behaviour. The achieved results will be used for further analysis of GAs behavior and for improvement of the GAs power in the solving of prominent engineering applications.

References

1. Goldberg, D.E.: Genetic Algorithms in Search, Optimisation and Machine Learning. Addison-Wesley (1989)
2. Blickle, T., Thiele, L.: A Comparsion of Selection Schemes Used in GA. TIK-Report (1995)
3. Matoušek, R., Popela, P, Karpíšek, Z.: Some Possibilities of Fitness –Value –Stream Analysis. In: Proc. 4th International Conference Mendel '98, Brno (1998) 69-73
4. Klir, G.J., Yuan, B.: Fuzzy Sets and Fuzzy Logic – Theory and Applications. Prentice Hall (1995)
6. Ackley, D.H.: A Connectionist Machine for Genetic. Kluwer, Boston (1987)
7. Bäck, T.: Evolutionary Algorithms in Theory and Practice, Oxford University Press, Oxford (1996)

Self-Learning Fuzzy Models and Neural Models in Environment Prediction

Gheorge M. Sandulescu, Mariana Bistran

Artificial Intelligence Laboratory - SIA at IPA SA
18 Mircea Eliade, Bucharest, Romania
sand@automation.ipa.ro

Abstract. Based on a special neural network approach we show efficient ways to tackle non-linear processes in air pollution problems.

Introduction

At present there are four trends in the advanced modelling for forecasting air pollution:

1. By using of the parallel computers (approx. 100 computers) in USA;
2. By using of the parallel computers (approx. 6-8 computers) in West European countries;
3. By using only one computer, but with so refined models of the evolution of air pollution that to be able at the reduced computing level (only one computer), in East European countries;
4. By using of the neural and Fuzzy models, especially for solving some punctual problems of the air pollution evolution;

For example, in 1998, Hass (Germany) based on parallel computing has developed some models like "Modal Aerosol Dynamics for Europe (MADE)", based on the "Regional Particulate Model" that was applied within the "EURAD model system to the simulation of tropospheric aerosols over Europe" [5].

In the same time, Kallos (Greece) has used a number of 16 RISC nodes for running weather forecasting models covering the regional and mesoscale portion of the spectrum of the atmospheric disturbances [7].

In the present times the intensive development of the Fuzzy, Neural and other procedures and systems belonging to the Artificial Intelligence fields are focused on the modelling of the very complex, non-linear, multivariate, uncertain processes.

Fuzzy models may represent today a very interesting alternative in the fields of the means for forecasting the evolution of air pollution and of pollution in general.

Now, there are many different procedures with the view of the accomplishment of models . So, it is well known that the fuzzy models are classified in 2 categories:

A) Linguistic Models (LMs) based on the collections of IF-THEN rules with vague predicates and fuzzy reasoning . In this case the models are qualitative expression of the systems;

B) Models based on the Takagi-Sugeno-Kang (TSK) method, models which are formed by logical rules that have a fuzzy antecedent part and functional consequent, or, with other words, which combine fuzzy and non-fuzzy models. TSK models integrate the LMs possibility of the qualitative knowledge representation with the possibility of processing quantitative informations .

The TKS fuzzy models solve the disadvantage of the LMs fuzzy models, at which, in principle, the incorporation of the knowledge about the system "it is not possible", if such knowledge cannot be expressed into the fuzzy set framework.

Taking into consideration that the training process may be viewed as a "game of the weights", the modelling processes of the fuzzy system may be focused in the direction of the optimum weights accomplishment . For an precise architecture the specificity of the model is included only in the weight matrix.

The application of the neural networks for the air pollutants evolution forecasting is based on :

A) the possibilities and features of these procedures and systems to extract the deterministic patterns and relations, hidden in the air pollution evolution;

B) on the use of
- the neural intrinsic interpolation features;
- of the neural generalisation intrinsic features .

Going this way based on the learned similarities, the extraction of the hidden, deterministic relations and rules (as patterns) which describe the future possible evolution become possible.

Moreover, if the use of the time series of the air pollution processes evolution, aimed at forecasting and assessing the pollution evolution trends and levels, may offer good results, the implication of the fuzzy multivariate procedures may, potentially, improve drastically the prediction quality.

Application to Air Pollution

The application of neural nets for modelling with the view of forecasting of the air pollution evolution has an interesting way. As an example of the accomplishment of neural models we present in Fig.1 one example of neural network, of the type "feedforward with BKP training", achieved from 4 levels including 10 hidden levels used in neural modelling of the air pollution. It was obtained the performance MSE=0.002926.

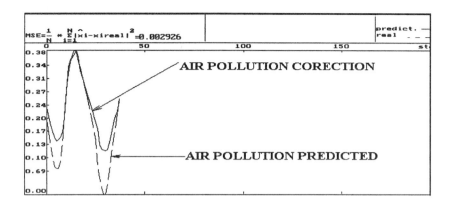

Fig. 1. Neural network in air pollution modelling with training type BKP

The information related to the specific model is captured and stored, in the course of the training process, in the weights of the neural networks, weights which consti- tute the connections between the neural nets nodes (central elements of the neu- rones). Practically, fuzzy, neural and neuro-fuzzy models are, in fact, arrangements and "a game" of the weights, weights which are combined inside the specific archi- tecture, configuration, type of neural networks used, and weights which are combined by the non-linear function of each neurone.

The output signal, from the neural model, is offered by the combination between the new input signals, the weights between the nodes and the non-linear, for instance sigmoidal characteristic, of the neural nodes.

The information at the output of the one artificial neurone may be of the form

$$y_j = \frac{1}{1+e^{-\sum [V_i W_{ij}]}}$$

with V_i the data at each input i of the respective neuron and W_{ij} the weights between the input i and the neurone (node) j.

Related to these, we would like to mention, for instance, the results accomplished by the Italian researchers and presented in [1]. They have found that the neural net- works forecasting works better than the linear predictors, and based on the neural forecasting and neural interpolation they have minimised the number of monitoring stations.

Some important results accomplished by German researchers in the field of air pollution forecasting based on the neural networks are presented in [2].

Other researchers [3] present the application of the neural nets in the evaluation of the relative contribution of various pollution sources to the pollution of certain eco- system. The basic idea behind the approach proposed here is the simulation of the ecosystem by a neural network.

316

In Fig.2 the dynamic demonstrative accomplishment of the fuzzy, genetically trained air pollution model is illustrated with performance MSE = 0.001273.

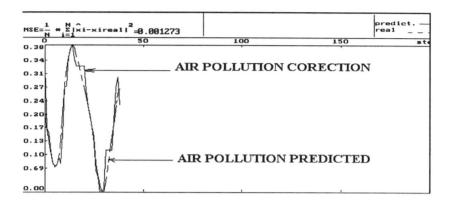

Fig. 2. Fuzzy auto-learning system in air pollution modelling with genetical algorithm for training

With the view of the application of the neural nets, especially in the fields of the Pollution Modelling, in Romania, at IPA SA, Research Institute of Bucharest, many neural models are developed and experimented, in the frame of the new complex system called "The Time Machine".

The fuzzy and neural modelling system "The Time Machine" is an innovative scientific and applicative flexible modelling configuration based on artificial fuzzy and neural (with the possibilities of the Neuro-Fuzzy and Genetically Algorithm applications). The system also includes means for the assessment of the modelling performance "The Neuro-Evaluator", which automatically offers the results of the application, based on many criteria and a mobile illustrative 3 D system.

"The Time Machine" - an intelligent, fuzzy and neural-based system has been also experimented with fuzzy system trained by genetic algorithms procedures.

At the input of the Fuzzy modelling system was used a set of samples of a pollutant with the sampling values accomplished based on the measurements offered by the monitoring stations.

The results of the modelling and forecasting with the "Time Machine" of (deduced) evolution of the NMHC in urban area are very promising. Similarly it is possible to forecast the evolution of SO2, NO2, CO.

Conclusion

Our experiments, based on "The Time Machine" artificial intelligence, fuzzy and neural based modelling system have indicated promising possibilities in the field of the fuzzy logic systems and the neural networks proving that they are very powerful tools for modelling arbitrary non-linear processes of air pollution.

In our experiments, the Fuzzy models with auto-training have offered error values smaller than the neural net models. Moreover, they require less processing power and provide robust performance.

Taking into consideration the experience and the procedures developed on "The Time Machine", neural system based, research developments, experiments and activities, we are open for finding new possibilities of co-operation focused on the achievement, experiment and assessment of the neural models in the pollution activities.

References

1. Ando, B., Cammarata, G., Fichera, A., Graziani, S., Pitrone, N.: Neural Networks for the Analysis of the Air Pollution in Urban Areas. In: Proc. EUFIT'95 Aachen (1995)
2. Schuurman, G., Muller, E.: Back Propagation Neural Networks - Recognition vs. Prediction Capability. Environ. Toxicol. Chem . 13 (1994) 2061-2077
3. Karayiannis, N.B., Venetsanopulos, A.N.: Applications of Neural Networks to Environmentual Protection. Conference
4. Sandulescu, G.M., Bistran, M.: The Self - Learning Fuzzy Models in Competition with the Neural Networks. In: Proc. 5th Zittau Fuzzy-Colloquium, Zittau, Germany (1997)
5. Hass, H., Ackermann, I.J.: Aerosol Modeling within the EURAD Model System: Development and Applications. In: Proc. NATO Advanced Workshop "On Large Scale Computations In Air Pollution Modelling", Sofia, Bulgaria (1998)
6. Sandulescu, G.M., M.Bistran, M.: Neural, Fuzzy Modelling in Pollution, an Interesting Alternative to Other Procedures: An Overview and PC Illustrations of Some Experiments. In: Proc. NATO Advanced Workshop "On Large Scale Computations In Air Pollution Modelling", Sofia, Bulgaria (1998)
7. Kallos, G., Nickovic, S.: The Regional Weather Forecasting Models as Predictive Tools for Major Environmental Disasters and Natural Hazards. In: Proc. NATO Advanced Workshop "On Large Scale Computations In Air Pollution Modelling", Sofia, Bulgaria (1998)

Fuzzy Stochastic Multistage Decision Process with Implicitly Given Termination Time

Klaus Weber[1], Zhaohao Sun[2]

[1] Lufthansa Systems Berlin GmbH, Fritschestraße 27-28, D-10585 Berlin, Germany
Klaus.Weber@LHSystems.com
[2] Bond University, School of Information Technology,
Gold Coast, QLD 4229, Australia
Zhaohao_Sun@bond.edu.au

Abstract. One of the most important goals in marketing is to realize the highest profit by applying appropriate means to optimize the process of acquiring customers. In order to assist the marketer in making marketing decision, this paper introduces a stochastic dynamic programming model for the process of acquiring customers. It is actually a stochastic multistage decision process, whose state space consists of granularized information on customers and whose transitions are controlled by marketing actions. Then it shows how to control this process using fuzzy constraints and how to characterize the goal of maximizing profit by a fuzzy set. After an introduction to dynamic programming under fuzziness this paper further presents a new model of fuzzy dynamic programming to solve the decision problem for a stochastic system with implicitly given termination time.

1 Introduction

This paper is related to marketing. One of the most important goals in marketing is to realize the highest profit by applying appropriate means to acquire customers, such as mailings, letters, telephone calls, offers, presentations, and profiles of a company. In the past this business was the realm of men and women who "had a nose" for selling goods or services. More recently it has also attracted computer scientists, who apply various intelligent techniques such as data mining, decision support system, and intelligent agent technologies [6] to assist the marketer with marketing decision. The goal of these efforts is the one-to-one ideal of understanding each customer individually [4]. For instance, customer segmentation takes advantage of the large amount of data usually stored in customer databases in order to find attributes which characterize groups of customers and so allows to choose most effective marketing measures [10][12][13][14].

Although sharing the common goal, we follow another direction in this paper and present a new method of fuzzy stochastic dynamic programming for optimizing the process of acquiring customers. As an extension and a complete review of [16], we first introduce a stochastic dynamic programming model for the process of acquiring customers (section 2). It is actually a stochastic multistage decision process, whose state space consists of granularized information on customers and state transitions are controlled by marketing actions. Section three gives an introduction to multistage decision-making under fuzziness, in particular the extension of the dynamic programming formulation given by Bellman and Zadeh [3]. This approach is the basis of a new method, which is discussed in section four and used to solve the decision problem for a stochastic system with implicitly defined termination time. It is also shown that an optimal decision is the solution of a functional equation which can be solved by iteration. Comparing with [16], this paper has also revised the proof of Theorem 3 completely.

2 Optimization of Acquiring Customers

Due to limitations of space we cannot here describe the process of acquiring customers step by step, for a detailed description see [16][17]. Instead, we proceed straightly to the mathematical model.

The process of acquiring customers is, in fact, a stochastic multistage decision process. Besides, Bellman's principle of optimality [1] obviously holds for this problem, i.e.

"An optimal policy (= decision strategy) *has the property that whatever the initial state and initial decision are, the remaining decisions must constitute an optimal policy with regard to the state resulting from the first decision."*

Thus it is of significance to use the formalism of dynamic programming to model the process of acquiring customers [1][2]. In other words, we first define control space, state space, state transition, return function, constraints, and policy.

Definition 1. The *control space* U is a finite, non-empty set of actions used in the process of acquiring customers, $U = \{\alpha_1, \alpha_2, ..., \alpha_j, ..., \alpha_m\}$, α_j is *action* j, and $u \in U$ is the *control variable* of the process.

Definition 2. The *state space* X is a finite, non-empty set of customer datasets, $X = \{\sigma_1, \sigma_2, ..., \sigma_n\}$ and $x \in X$ is the *state variable* of the process.

Each customer dataset, σ_i, is represented by a tuple of customer attribute values, $\sigma_i = (a_1^i, a_2^i, ..., a_p^i)$, where a_j^i is the value of attribute a_j for customer i, with $i = 1, ..., n$, $j = 1, ..., p$. The attributes are, for instance, income, age, education, etc. The attribute p of every customer dataset indicates his interest on the product or business service offered to him. For the sake of easy interpretation and practicability, we assume that the number of values which an attribute (variable) can take is finite and relatively small, i.e. the granularity of information is high [20], e.g. income is small, middle or high; age is teenage, young, middle-aged or old. The default value of each attribute is "unknown". Possible values of attribute p could be "final rejection", "declining", "indifferent", "offer wish", "final acceptance" and "unknown" by default.

The process of acquiring customers is stochastic by nature. If an action u_t is imposed on a customer characterized by dataset x_t, then his reaction, and thus the marketer's knowledge about him at time $t + 1$, i.e. x_{t+1}, is usually not uniquely determined. However, for given x_t and u_t we can obtain a probability distribution of x_{t+1} by means of statistical experience or from an expert by rule of thumb. So the process is considered a time-invariant stochastic system, which leads to

Definition 3. The *state transition* in a process of acquiring customers is governed by a conditional probability function $p(x_{t+1}|x_t, u_t)$, which specifies the probability of reaching state x_{t+1} from state x_t under control u_t.

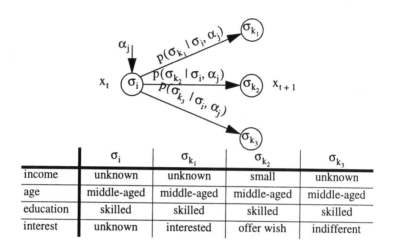

	σ_i	σ_{k_1}	σ_{k_2}	σ_{k_3}
income	unknown	unknown	small	unknown
age	middle-aged	middle-aged	middle-aged	middle-aged
education	skilled	skilled	skilled	skilled
interest	unknown	interested	offer wish	indifferent

Fig. 1. State transition from state $x_t = \sigma_i$, given input α_j to state x_{t+1}, which can be σ_v with probability $p(\sigma_{k_v} | \sigma_i, \alpha_j)$, $v = 1, 2, 3$.

An example of state transitions is given in Fig. 1. Let state x_t be σ_i = (unknown, middle-aged, skilled, unknown) and control α_j a letter with product information and a reply coupon on which a customer can select either the request for further information or that for a concrete offer. In the latter case, the customer has to give his income figures in order to obtain a tailored offer. Now, a customer can react in one of three different ways as follows:

1. He could send back and ask for more detailed information. Then his income is still unknown, but his up-to-now "unknown" interest on the product has been changed into "interested" (state σ_{k_1}).
2. He requires an offer and quotes his income. Therefore he now belongs, for instance, to state σ_{k_2}, where income is "low" and interest is "offer wish".
3. He may not answer at all. So state arrives at σ_{k_3}, where income is still "unknown" and interest "indifferent".

Transition probabilities of these cases are indicated by $p(\sigma_{k_v} | \sigma_i, \alpha_j)$, $v = 1, 2, 3$.

The goal of the process is to realize the highest profit, which can be achieved only after the customer has signed a contract.

Definition 4. The *positive termination set* T_{pos} is the set of states $\sigma_i = (a_1^i, a_2^i, ..., a_p^i) \in X$, $i = 1, ..., n$, where a_p^i is "final acceptance", the *negative termination set* T_{neg} is the corresponding set where a_p^i is "final rejection". Their union $T = T_{pos} \cup T_{neg}$ constitutes the *termination set*. $\bar{T} = X \setminus T$ is the *set of non-*

terminated states.

The process of acquiring customers will stop if it reaches the termination set. Then - in the positive case - the marketer earns income, whose amount depends on the specific product or business service and on customer attributes. So the expected income is a function of the terminating state in the termination set.

Definition 5. The *return function* $r : X \to R$ maps each state in the positive termination set, $x \in T_{pos}$, to an expected income value, which usually lies in the closed interval of possible income values R. For $x \in \overline{T}$ and $x \in T_{neg}$ we set $r(x) = 0$.

Using normalization r is transformed into *fuzzy goal* G with membership function $\mu_G(x) = g(r(x)/r_{max})$, where $r_{max} = \max R$, and g is monotone non-decreasing with $g(0) = 0$ and $g(1) = 1$ [3].

All actions are subject to constraints resulting from costs, customer characteristics (preferences), or supplementary marketing knowledge. For example, the total cost of actions should be low in order to reduce the potential income as little as possible. Actions should also fit the customer characteristics and preferences as well as possible with the aim to reach the termination set finally. Since fitness and preference are a matter of degree, we use fuzzy sets to represent constraints.

Definition 6. *Constraints* are state-dependent fuzzy sets $C(x_t)$, $x_t \in X$ over the control space U, i.e. $x_t \in X$ is a parameter and U is the fuzzy set's universe of discourse. The membership function of these sets will be denoted by $\mu_C(u_t \mid x_t)$, $u_t \in U$.

Finally, we turn our attention to finding the sequence of controls that maximizes the return function subject to the given constraints. It is denoted *optimal policy* [2] (or *optimal decision strategy*). Since the termination time is unknown in advance, it is convenient to express the controls by a stationary policy function [8].

Definition 7. A stationary *policy function* $\pi : \overline{T} \to U$ associates with each state $x_t \in \overline{T}$ an input u_t, which should be applied to the system when it is in state x_t, i.e. $u_t = \pi(x_t)$, $t = 0, 1, \ldots, x_t \in \overline{T}$. The set of policies π constitutes *policy space* Π.

3 Dynamic Programming under Fuzzy Environment

The extension of dynamic programming to multistage decision making under fuzziness was established initially by Bellman and Zadeh [3]. They introduced the concepts of fuzzy goal, fuzzy constraint and the following idea of fuzzy decision.

Definition 8. Given n fuzzy goals $G_1, \ldots, G_n$ and m fuzzy constraints $C_1, \ldots, C_m$ in a space of alternatives X. Then, the resultant *decision* is the fuzzy set D with the membership function $\mu_D(x) = \mu_{G_1}(x) * \ldots * \mu_{G_n}(x) * \mu_{C_1}(x) * \ldots * \mu_{C_m}(x)$ for each $x \in X$, where "$*$" is an aggregation (confluence) operator. The *maximizing decision* is defined as $x^{opt} \in X$ such that $\mu(x^{opt}) = \max_{x \in X} \mu_D(x)$.

An example for case $m = n = 1$ and minimum aggregation operator is given in Fig. 2. Obviously, in terms of optimization, a fuzzy decision serves the purpose of an objective function introducing an order in the space of alternatives, where the

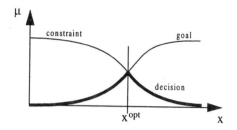

Fig. 2. Decision = confluence of goal and constraint

performance of the decision making is given by $\mu_D(x)$.

Multistage decision-making in a fuzzy environment is usually classified with respect to three aspects [8][9]:

- the type of termination time: fixed and specified in advance, implicitly given by entering a termination set of states, fuzzy, and infinite;
- the type of system under control: deterministic, stochastic, and fuzzy;
- the type of aggregation operator: min, product, weighted-sum, max, etc.

For a detailed introduction to decision-making under fuzziness we refer to the article of Bellman and Zadeh [3]. Other approaches can be found in [5][8][9].

4 Fuzzy Dynamic Programming for a Stochastic System with Implicitly Defined Termination Time

We turn now to the case of controlling a stochastic system with implicitly defined termination time using min type aggregation operator, which fits the optimization of acquiring customers stated in section 2.

Given $X = \{\sigma_1, ..., \sigma_k, \sigma_{k+1}, ..., \sigma_n\}$, we assume that termination set T consists of $\sigma_{k+1}, ..., \sigma_n$ and the initial state x_0 belongs to $\bar{T}$. From Definition 4 and Definition 6 we derive easily that $\mu_C(u_t | x_t) \geq 0$ for $x \in \bar{T}$ and $\mu_C(u_t | x_t) = 0$ otherwise, for arbitrary $u_t \in U$ respectively. Furthermore, for the conditional probabilities of a state transition under a control $u_t \in U$ holds

$$p(x_{t+1} | x_t, u_t) \begin{cases} = 1 & \text{if} \quad x_{t+1} = x_t \in T \\ \geq 0 & \text{if} \quad x_{t+1} = x_t \in \bar{T} \\ = 0 & \text{if} \quad x_t \in T, x_{t+1} \neq x_t \end{cases}$$

for every $u_t \in U$ respectively.

For a stochastic system with implicitly defined termination time the number of stages from initial state to final one is not predetermined. Now we suppose that the sequent inputs $u_0, u_1, ..., u_t, ...$ are determined by a stationary (time-invariant) policy function $\pi : \bar{T} \to U$ (see Definition 7). So for a certain policy π, in a given state

$x_t \in \bar{T}$ the decision $D_\pi(x_t)$ with respect to the policy π is the confluence of the constraint in the transition from x_t to x_{t+1}, $C_\pi(x_t)$ and the next decision $D_\pi(x_{t+1})$ with respect to π (see Definition 8). If $x_{t+1} \in T$ we set $D_\pi(x_{t+1}) = G(x_{t+1})$ for arbitrary π. Given x_t, the state x_{t+1} is a random variable characterized by the conditional probability function $p(x_{t+1} \mid x_t, u_t)$. Thus

$$D_\pi(x_t) = C_\pi(x_t) \cap ED_\pi(x_{t+1}) \text{ or } \mu_D(x_t \mid \pi) = \mu_C(\pi(x_t) \mid x_t) \wedge E\mu_D(x_{t+1} \mid \pi),$$

where
$$E\mu_D(x_{t+1} \mid \pi) = \sum_{x_{t+1} \in \bar{T}} p(x_{t+1} \mid x_t, \pi(x_t)) \cdot \mu_D(x_{t+1} \mid \pi)$$
$$+ \sum_{x_{t+1} \in T} p(x_{t+1} \mid x_t, \pi(x_t)) \cdot \mu_G(x_{t+1})$$

and hence

$$\mu_D(x_t \mid \pi) = \mu_C(\pi(x_t) \mid x_t) \wedge \left[\sum_{x_{t+1} \in \bar{T}} p(x_{t+1} \mid x_t, \pi(x_t)) \cdot \mu_D(x_{t+1} \mid \pi) \right. $$
$$\left. + \sum_{x_{t+1} \in T} p(x_{t+1} \mid x_t, \pi(x_t)) \cdot \mu_G(x_{t+1}) \right] \quad (1)$$

which is, in effect, a system of n equations (one for each value of x_t).

Let $\pi = (\pi(\sigma_1), ..., \pi(\sigma_k))^T$ be a policy, its i-th component is a control in state of system σ_i. $\mu_C(\pi) = (\mu_C(\pi(\sigma_1) \mid \sigma_1), ..., \mu_C(\pi(\sigma_k) \mid \sigma_k))^T$ is a constraint vector, $\mu_D(\pi) = (\mu_D(\sigma_1 \mid \pi), ..., \mu_D(\sigma_k \mid \pi))^T$ is a control vector, while a goal vector is $\mu_G = (\mu_G(\sigma_{k+1}), ..., \mu_G(\sigma_n))^T$. The components of $\mu_C(\pi)$ and $\mu_D(\pi)$ are the values of the membership function of C and D at $\sigma_1, ..., \sigma_k$ respectively. The components of μ_G are the values of the membership function of the fuzzy goal G at the states in the termination set T. We further introduce the $k \times k$-transition matrix

$$P_{\bar{T}}(\pi) = (p_{ij}(\pi))_{i, j = 1, ..., k} \quad (2)$$

and the $k \times (n - k)$-transition matrix

$$P_T(\pi) = (p_{ij}(\pi))_{\substack{i = 1, ..., k \\ j = k+1, ..., n}}, \quad (3)$$

whose elements are the conditional probabilities for the state transition

$$p_{ij}(\pi) = p(\sigma_j \mid \sigma_i, \pi(\sigma_i)) \quad (4)$$

provided policy π. With these definitions the system of equations in (1) can be converted into a more compact form:

$$\mu_D(\pi) = \mu_C(\pi) \wedge [P_{\bar{T}}(\pi) \cdot \mu_D(\pi) + P_T(\pi) \cdot \mu_G] \quad (5)$$

where the minimum operator $\wedge$ is applied componentwise to the vectors.

After having obtained functional equation (5), three questions immediately arise:

1. Are there policies π which transfer the system to a terminated state?
2. What is an optimal policy π^{opt}? Does it exist?

3. How can π^{opt} be calculated from (5)?

In the remaining section we give answers to these questions. In the following theorems and definitions we always presuppose a multistage decision process with stochastic state transitions and fuzzy constraints and goals as described in the first part of this section.

Theorem 1. The termination set T is reachable from any initial state in $\overline{T}$ iff there exists a policy π and a natural number K such that $\left\|P_{\overline{T}}^K(\pi)\right\|_\infty < 1$, $1 \le K \le k$, where k is the number of states in $\overline{T}$ and the matrix norm is defined by $\|A\|_\infty := \max_i \sum_j a_{ij}$.

Proof. *Necessity.* Let π be a policy such that $\left\|P_{\overline{T}}^K(\pi)\right\|_\infty < 1$ for some $1 \le K \le k$. From the definition of $P_{\overline{T}}(\pi)$ and $P_T(\pi)$ in (1), (2), (3) and (4) follows that matrix

$$P = (p_{ij}(\pi))_{i,j} := \begin{pmatrix} P_{\overline{T}}(\pi) & P_T(\pi) \\ 0 & E \end{pmatrix}$$

is a stochastic matrix, i.e. each coefficient lies in the interval $[0, 1]$ and the coefficients of a row add up to one for every row. It can be shown easily by induction

that $$P^K = \begin{pmatrix} P_{\overline{T}}^K(\pi) & B(\pi) \\ 0 & E \end{pmatrix} = (p_{ij}^{(K)})_{i,j}$$

where $B(\pi)$ is a function of $P_{\overline{T}}(\pi)$ and $P_T(\pi)$.

In accordance with the Chapman-Kolmogorov equations [15] $p_{ij}^{(K)}(\pi)$ is the K-step transition probability $p(x_{t+K} = \sigma_j \mid x_t = \sigma_i, \pi)$. If $\left\|P_{\overline{T}}^K(\pi)\right\|_\infty < 1$, then $\forall i = 1, ..., k : \sum_{j=1}^k p_{ij}^{(K)}(\pi) < 1$, and because each power of a stochastic matrix is still a stochastic matrix, $\forall x_t \in \overline{T} \; \exists x_{t+K} \in T : p(x_{t+K} \mid x_t, \pi(x_t)) > 0$, i.e. for any initial state in $\overline{T}$ a termination state is reachable after K steps at most, q.e.d.

Sufficiency: Let the termination set T be reachable from any initial state in $\overline{T}$. Then for every state $x_0 \in \overline{T}$ exist a common policy π which induces a chain of states $x_0 \to x_1 \to ... \to x_{N-1} \to x_N$, such that $x_0, ..., x_{N-1} \in \overline{T}$ and $x_N \in T$, where N depends on x_0, i.e. $N = N_j$ for $x_0 = \sigma_j$, $\sigma_j \in X$, and the N-step transition probability $p(x_N \mid x_0, \pi)$ is positive. So, for every initial state $\sigma_i \in \overline{T}$ and every state $\sigma_j \in \overline{T}$, $i, j = 1, ..., k$ holds $\sum_{j=1}^k p_{ij}^{(N_{max})}(\pi) < 1$, where $N_{max} = \max_{i = 1, ..., k} N_i$.

Hence, $\left\|P_{\bar{T}}^{N_{max}}(\pi)\right\|_{\infty} < 1$. Now, it remains to prove that for N_{max} holds $N_{max} \le k$, which can be done indirectly. Let us assume that for initial state σ_i $N_i > k$, i.e. when starting at state σ_i the termination set cannot be reached in k steps. Thus $\forall \sigma_j \in T : p_{ij}^{(k)} = 0$, and so

$$\left\|P_{\bar{T}}^{k}(\pi)\right\|_{\infty} = 1. \tag{6}$$

Since the number of states in $\bar{T}$ is k, k consecutive state transitions have to obtain a circuit, which has the transition probability one, due to (6). Hence, termination set T is not reachable which contradicts the assumption. Therefore, there cannot exist an $N_i > k$, and so $N_{max} \le k$, q.e.d.

Definition 10. A policy $\pi \in \Pi$ is *proper* if the termination set T is reachable from every initial state $x_0 \in \bar{T}$. The set of proper policies is indicated Π_p.

From Theorem 1 we derive the following

Corollary 1. A policy $\pi \in \Pi$ is proper iff there exists a natural number K such that $\left\|P_{\bar{T}}^{K}(\pi)\right\|_{\infty} < 1$, $1 \le K \le k$, where k is the number of states in $\bar{T}$.

Remark. Since the number of states in $\bar{T}$ and the number of controls in U are finite, the number of policies in Π_p is also finite, i.e. $|\Pi_p| = |U|^{|X|}$, where $|\cdot|$ indicates the number of elements.

In order to answer question 2 we introduce

Definition 11. A policy $\pi' \in \Pi_p$ is said to be *better or not worse* than a policy $\pi'' \in \Pi$ ($\pi' \ge \pi''$) iff the corresponding decision vectors $\mu_D(\pi')$ and $\mu_D(\pi'')$ satisfy

$$\forall i = 1, ..., n : \mu_D(\sigma_i \mid \pi') \ge \mu_D(\sigma_i \mid \pi'') .$$

A policy $\pi^{opt} \in \Pi_p$ is said to be *optimal*, i.e. $\pi^{opt} = \max\limits_{\pi \in \Pi_p} \pi$ iff $\pi^{opt} \ge \pi$ for any other policy $\pi \in \Pi_p$, $\pi \ne \pi^{opt}$.

Theorem 2. There always exists an optimal policy in the set of proper policies Π_p.

Proof. The existence of an optimal policy can be proved in a formal way by showing that Π_p is a complete lattice and such a supremum exists [11]. A more illustrative and constructive way is provided by the *Alternation Principle* [19].

Now we can derive the following equation for the optimal policy π^{opt} from (5):

$$\mu_D^{opt} = \max_{\pi}\{\mu_C(\pi) \wedge [P_{\bar{T}}(\pi) \cdot \mu_D^{opt} + P_T(\pi) \cdot \mu_G]\} \tag{7}$$

where

$$\mu_D^{opt} := \mu_D(\pi^{opt}) = \max_{\pi}\mu_D(\pi) . \tag{8}$$

Since $\Pi_p = \{\pi^1, ..., \pi^r\}$, using $\vee$ in place of "max", (7) becomes

$$\mu_D^{opt} = \bigvee_{v=1}^{r}\{\mu_C(\pi^v) \wedge [P_{\bar{T}}(\pi^v) \cdot \mu_D^{opt} + P_T(\pi^v) \cdot \mu_G]\} , \tag{9}$$

where $\vee$ is defined componentwise. For transparency in solution, we set $\omega := \mu_D(\pi)$,

$A_v := P_{\bar{T}}(\pi^v)$, $b_v := P_T(\pi^v) \cdot \mu_G$, and $c_v := \mu_C(\pi^v)$. Then (9) can be rewritten as

$$\omega^{opt} = \bigvee_{v=1}^{r} \{c_v \wedge [A_v \omega^{opt} + b_v]\} \ . \tag{10}$$

From (10) it can be concluded that μ_D^{opt} is a fixed point of the transformation

$$T(\omega) := \bigvee_{v=1}^{r} \{c_v \wedge [A_v \omega + b_v]\} \ . \tag{11}$$

The last one of the above-mentioned three questions is answered by

Theorem 3. The optimal policy $\pi^{opt} \in \Pi_p$ given by (7) and (8) can be obtained from the solution of fixed point problem $T(\omega) = \omega$, where T is defined in (11). The fixed point can be yielded through the iteration $\omega^{k+1} = T(\omega^k)$ for arbitrary $\omega^0 \in [0, 1]^k$.

The proof of theorem 3 requires the following two lemmata, whose proofs can be found in [18].

Lemma 1. Let $a \vee b$ denote the maximum of the two real numbers $a, b \in \mathbb{R}$. Then, for arbitrary $a_v, b_v \in \mathbb{R}$ and $r \in \mathbb{N}$ holds:

$$\left| \bigvee_{v=1}^{r} a_v - \bigvee_{v=1}^{r} b_v \right| \leq \bigvee_{v=1}^{r} |a_v - b_v| \ . \tag{12}$$

Lemma 2. For arbitrary $a, b, c \in \mathbb{R}$ and the minimum operator $\ddot{Y}$ holds:

$$|(a \wedge b) - (a \wedge c)| \leq |b - c| \ . \tag{13}$$

Proof of Theorem 3. We regard the subset $\Omega = [0, 1]^k$ of the Banach space $\mathbb{R}^k$ with maximum norm $\|x\|_\infty = \max_{i=1}^{k} |x_i|$ and the canonical metric $d(x, y) = \|x - y\|_\infty$ for all $x, y \in \mathbb{R}^k$. Obviously, T maps Ω on itself. Let $T_i(\omega)$ indicate the i-th component of the image of ω. Then for arbitrary $\omega', \omega'' \in \Omega$ holds

$$d(T(\omega'), T(\omega'')) = \max_{i=1}^{k} |T_i(\omega') - T_i(\omega'')|$$

$$= \max_{i=1}^{k} \left| \bigvee_{v=1}^{r} \left\{ c_i(v) \wedge \left[\sum_{j=1}^{k} a_{ij}(v)\omega_j' + b_i(v) \right] \right\} - \bigvee_{v=1}^{r} \left\{ c_i(v) \wedge \left[\sum_{j=1}^{k} a_{ij}(v)\omega_j'' + b_i(v) \right] \right\} \right|$$

$$\overset{(11)}{\leq} \max_{i=1}^{k} \bigvee_{v=1}^{r} \left| \left\{ c_i(v) \wedge \left[\sum_{j=1}^{k} a_{ij}(v)\omega_j' + b_i(v) \right] \right\} - \left\{ c_i(v) \wedge \left[\sum_{j=1}^{k} a_{ij}(v)\omega_j'' + b_i(v) \right] \right\} \right|$$

$$\overset{(12)}{\leq} \max_{i=1}^{k} \bigvee_{v=1}^{r} \left| [\sum_{j=1}^{k} a_{ij}(v)\omega_j' + b_i(v)] - [\sum_{j=1}^{k} a_{ij}(v)\omega_j'' + b_i(v)] \right|$$

$$\leq \max_{i=1}^{k} \bigvee_{v=1}^{r} \left| \sum_{j=1}^{k} a_{ij}(v)(\omega_j' - \omega_j'') \right| = \bigvee_{v=1}^{r} \max_{i=1}^{k} \left| \sum_{j=1}^{k} a_{ij}(v)(\omega_j' - \omega_j'') \right|$$

$$= \bigvee_{\nu=1}^{r} \|A(\nu)(\omega' - \omega'')\|_{\infty} \leq \bigvee_{\nu=1}^{r} \|A(\nu)\|_{\infty} \cdot \|\omega' - \omega''\|_{\infty} \, .$$

Since $\pi^{\nu} \in \Pi_{p}$, $\nu = 1, \dots, r$ is a proper policy $\|A(\nu)\|_{\infty} = \left\|P_{\bar{T}}(\pi^{\nu})\right\|_{\infty} < 1$ (see Corollary 1) and thus T is a contraction on Ω. Now, we can apply the Banach fixed point theorem [7] and obtain all statements of the theorem.

5 Conclusion Remarks

In this study, we examined the process of acquiring customers as a multistage decision process with stochastic state transitions, fuzzy constraints, a fuzzy goal, and an implicitly defined termination time. We presented a new model of fuzzy stochastic dynamic programming, based on Bellman and Zadeh's concept of fuzzy decision [3]. Since the state space in this model consists of granularized information on customers, it fits the requirements of customer-relationship marketing. The problem of how to choose appropriate marketing actions, in order to maximize profit, leads to solve the decision problem by iterative solution of a fixed point problem, which yields an optimal policy.

References

1. Bellman, R.E.: *Dynamic Programming*. Princeton University Press, Princeton, NJ (1957)
2. Bellman, R.E., Kalaba, R.: *Dynamic Programming and Modern Control Theory*. Academic Press, New York, London (1965)
3. Bellman, R.E., Zadeh, L.A.: Decision-Making in a Fuzzy Environment. Management Science 17 (1970) B-141 - B-164
4. Berry, M.J.A., Linoff, G.: *Data Mining Techniques for Marketing, Sales, and Customer Support*. Wiley, New York et al. (1997)
5. Esogbue, A.O., Fedrizzi, M., Kacprzyk, J.: Fuzzy Dynamic Programming With Stochastic Systems. In: Kacprzyk, J., Fedrizzi, M. (Eds.): *Combining Fuzzy Imprecision with Probabilistic Uncertainty*. Lecture Notes in Economics and Mathematical Systems. Vol. 310, Springer, Berlin et al. (1988)
6. Finnie, G., Sun, Z., Weber, K.: A Multiagent-based Intelligent Broker Architecture for Bargaining Processes. *Proceedings of the Australian Workshop on AI in Electronic Commerce*, December 6, 1999, Sydney (1999) 47-56
7. Heuser, H.: *Funktionalanalysis*. 2., neubearb. Aufl., B. G. Teubner, Stuttgart (1986)
8. Kacprzyk, J.: *Multistage Decision-Making Under Fuzziness*. Verlag TÜV Rheinland, Köln (1983)
9. Kacprzyk, J., Esogbue, A.O.: Fuzzy Dynamic Programming: Main Developments and Applications. Fuzzy Sets and Systems 81 (1996) 31-45
10. Klingsporn, B.: Teilmärkte bilden: Yuppie oder Skippie - wer ist ihr Kunde? Bank Magazin (1996) No. 7, 34-42
11. Kulisch, U.W., Miranker, W.L.: *Computer Arithmetic in Theory and Practice*. Academic Press, New York et al. (1981)
12. Küspert, A.: Bildung und Bewertung strategischer Geschäftsfelder. Die Bank (1991) No. 8, 425-434
13. Küspert, A.: Kundengruppenbildung im Privatkundengeschäft von Kreditinstituten - eine Fallstudie. Zeitschrift für Bankrecht und Bankwirtschaft (1992) No. 3, 184-201

14. Link, J., Hildebrand, V.: *Database Marketing and Computer Aided Selling*. Verlag Franz Vahlen, München (1993)
15. Viertl, R.: *Einführung in die Stochastik*. Springer-Verlag, Wien, New York (1990)
16. Weber, K., Sun, Z.: Fuzzy Stochastic Dynamic Programming for Process of Acquiring Customers. *Proceedings of the 7th Zittau Fuzzy Colloquium*, Sept. 8-10, 1999, Zittau, Germany (1999) 224-233
17. Weber, K., Sun, Z.: Fuzzy Stochastic Dynamic Programming for Marketing Decision Support. Submitted to International Journal of Intelligent Systems (1999)
18. Weber, K.: Database Marketing mit unscharfen Verfahren. Unpublished (2000)
19. Zadeh, L.A., Eaton, J.H.: An Alternation Principle for Optimal Control. Automation and Remote Control 24 (1963) 305-306
20. Zadeh, L.A.: Some Reflections on Soft Computing, Granular Computing and their Roles in the Conception, Design and Utilization of Information/Intelligent Systems. Soft Computing 2 (1998) 23-25

Diagnosis, Monitoring, and Decision Support Systems

Fuzzy Time Series Analysis

Steffen F. Bocklisch, Michael Päßler

TU Chemnitz, Professur für Systemtheorie
09107 Chemnitz , Germany
bocklisch@infotech.tu-chemnitz.de
mipa@hrz.tu-chemnitz.de

Abstract. A modeling method is suggested in this paper, which permits building multidimensional fuzzy models of time series consisting of fuzzy prototypes. These models have to be trained in a so-called period of learning and are suitable for short, medium and long range forecasts. The prediction of an incomplete time series is based on fuzzy classification to the prototypes. The results are grades of membership. In principle, these grades and further courses of prototypes are used to forecast the time series.

1 Introduction

In the last years at the chair of System Theory at the Chemnitz University of Technology several methods for fuzzy time series analysis and forecasting have been developed. These methods are based on the fuzzy pattern classification, which was developed and successfully used in various projects as well.

The basic method of time series analysis and forecasting by fuzzy pattern classification is explained herein. Furthermore, possible useful extensions are described and finally short descriptions and results of three different examples by using these methods are given.

2 The Fuzzy Time Series Concept

The main goal of the classic time series analysis is the building of mathematical models based on the known past in order to forecast time series in future. Functional trend models are used, assuming that these are able to generate future values [4], whereby this assumption is not suitable for medium or long range forecasts.

In the past few years some new methodologies were developed, based on created models in a comprehensive period of learning. Generally, this kind of models represents a well defined period of time, for example daily courses of time or batch processes of production. These models contain all types of known courses of time due to the fact that during a learning period the models have been trained with various time series. If such a model and course of time, known up to the point of prediction t_v, are on disposal, both short and long range forecasts are possible. One of these approaches is using neuronal networks and the other the fuzzy set methods.

Fuzzy time series are based on a set of elementary finite time series and composed of several significant representative courses. These courses of time are described in a fuzzy way. In our case the basis therefore is the fuzzy pattern classification [2,6].

2.1 Classification of Time Series

A course of time is described as a set Z of vectors $\underline{z}(t_i)$ over sample points of time t_i. An important assumption of fuzzy time series analysis is that all courses are sampled at the same points of time t_i. Furthermore, all vectors $\underline{z}(t_i)$ are defined in the same feature space.

So, an elementary time series Z can be described by feature vectors $\underline{z}(t_i)$ as follows

$$Z = \left\{ \underline{z}_{t_i} : i = 1, \ldots, n \right\} = \left\{ \underline{z}(t_1), \ldots, \underline{z}(t_n) \right\} \tag{1}$$

The times t_1 and t_n characterise so-called "trigger points" of time series and represent the boundaries of all comparable time series. Time series containing such trigger points are usually periodic but they can be aperiodic too, e.g. starting processes. Figure 1 shows the periodic course of traffic flow, recorded by a measuring station.

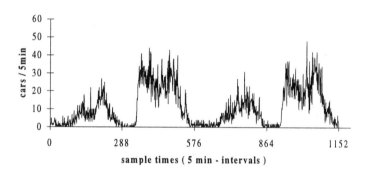

Fig. 1. Periodic course of time of traffic flow over 4 days

For this example, the time series can be subdivided in daily courses and the trigger points are fixed intuitive at 12.00am (points of time t_0, t_{288}, t_{576}, t_{864} and t_{1152}) of the respective day. Figure 2 shows such a subdivision. However, the trigger points could be also at 4.20am and 8.20pm[1]. In this case, the period from 8.20pm to 4.00am is not considered because it contains no relevant information for prediction.

The modelling method of fuzzy time series is based on finding subsets of similar time series in a set Z of single time series Z_j. These subsets are called classes of time series Z_k (k=1,...,K) and they are generated by clustering techniques or generally defined by experts with semantic contexts (workday, weekend, see Figure 2) [1,2,5].

---

[1] see Section 4.1: Analysis of traffic density

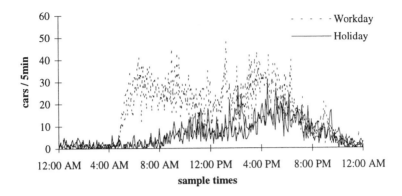

Fig. 2. Sets of daily courses of traffic flow for two different types (classes)

Consequently, these classes contain representative time series, in which variations were tolerated.

The fuzzy pattern classification represents the possibility to convert such classes of time series into fuzzy descriptions in time. They reflect the representative courses of the classes. Additionally, they contain the deviations between the time series and various measurement errors as fuzziness. In some special cases the fuzzy prototypes can be defined only with expert knowledge.

2.2 Construction of the Fuzzy Time Series Model

Given are a set Z of time series and a class structure $Z_k \subset Z$. A class of time series Z_k consists of N_k courses of time Z_j.

$$Z_j = \left\{ \underline{z}_{j,t_i} ; i = 1,...,n \right\} \forall \, j = 1,...,N_k \tag{2}$$

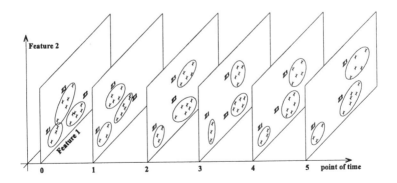

Fig. 3. Classes Z_k of time series

All observations with the same index i are from same sample point t_i. Consequently, these sample points can be considered separately for construction the fuzzy model. For every point of time, there are K several sets with N_k vectors

$$Z_{k,t_i} = \left\{ \underline{z}_{j,t_i} : j = 1,...,N_k \right\}; k=1,...,K. \tag{3}$$

These sets represent classes for each sample point t_i and can be converted in a fuzzy description P_{k,t_i} based on the theory of fuzzy pattern classification [6] (see Figure 4).

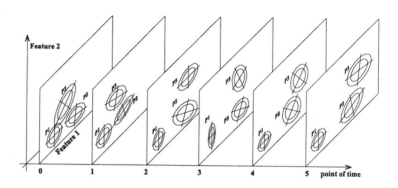

Fig. 4. Two-dimensional fuzzy time series model

As calculation results fuzzy classes Z_k over t_i lead to prototypes P_k

$$P_k = \left\{ P_{k,t_i} : i = 1,...,n \right\} \tag{4}$$

Fuzzy classes were described by multidimensional membership functions based on the potential function of Aizerman [1]. Vectors $\underline{z}_{0;t_i}^{P_k}$ give the representatives of prototype and are included in fuzzy description by membership functions.

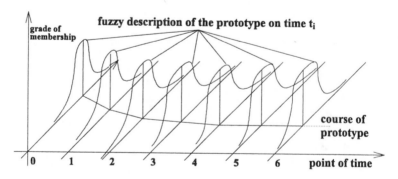

Fig. 5. One-dimensional fuzzy prototype

By using the fuzzy model a sufficient data compression is realised. Simultaneously, errors (e.g. errors in measurement or rounding errors) can be inserted by the elementary fuzziness $c_{e,m}$ for every feature m of the model [6].

The construction of fuzzy prototypes from a set of time series is practicable by the program ZR_BUILD of the FX-Software-System [3].

3 Forecasting

3.1 The General Attempt of Forecasting

The precondition for time series prediction with fuzzy pattern classification is the fuzzy model. Furthermore, the incomplete time series being predicted must correspond to the model, i.e. context (measuring conditions, etc.) and the sample points t_i have to be the same. So the course can be assigned to one or several representative prototype and it is forecasted by this classification, starting after the prediction point t_v. That is the last known of the actual time series. The forecasting of time series by fuzzy methods is subdivided in two steps: the period of identification and the period of defuzzification.

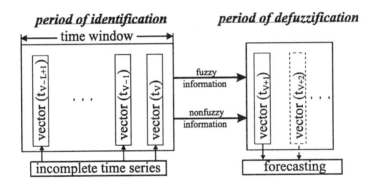

Fig. 6. Schematic description of the forecast by using fuzzy methods

The period of identification contains all algorithms for classification of time series to the prototypes. The classification of time series is done only in a time window of length L. This time window contains the last sample times in the past, which will be used for identification of time series. The length of time window can be different for various prediction points. The results of the period of identification are grades of membership to every prototype and vectors of direction[2] [6].

In the period of defuzzification the identification results will be transferred to the future for building new vectors of time series with algorithms as described in this paper. These new vectors represent the forecast for the corresponding time series.

---

[2] see also Section 3.3

In principle, there are two kinds of forecast:

1. global forecast

   With the results from the period of identification at the prediction point t_v the time series will be predicted for all points of times in the future.

2. recursive forecast

   With the identification results at t_v only the value at the next point of time t_{v+1} will be forecasted. Then the time t_{v+1} will be regarded as the "new time of prediction" and the algorithm starts again.

3.2 The Period of Identification

In the period of identification the vector of grades of membership $\underline{\mu}$ is calculated first. It describes in a fuzzy way the position of the course of time which have to be forecasted. For each prototype P_k a grade of membership exists in this vector.

The calculation of the grades of membership is carried out in the time window with length L. I.e., the feature vectors of the course must be identified for all sample times within the time window, which results to L different vectors of membership

$$\underline{\mu}_{t_{V-i}} = \left(\mu_{t_{V-i}}^{P_1}, ..., \mu_{t_{V-i}}^{P_K} \right)^T ; i = 1,...,L\text{-}1 . \tag{5}$$

They are obtained as follows:

$$\mu_{t_{V-i}}^{P_k} = \left(1 + \frac{1}{M} \sum_{m=1}^{M} \left(\frac{1}{b_{m,t_{V-i}}^{P_k}} - 1 \right) \left(\frac{\left| u_{m,t_{V-i}}^{P_k} \right|}{c_{m,t_{V-i}}^{P_k}} \right)^{d_{m,t_{V-i}}^{P_k}} \right)^{-1} ; i \in \{0,...,L-1\} \tag{6}$$

where $\mu_{t_{V-i}}^{P_k}$ is the vector of membership to the prototype P_k for time t_{v-i}.

The $b_{m,t_{V-i}}^{P_k}, c_{m,t_{V-i}}^{P_k}, d_{m,t_{V-i}}^{P_k}$ are the fuzzy parameters of the corresponding prototype P_k for time t_{v-i} and feature m. $u_{m,t_{V-i}}^{P_k}$ is the feature vector of the time series in the class space [6].

The L different vectors $\underline{\mu}_{t_{V-i}}$ must be summarized to a resulting vector $\underline{\mu}$ of membership with adequate methods. In this paper there will be used weighting functions or a fuzzy operation.

The weighting functions are defined over the time window as follows

$$\mu^{P_k}(t_V) = \frac{\sum\limits_{i=1}^{L} g_i \cdot \mu^{P_k}_{t_{V-L+i}}}{\sum\limits_{i=1}^{L} g_i} \quad , \tag{7}$$

where $\mu^{P_k}(t_V)$ is the resulting grade of membership to the prototype P_k. The g_i are the weighting parameters for the points of time t_{V-L+i}.
Here three weighting functions are used
  1. arithmetical mean

$$g_i = 1 \; ; \, i = 1,...,L \tag{8}$$

  2. linear rising:

$$g_i = i \; ; \, i = 1,...,L \tag{9}$$

  3. exponential rising

$$g_i = e^i \; ; \, i = 1,...,L \tag{10}$$

Another kind of summarizing the grades of membership over time is based by a fuzzy operation, based on the potential function of Aizerman. For that operation all points of time will be regarded equal.[6]

3.3 The Period of Defuzzification

The results of the period of identification are transferred to the period of defuzzification. Generally, the results are a vector of membership and information about the non-fuzzy position. These will be used for generating the new vector of the incomplete time series at the time t_{V+1} (recursive forecasting). In case of global forecasting this data will be used for all remaining time points.

During forecasting, the grades of membership μ^{P_k} are transferred to the sample time t_{V+1}, i.e. the forecast vector $z^{P_k}_{t_{V+1}}$ must possess this grade of membership to the corresponding prototype P_k.
There is a set of feature vectors having this grade of membership of prototype for the time t_{V+1}. This set is described by a hyperelliptic surface based on the potential function of Aizerman (Figure 7).

By the nonfuzzy information of position exactly one feature vector can be selected out of this set. Generally, it is a vector of direction $\underline{r}^{P_k}$. This vector is calculated at the time of prediction and describes the position of the vector of time series related to the representative vector $z^{P_k}_{0,t_{V+1}}$ of the corresponding prototype. It will be added to the representative vector at time t_{V+1} and there will be determined exactly one feature vector out of the hyperelliptic surface.

338

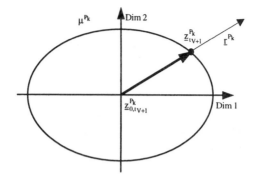

Fig. 7. Determination of the forecast vector

All of these vectors will be connected as follows

$$
\underline{z}_{t_{v+1}} = \frac{\sum_{k=1}^{K} \underline{z}_{t_{v+1}}^{P_k} \cdot \mu^{P_k}}{\sum_{k=1}^{K} \mu^{P_k}}
\tag{11}
$$

The result is exactly one feature vector $\underline{z}_{t_{v+1}}$. Simultaneously, it represents the prediction at the point of time t_{v+1}. For the next step the point t_{v+1} is considered as the new time of prediction and the algorithm starts again.

3.4 Implemented Methods for Forecasting

At present, there are three various kinds for forecast by fuzzy methods. In principle, the prediction algorithms are described in Sections 3.2 and 3.3., however for those three methods there are several extensions.

The first algorithm is called **"forecasting by vector of direction"**. This algorithm is exactly the method from Sections 3.2 and 3.3.

The second algorithm is derived from the "forecasting by vector of direction". This forecast algorithm is carried out for each feature, i.e. the multidimensional problem will be decomposed into several one-dimensional problems. Therefore it is called **"method of one-dimensional forecasting"**.

The third algorithm is named **"method of the weighted gradients"**. Unlike the two other methods, the gradients of the representative vectors are implemented for forecasting. A second classificator must be used for this method. It contains several prototypes of gradients which are connected to the prototypes of the time series[3]. There is a prototype of time series for every prototype of gradients. The incomplete course of time will be classified to both prototypes and there two grades of member-

---

[3] Generally, the gradients of the course of prototype are equal to the representative vectors of the prototypes of gradients.

ship corresponding to every prototype of gradient emerge. In the period of prediction these two grades will be combined by multiplication (based on the fuzzy AND-operation). The results are the weights for the gradients associated by the weighted average defuzzification method, i.e. the prediction consists of a vector of gradients continuing the incomplete time series up to the point of time t_{v+1}.

3.5 Extensions of Forecast for Fuzzy Methods

In this section several possibilities are suggested for extending the fuzzy methods of forecasting in a useful way and for eliminating errors of prediction. Usually, not all of the presented methods will be used for solving a problem.

1. Optimization of time windows

   For all methods of forecast the length of time window at the time of prediction must be chosen. For the recursive methods that has to be done for all sample times. The length of time windows can influence the results of forecasting significantly. Particularly, for models of time series with several structures (e.g. crossing or overlapping prototypes) it becomes enormously important.
   The time series built the fuzzy models can be used for optimizing the time window.
   In order to achieve that, a prediction error for every length of time window will be calculated for the corresponding sample time as a sum of forecasts. In principle, the time window with the minimum error will be used for prediction.

2. Rejection of time series

   The potential function of Aizerman is defined over the whole feature space. Therefore, the possibility exists that prototypes with low grades of membership affect the forecast. Under the condition that there are many prototypes with this property the forecast could be wrong. Another example are crossing prototypes. At the time of crossing the influence of a "false" prototype is getting higher , i.e. the grade of membership is increasing. Generally, the prediction after this time could be wrong[4]. In these cases, the incomplete time series can be rejected by prototypes within the time window at time of prediction. Consequently, the grades of membership are set to zero for the remaining time.

3. Masking of prototypes

   For solving problems it could be useful to create several fuzzy models of time series based on certain verified supplementary conditions. This can result in a more exact description of reality. Especially, the mutual influence of prototypes without context can be avoided. The supplementary conditions are mostly non-fuzzy information like season[5], weekday, type of machine or kind of production.

---

[4] This effect will be preferred by several models. Therefore the rejection of time series depends on specified problems.

[5] See also Section 4.2 Forecasting of load curves

The incomplete time series is classified to one of the models according such a condition[6] and thus makes possible sophisticated forecast.

4. Additional features

For some problems it is useful to possess further forecasting information. These could be available as values of features, which are not going to be forecasted. In principle, the problem can be divided in two parts: the fuzzy model of time series and the further fuzzy model "of correction". Similar to the algorithm of prediction from the original model, the model of correction generates a feature vector of offsets which have to be added to the results of prediction at the sample time t_{v+1}. Only at this time offsets are added. Examples for such features are brightness and temperature at forecast load curves of power or numbers of sell cars for forecasting the demand of spare parts.

4 Examples of Projects

In the following several projects processed at the Professorship of System Theory will be introduced.

4.1 Analysis of Traffic Density

Analysis and forecasting of traffic flow was the topic of Warg [10]. At a federal road the traffic flows have been measured over several weeks (fixed measurement position). The trigger points were set at 4.20am and 8.20pm.

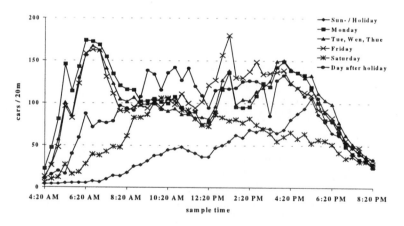

Fig. 8. Fuzzy model of traffic flows; The meanings of the prototypes are shown in the legend.

---

[6] It is conceivable that more than one of the models can be selected according to conditions. But this idea tends towards networks of fuzzy prototypes and is not an issue in this paper.

The model of time series could be considered as different time slot patterns. Here only a time interval of 20 minutes is presented. Furthermore, the classes of time series were built based on semantic information. Figure 8 shows the corresponding fuzzy model.

The forecast of incomplete traffic flows leads to very good results. Three random selected traffic flows are shown in the following figure. The time of prediction was at 6.20am and forecasts were calculated to the end of time of the model (8.20pm). The method of weighted gradients was used. As shown, the forecasted courses of time describe the original courses of the corresponding time series very well.

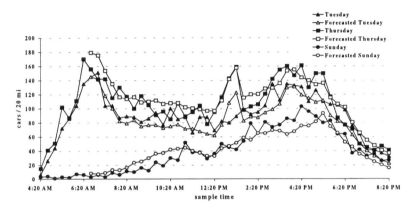

Fig. 9. Forecast of traffic flows; At several sample times the forecasted course does not correspond to the original course. E.g. at 1.00pm and 1.20pm there does not exist a peak at Tuesday course such as in the model. That is not foreseeable and errors of forecasting are higher than usual.

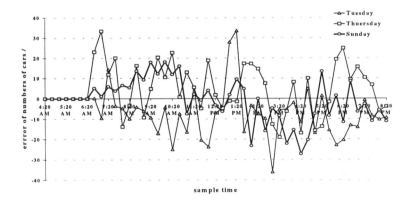

Fig. 10. Forecasting errors of traffic flow. It can be noticed that the forecasting errors do not enlarge in principle.

342

4.2 Forecasting of Load Curves

The assignment of this study was the implementation of a forecast method for load curves of energy [9]. There were 100 daily load curves at disposition. Naturally, the trigger points were at 12.00am. Two days had to be rejected because they had too much measuring faults. All other faults could be recovered by linear regression.[7]

Four fuzzy models of time series were built by masking the prototypes[8]. The distinctive feature was the date. By date the periods of summer and winter time and furthermore the workdays and the holidays are distinguished.

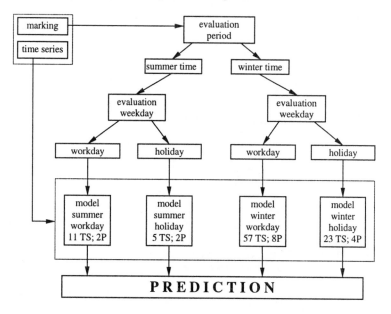

Fig. 11. Allocation of time series for several models by dates. In this figure there are shown the numbers of time series (TS) which built the models and the numbers of corresponding prototypes (P). Incomplete time series are at first assigned to one model and the forecasting is calculated within the model.

The forecast of the load curves should fulfill the following restriction: The errors of prediction of the next point of time (15min) must be lower than 3% of the original value at 98% of all forecasts. Figure 12 shows the results for the "workday-winter time" model for 7 daily courses, which were not used for building the prototypes. In addition, there are shown the results for the one and two hour forecasts.

---

[7] There were maximal two faults in turns.
[8] See also Section 3.5

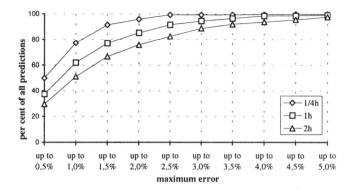

Fig. 12. Error distribution for the model of load curves (workday – winter time) – by 7 time series; The numerical values of the errors up to 3% are 99.26% for the 1/4h, 94.49% for the 1h and 88.84% for the 2h prediction.

Table 1 shows the quality of prediction of all time series with the condition "workday – winter time". The error distribution is nearly the same as in Figure 12. For all other models the results are similar.

Table 1. Error distribution of the model of load curves (workday – winter time)

	1/4h	1h	2h
up to 0,5 %	54,53 %	41,31 %	32,37 %
up to 1,0 %	82,09 %	68,35 %	56,80 %
up to 1,5 %	92,70 %	82,28 %	72,35 %
up to 2,0 %	96,71 %	89,43 %	82,01 %
up to 2,5 %	98,40 %	93,45 %	87,74 %
up to 3,0 %	99,05 %	95,59 %	91,56 %
up to 3,5 %	99,42 %	97,01 %	93,86 %
up to 4,0 %	99,70 %	98,13 %	95,39 %
up to 4,5 %	99,82 %	98,68 %	96,48 %
up to 5,0 %	99,91 %	99,08 %	97,40 %

4.3 Analysis and Prediction of Ecological Data

The task of the project "Time series analysis and prediction of ecological data" [8] was to transfer ecological data of sewage works in a fuzzy model of time series and to test this model for forecasts up to 10 hours. There were 5 different measurement variables. In this paper only the "pH-value" will be discussed.

The time series consist of daily courses of 96 measurement values (15min intervals). Altogether, there were 69 time series for building the model. The first step was the clustering of time series. After evaluation by experts from the 8 emerged classes only 4 prototypes remained (Figure 13).

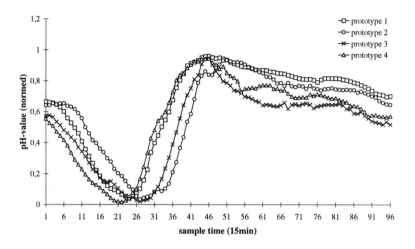

Fig. 13. Fuzzy model of time series of pH-values (daily courses)

Forecasting results of incomplete courses of pH-values are really good. The prediction up to 10 hours has a mean relative error at about 10%. The next figure shows the forecast of a time series from 7.45am and another from 12.45pm.

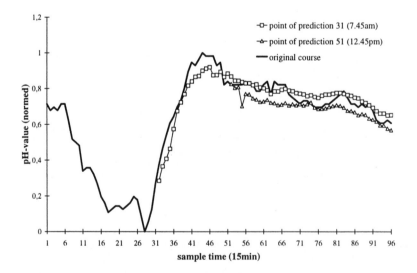

Fig. 14. Examples for the forecasting of ecological time series (pH-value); The mean relative error of forecasting was 9% for both forecasts.

References

1. Bock, H.H.: Automatische Klassifikation. Vandenhoeck & Ruprecht, Göttingen 1974
2. Bocklisch, S. F.: Prozeßanalyse mit unscharfen Verfahren. Verlag Technik Berlin (1987)
3. FX-System, Nutzerhandbuch. TU Chemnitz, Professur für Systemtheorie (1999)
4. Harvey, A. C.: Zeitreihenmodelle. R. Oldenburg Verlag, München Wien (1995)
5. Päßler, M.: Automatische Klassifikation, Praktikumsarbeit. TU Chemnitz, Professur für Systemtheorie (1995)
6. Päßler, M.: Mehrdimensionale Zeitreihenanalyse und -prognose mittels Fuzzy Pattern Klassifikation, Diplomarbeit. TU Chemnitz, Professur für Systemtheorie (1998)
7. Päßler, M.: Zeitreihenanalyse und -prognose mittels Fuzzy Pattern Klassifikation. Information 06/99, TU Chemnitz, Professur für Systemtheorie
8. Bocklisch, Buchholz, Lindner, Stephan: Zeitreihenanalyse und -prognose umweltrelevanter Meßverläufe. Information 05/99, TU Chemnitz, Professur für Systemtheorie (1999)
9. Kurzbeschreibung der Zeitreihenanalyse und -prognose mittels Fuzzy Pattern Klassifikation für die Problematik der Energielast- und -bezugsprognose. TU Chemnitz, Professur für Systemtheorie (1998)
10. Warg, S.: Verkehrsstromanalyse und -prognose mit Fuzzy Pattern Klassifikation, Diplomarbeit. TU Chemnitz, Professur für Systemtheorie (1997)

Fuzzy Modelling of Heart Rate Variability

J. Bíla, P. Zítek and P. Kuchar

Institute of Instrumentation and Control technology, Faculty of Mechanical Engineering, The Czech Technical University in Prague, Technická 4, 166 07 Prague 6, Czech Republic
{bila,zitek,kuchar}@fsid.cvut.cz

Abstract. Heart rate, actual blood pressure, blood flow rate and other cardiovascular parameters fluctuate in their values even in quite steady physiological state. Unlike the earlier conception has been proved that these fluctuations are to be viewed as specific symptoms of heart performance control. Based on this idea a model of brain interventions into control strategy of heart is presented (consisted of fuzzy blocks, integration blocks and delays) performing "beat by beat" control principle and resulting in Heart rate variability signals with features of deterministic chaos. The model is implemented in MatLab/Simulink environment.

1 Introduction

In the proposed paper there is presented a novel concept of modelling the blood flow rate control (with reference to publications [1,9,10] and to own measurements). This model allows explaining the heart rate variability as consequence of a beat-by-beat control of heart performance. The development of the model presented has been based on the assumption the heart rate variability might appear as an effect of multi-level feedback control functions. This assumption results from experience in technical applications, where this control principle leads to more or less quasi-periodical control system behavior. Before starting the model design all available published resources from the field of cardiovascular physiology have been collected and relevant system relationships among heart rate, stroke volume, arterial blood pressure, peripheral resistance, venous preload, respiration frequency etc. have been summarized. Nevertheless, the resulting analysis of the collected knowledge has revealed significant gaps, i.e. missing data concerning the control mechanisms.

Primarily, with reference to models in [4] and [6] it was decided first to include only the control mechanisms providing the basic and rather fast influences into the model structure, particularly those which can manage to modify the heart action from beat to beat and thus to contribute to the heart rate variability. As the dominant control mechanism the baroreflex was considered in the model. The signals from aortal and carotic baroreceptors are transmitted to the vasomotoric and cardioinhibitting centers, representing thus sympathetic and parasympathetic (vagal) feedback influence on the heart rate, stroke volume, and peripheral resistance. In this manner the level of both arterial blood pressure and blood flow rate needed to provide the appropriate oxygen supply can be met. Next - feedbacks included into the model influencing the heart action and blood pressure analogically are the reflexes mediated by atrial receptors (type B) corresponding to the increased tension resulting from instantaneous atrial blood pressure involving the change in venous preload. Both the feedback mechanisms are interconnected into one control loop by means of Frank - Starling law expressing a nonlinear relationship between atrial filling and myocardial ejection force. Although the model is designed to give a comprehension how the controlling mechanisms result in heart rate variability, its structure respects all available knowledge of physiological relationships in cardiovascular system. Its structure depicted in Fig. 1 consists of two principal parts, namely of the heart action and blood circulation model as the first part, and the control feedback mechanisms as the second one.

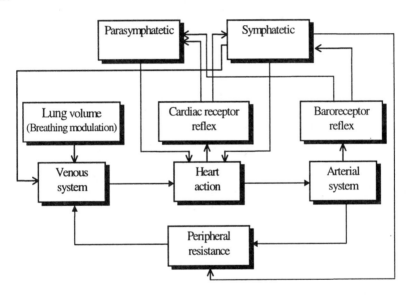

Fig. 1. Basic structure of the model

2 Model implementation

On the basis of the above explained model structure a computer model has been completed having applied the simulation program Matlab - Simulink. Primary model part is a special "saw-tooth" signal generator, Block 2, which serves for timing the beats as well as adjusting their amplitudes using repeatedly launched and interrupted integration.

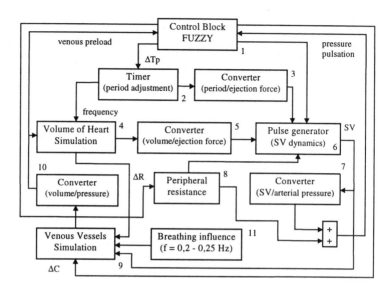

Fig. 2. Structure of model implementation

The generator function corresponds to cyclic polarization and depolarization of sineatrial node [2], in this way adjusting the duration of heart action cycle. Like in the reality the cycle timing is subject to both neural activities - parasympathetic (vagal, prolonging the depolarization in case of increasing pressure and venous backflow) and sympathetic (shortening the depolarization in the opposite case). However, unlike the real adjusting the depolarization velocity an artificial threshold is moved finishing the running „saw tooth" and thus timing the next coming short impulse representing the systolic ejection. Impulse amplitude is adjusted too, by the Block 6, being derived from the heart volume on the end of diastole (Frank - Starling law), and from the instantaneous heart rate (inotropy effect). Simultaneously the ventricle filling is simulated by the Block 4, satisfying the following equation

$$V_{heart} = V_{end\_systolic} + k \cdot \int_{0,26}^{t} \left(P_{vessels} - P_{in\_heart} \right) dt . \tag{1}$$

On the other hand, this positively influencing the stroke volume is compensated by adapting effective impedance of the blood vessel system. Hydraulic resistance changes, Block 8, are provided by a fuzzy inference block simulating the sympathetic control (ΔR). The stroke volume ejection in the systolic phase is followed by a specific pressure wave given not only by the stroke volume amplitude but also by the instantaneous impedance, just characterizing the blood vessel system. The pressure wave is shaped and adjusted by a filter, Block 7, given by the transfer function

$$\frac{\bar{p}}{\bar{V}} = \frac{1{,}01s}{0{,}0065s^2 + 0{,}1s + 0{,}45} . \qquad (2)$$

The model is built up on the assumption baroreceptors take continuously the arterial pressure changes and promptly activate the baroreflex neural control, which is adjusting the depolarization velocity in the reality. Hydrodynamic resistance properties of blood vessel system are represented by the Block 8 (sympathetic vasoconstriction) while the venous capacity influence is simulated by the Block 9 (ΔC). The blood volume contained in capacity vessels is adjusted according to the following formula

$$V_{venous} = V_{konst} + \int_0^t (SV - V_{flow\_in\_hart})dt \qquad (3)$$

Besides, its final value is affected also by the breathing rhythm, i.e. by the lung ventilation, which is simulated by the Block 11. The venous preload together with pressure changes and other relevant combined effect of the Blocks 8 and 9 determines finally the venous preload, i.e. the total blood flow cardiovascular variables are transmitted to rule-based Block 1 where actuating signals are generated adjusting the heart rate timing and amplitude

2.1 Fuzzy control block

Control Fuzzy Block consists of *three fuzzy blocks* which influence on variables HR, R and SV (Heart Rate, Resistance, Stroke Volume). Each of these introduced fuzzy blocks has the same basic structure of rules

$$\langle \textbf{IF}\langle(B_A \text{ is } Q) \textbf{ AND } (B_V \text{ is } S)\rangle \Rightarrow \langle\textbf{THEN}\langle(U \text{ is } W)\rangle\rangle\rangle, \qquad (4)$$

where B_A and B_V are signals which respresent activities of arterial (i.e. aortal and carotic) baroreceptors and arterial receptors respectively. U is the action variable which effects to the values of variables HR, R, SV. Symbols Q, S, and W denote fuzzy values (of the type Large Positive, Middle Positive, ..., etc.) of fuzzy variables B_A, B_V and U. Outputs of fuzzy inference blocks serve to set up correcting signals which adjust (beat by beat) appropriate deviations of heart rate (ΔHR), stroke volume (ΔSV), peripheral resistance (ΔR). The primary inputs B_A and B_V are generated by means of non-linear pressure-frequency characteristics of this activity, [4] describing the operation of arterial baroreceptors, atrial receptors driven by the venous preload (ZN) and chemoreceptors alarming the heart rate if aortal pressure has dropped under the threshold of 80 mmHg. The low-pass filter linked in the sympathetic branch is adjusted on a cross-over frequency 0.1 Hz.

Fuzzy inference blocks are built-up on three membership functions in the input and on five in the output, with symmetrical distributions. The inference outputs are delayed in correspondence to actual delays in neural transfer properties and then, via non-linear characteristics (look-up tables) they are converted to deviation signals of heart rate, stroke volume, peripheral resistance and venous volume respectively. These signals represent control variables of the model. The both sympathetic and vagal branches of control generate the heart rate deviation, while the vagal is approximately five times stronger approximately. On the contrary, in the remaining deviations the sympathetical influence is dominant as much, that the vagal contributions in these feedbacks have been neglected in the model. After a proper tuning the parameter influence of the above specified non-linear and delayed feedbacks result in moderate irregular oscillations of aortal pressure, heart rate, venous preload and other variables in a vicinity of their ideal steady state points. Notice, this type of behaviour is not conditioned by the breathing influence. It keeps its chaotic character even if the periodical respiration model is stopped. The main part of Fuzzy Control Block is illustrated in Fig. 3.

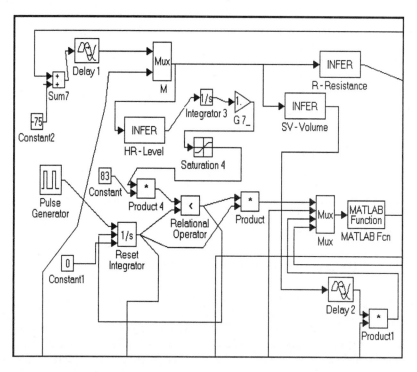

Fig. 3. Main part of Fuzzy Control Block

3   Results

Selection of experiments with the model proved that the model could explain the main control mechanisms of blood circulation, particularly the relationships among the instantaneous heart rate, stroke volume and arterial blood pressure values and to demonstrate some of special phenomena of heart system behaviour such as deterministic chaos. Figures 4. and Fig. 5. illustrate features of deterministic chaos in Heart Rate Variability (time and phase diagrams) in conditions of a constant time delays. Figures 6. and 7. illustrates the run of Arterial Pressure after beginning phase and the run Ventricular volume at the end of the observed time interval.

Turning back to the first section of this paper - it is interesting that such a complex behavior can be modeled by *deterministic system controlled by not very large set of rules*. The type of HR run (illustrated in Fig. 4.) corresponds to image of - "beat by beat control" provided by heart system - which had been introduced e.g. in [1] and proved by appropriate data arguments. Nevertheless - it is fair to add that

Heart rate [beat/min]

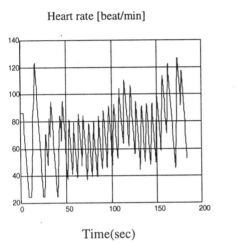

Time(sec)

Fig. 4. Heart rate in time segment

Heart rate derivation

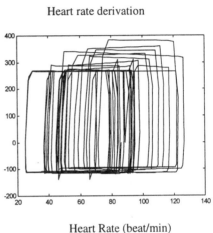

Heart Rate (beat/min)

Fig. 5. Phase diagram of heart rate

Arterial pressure [mmHg]

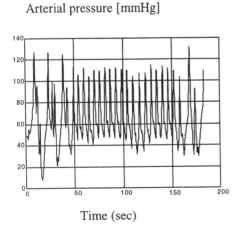

Time (sec)

Fig. 6. Ventricular pressure

Ventricular volume [ml]

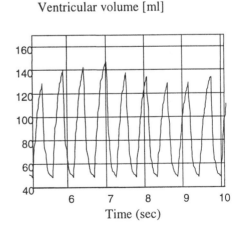

Time (sec)

Fig. 7. Ventricular volume

chaotic behavior emerged in our case as a consequence of effects of *delayed signals in interconnections between fuzzy blocks*. (This fact has already been observed and verified in some works, e.g., in [7,8].)

Delays represent a creative behaviour of brain as the ability to combine data acquired in different time intervals (moments). Seems to be important to model brain interventions into heart control strategy, especially the delayed bonds between fuzzy blocks which could represent unhealthy activities of the brain.

5. Conclusions

By the present version of the model all basic cardiovascular control mechanisms are represented and the heart rate and arterial pressure control behavior generated by the model can be compared with the measured ones. Systematic measurements are performed at the cooperating Institute of Physiology of the 1st Faculty of Medicine and also at the Faculty of Medicine Motol, both from Charles University in Prague. The authors are aware of simplifications made in model development, but have become convinced of its value in explaining the connection between the heart rate variability and principles of blood flow rate control. So far the primary assumption of model design is proved by the model results and that is why the model is being further developed and supplemented by additional control mechanisms (chemoreceptors, renin-angiotension reflex etc.) as well as by a more elaborated description of blood vessel system behavior.

Acknowledgements

This research has been conducted at the Institute of Instrumentation and Control Technology as a part of grant project GACR 101/98/0336 and as a part of research grant CTU/FME/J04/98:212200008.

References

1. Šrámek, B.B., Valenta, J., Klimeš, F: Biomechanics of the Cardiovascular System. Czech Technical University Press, Prague, CR (1995)
2. Trojan, S. et al.: Physiology (in Czech). Grada, Prague, CR (1996)
3. van Wijk, R. and van Brieving, P.: Biomedical Modelling and Simulation on a PC. Springer-Verlag, New York (1993)
4. Malliani, A., Pagani, M., Lombardi, F., Cerrutti, S.: Cardiovascular Neural Regulation Explored in the Frequency Domain. Circulation, Vol. 84, No.2 (1991) 482-485
5. Akselrod, S., Eliash, S., Oz, O., Cohen, S.: Hemodynamics regulation in SHR: Investigation in Spectral Analysis. American Physiological Society (1987) 176-182
6. Appel, M.L., Berger, R.D., Saul, S.J., Smith, J.M.: Beat to Beat Variability in Cardiovascular Variables: Noise or Music? JACC 14 (1995) 1139– 1143

7. Teodorescu, H.N.L., Belous, V., Mlynek, D., Forte, M.W.: Creativity Models and Chaos in Networks of Fuzzy Systems. In: Gonzáles,E.L. (ed.): Proc. of Int. Conf. on Intelligent Technologies in Human-Related Sciences - ITHURS 96, Leon Spain (1996) 317-324

8. Teodorescu, H.N.L., Brezulianu, A., Yamakawa, T.: Temporary and permanent Self-Organization Phenomena in Chaotic Fuzzy Systems Networks (CFNS). In: Gonzáles,E.L. (ed.): Proc. of Int. Conf. on Intelligent Technologies in Human-Related Sciences - ITHURS 96, Leon Spain (1996) 283-288

9. Malik, M.: Heart Rate Variability. Standarts of Measurement, Physiological Interpretation and Clinical Use. Special Report of Task Force of the European Society of Cardiology and North American Society of Pacing and Electrophysiology. American Heart Association, Inc. (1996)

10. Zítek, P., Bíla, J. and Kuchar, P. : Blood Circulation Control Model based on Heart Rate Variability Syndrome. In: Stefan, J. (ed.), Proc. of 32nd Spring Int. Conf. On Modelling and Simulation of Systems – MOSIS 98, C.R. (1998) 297-302

Fuzzy Approach to Combining Parallel Experts Response

Andrzej Marciniak

Department of Robotics and Software Engineering
Technical University of Zielona Gora
65-246 Zielona Gora, ul.Podgorna 50, Poland
A.Marciniak@irio.pz.zgora.pl

Abstract. This paper gives description of a parallel neural expert system for solving complex diagnostic problems. The idea of this paper is to propose an ensemble composed of independently trained neural networks. The aim is to build a classifier, which does not need a heavy training process and has good generalisation properties. This work contains a comparison of a few basic methods for combining multiple classifiers. Two classification problems are considered: breast cancer diagnosis and hand-written digit recognition.

1 Introduction

Recently, the concept of using parallel experts in pattern recognition has been actively exploited, mainly for applications in highly reliable character recognition systems, but also in avionics, medical diagnostic or for the safe control of nuclear reactors [1]. The application of redundancy to hardware components has long been established as an effective methodology for increasing reliability. Its application to software is a relatively new technology motivated by the need for high reliability in life-critical applications.

This concept seems to be especially apt in case of using a feed-forward neural network as an individual expert [2]. It is common knowledge, that supervised learning of neural nets is implemented using a database, which consist of an input pattern set together with the corresponding classifications. The objective is to let the trainee extract relevant information from the database in order to classify patterns from behind, i.e. to reach the best generalisation ability. The training process is critical for the classification results. Its efficiency depends on selection of network structure, learning algorithm and its parameters, and also on quality of representative samples in database. Therefore, in case of using many individual neural experts, each of them could attain a different degree of success, but maybe none of them is as good as expected for practical applications.

The other aspect is that for specific recognition problem, usually numerous types of features could be used to represent and recognise patterns [6]. For example in waveform analysis and recognition, the usable features may come from power spectrum, AR modelling, function approximation, hidden Markov modelling, and many types of structural line segments. These features are represented in various forms, so it is very difficult to lump them into one classifier. Even, if the features are not drastically different and even if normalisation is not required in lumping a lot variables into one vector, it may still be favourable to divide this high-dimensional vector into several vectors with lower dimensions. It may decrease computational complexity and also produce less implementation and accuracy problems.

The solution for above mentioned problems can be reached using a parallel neural networks classifier, composed of the independently trained neural networks, that are trained by the gradient methods. The basic idea is to get classification from each neural network and then using a consensus scheme to obtain the collective response. Because of the different weights corresponding to different ways of forming generalisations in training set, one can argue that the collective decision produced by the ensemble is less likely to be erroneous than the decision made by any of the individual networks [1].

2 Parallel Structure of Neural Experts

A common neural network of a finite size does not often approximate a particular mapping completely or it generalises poorly. Increasing the size and number of hidden layers most often does not lead to any improvements. Furthermore, in complex problems both the number of available features and the number of classes are large. Thus, if we can build the network, which considers the only characteristic part of the complete mapping, it should fulfil its task better.

The basic idea of the presented parallel network scheme is to develop n independently trained neural networks with relevant features, and to classify a given input pattern by utilising the combination method to decide the collective classification, as is shown in Fig.1. Each of the independently trained neural networks may have an optional structure and should be learned with a convenient algorithm. The only condition is that the number of outputs for each network must be equal to the number of classes. The collective decision is taken from the conditional probabilities which are returned by the network outputs of all classifiers. For better efficiency, the outputs of individual networks can have a weights.

In the event of neural networks training process, standard practice dictates that the cross-validation should be used for preparing learning and testing sets. It gives the best measure of expected generalisation ability, if both training and testing set fully reflect the probability distribution of the data in input space.

3 Methods of Combining Multiple Classifiers

The task of a single classifier is to assign the input pattern x the index $j \in \Lambda \cup \{M+1\}$ as a label to represent that x is regarded as being from the class C_j

if $j \neq M+1$, where Λ is the label set. Although, in the case of single classifier, j is the only output information that we need, but in case of many classifiers, it could be proper to keep other related information.

It is well known, that neural networks trained to classify the n-dimensional input x in one out of M classes can actually learn to compute the *a posteriori* probabilities that the input x belongs to each class (it can be proved, that ideally trained feed-forward neural network with sigmoid activation function is a Bayes classifier) [4]. It can be denoted as $P(C_i|x)$, $\forall i \in \Lambda$.

For the single neural classifier, the final label j may be obtained as a result of maximum selection from the M values

$$P(C_j|x) = \max_{i \in \Lambda} P(C_i|x), i = 1, \ldots, M, \tag{1}$$

and this selection rejects some information, which could be useful for many classifiers combination.

In regard to the output information that various classification algorithms are able to supply, the methods combining many classifiers can be divided into three groups.

The first group contains methods which use the information on the abstract level, i.e., the j values. The best known are the voting methods based on *majority* and *plurality* rules [1]. Majority rule chooses the classification made by more than half networks, next plurality rule chooses classification reached by more networks than any other. These methods are simple and powerful enough in many applications.

The next group of methods contains the ones that use the information on the rank level, i.e., every classifier ranks all labels or a subset in a queue with label at the top being the first choice. These algorithms are suitable for problems with many classes, for instance in some medical diagnostic problems.

The last group are methods which combine classifiers on the measurement level, i.e., in case of neural networks output, on the level of returned member-

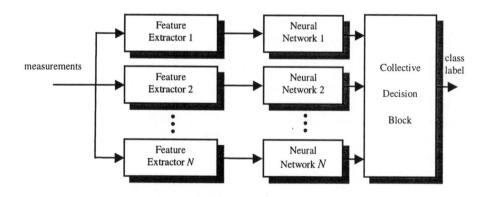

Fig. 1. The parallel neural networks scheme.

ship values (estimated probabilities). The simplest method is *averaged Bayes classifier* that use the following average value as a new estimation of combined classifier:

$$P(C_i|x) = \frac{1}{N}\sum_{n=1}^{N} P_n(C_i|x), i = 1,\ldots,M. \tag{2}$$

The final decision made by whole classifier is given by (1). The more complicated method is *weighted averaged Bayes classifier*, given by

$$P(C_i|x) = \sum_{n=1}^{N} w_n P_n(C_i|x), i = 1,\ldots,M, \tag{3}$$

with

$$\sum_{n=1}^{N} w_n = 1. \tag{4}$$

The weight values can be chosen with use of the correlation matrix of the error residuals on the testing set. Let the error matrix $\mathbf{E}$ have N rows and M columns, and each its component is a quadratic error after all training samples. The quadratic error of the ensemble is then

$$Q = \frac{1}{N^2}(w_1, w_2,\ldots,w_N)\mathbf{E}\mathbf{E}^T(w_1, w_2,\ldots,w_N)^T. \tag{5}$$

The minimum of this expression can be found by differentiating with respect to the weight vector and setting the result to zero. Because of the condition (4), the Lagrange multipliers method should be used [4]. Excepting the transformations, the final optimal set of weights is

$$\mathbf{w} = \frac{1(\mathbf{E}\mathbf{E}^T)^{-1}}{1(\mathbf{E}\mathbf{E}^T)^{-1}1^T}. \tag{6}$$

This method is useless if the matrix $\mathbf{E}\mathbf{E}^T$ is ill-conditioned or if its computation requires too many operations. In that case the weights can be chosen arbitrary or an adaptive method can be used.

The other approach to combining multiple classifiers is a method based on the fuzzy measure and fuzzy integral [5]. To present this method a few definitions have to be introduced. A set function g converting class subset Ω of space Y to interval [0,1] is called a fuzzy measure if the following conditions are satisfied:

1. $g(0) = 0$, $g(Y) = 1$,
2. if $g(A) \leq g(B)$, then $A \subset B$, for $A, B \in \Omega$,
3. if $\{An\}$ is a monotonic sequence on Ω, then

$$\lim_{n\to\infty} g(A_n) = g(\lim_{n\to\infty} A_n).$$

The fuzzy measure g_λ is called λ-additive if

$$g_\lambda(A \cup B) = g_\lambda(A) + g_\lambda(B) + \lambda g_\lambda(A)g_\lambda(B), \tag{7}$$

where $A, B \in \Omega$, $A \cap B = \emptyset$, $A \cup B \in \Omega$, and for some $\lambda > -1$ and $\lambda \neq 0$, given by solving the following equation

$$\lambda + 1 = \prod_{i=1}^{n}(1 + g^i). \tag{8}$$

Let Y be a finite set and $h : Y \to [0,1]$ a fuzzy subset of Y. The fuzzy integral over Y of the function h with respect to a fuzzy measure g is defined by

$$e = \max_{i=1}^{n}[\min(h(y_i), g_\lambda(A_i))], \tag{9}$$

where $A_i = \{y_1, y_2, ..., y_n\}$.

The classifier response is calculated from relation (9). Finally, the class with the largest integral value is chosen as the output class. The fuzzy measures can be obtained from (6) or can be chosen arbitrary.

4 Simulation Results

4.1 Breast Cancer Diagnosis

This application utilises characteristics of individual cells, obtained from a minimally invasive needle fine aspirate, to discriminate benign from malignant breast lumps. This allows an accurate diagnosis without the need for a surgical biopsy [3]. Ten features are computed for each nucleus from the microscope image. The mean value, extreme value and standard error of each these cellular features are computed for each image and resulting in a total of 30 features. The database contains 569 vectors with 212 malignant and 357 benign points, confirmed by biopsy or subsequent periodic medical examinations.

To obtain the best estimate of generalisation ability, the learning set is smaller than the testing set, and contains only 206 samples, randomly chosen

Classifier	Structure	Error (%)
1	8-3-2	13,008
2	8-3-2	11,382
3	7-3-2	12,737
4	30-3-2	7,317
5	30-3-2	8,130
6	30-2	4,336
7	30-2	2,439

Table 1. Results of individual classifiers.

Classifier	Voting(%)	Averaged (%)	Weighted (%)	Fuzzy(%)
$C_{1,2,3,4,5,6,7}$	3,794	3,794	2,168	2,168
$C_{2,3,4,5,6,7}$	2,439	3,523	2,168	2,168
$C_{2,3,4,6,7}$	2,168	2,168	2,168	2,439
$C_{2,4,6,7}$	1,355	2,168	2,168	2,439
$C_{4,6,7}$	3,252	3,252	2,439	2,439

Table 2. Results of combined classifiers.

from the database. The different neural structures have been trained (with or without pre-processing of Principle Component Analysis) and the best classifiers after selection have been chosen. The selection of the multiple classifier is done with regard to the equation (6), i.e., when the weight for given classifier is negative, the classifier is rejected.

The results for individual classifiers are presented in Table 1, where the structure determines respectively the number of: input nodes, hidden neurones (if they are) and output neurones. The error is computed on the training set (363 samples). In Table 2, the results for the multiple combination classifier with use, adequately: plurality rule (1), averaged Bayes classifier (2), weighted Bayes classifier (3) and fuzzy integral (9), are presented. In case of even number of classifiers, a ties were settled to the second class advantage, as more represented.

4.2 Handwritten Digit Recognition

The digit image was obtained with a hand scanner, which was working in B&W mode and resolution up to 400 dpi. The image was transformed and scaled to the binary matrix with size 19×19 pixels. Acquired database contains 750 samples, i.e., 600 in the training set and 150 in the testing set. The multiple classifier contains 5 neural networks. The raw binary matrix is introduced directly to the classifier NN_1. The classifiers NN_2 - NN_5 use the Kirsch masks for extracting directional features, i.e., in turn: horizontal, vertical, right diagonal and left diagonal line detectors. The recognition rates for all classes and all classifiers are given in Table 3.

5 Conclusions

Two different recognition problems were presented and solved using the multiple combination classifier. In both, one can observe that the response of the multiple classifier is better than the response of the best single neural network, if the proper method for combining is used. The first experiment shows, that for some problems, the best results can be obtained with simple methods, like

Classifier	0	1	2	3	4	5	6	7	8	9	All
NN_1	87	93	80	100	87	100	93	93	87	93	91,3
NN_2	73	100	53	93	100	73	80	93	93	93	85,6
NN_3	80	87	60	93	87	80	87	73	73	100	82,0
NN_4	87	100	73	87	80	80	93	93	80	73	85,6
NN_5	87	100	67	73	80	73	87	93	53	80	79,6
Weighted average	87	100	73	100	93	100	93	93	87	100	92,6
Fuzzy integral	87	100	73	100	93	100	93	93	93	100	93,3

Table 3. Recognition efficiency for individual and collective classifiers.

the voting or weighted Bayes classifier. In the other experiment, where the classifiers get a different feature vectors and have many classes, the best method is using the fuzzy integral.

In practice, increasing the number of layers and neurones is a method frequently applied for improvement of recognition quality. In this paper, an alternative technique for improving the results in feed-forward networks was described. Considering the results obtained, we can see that individual networks can complement each other, i.e., some of them can map better some patterns, whereas others can do it for another patterns. The parallel experts systems seems interesting for increasing the quality of neural classifiers, because it does not call for a large increase in calculations complexity and allows every network to be trained independently. Taking the generalisation ability into consideration, they could be interesting solution in cases, where the possible number of input values combinations is practically infinite.

References

1. Hansen, L.K., Salamon P.: Neural Network Ensembles, IEEE Trans. on PAMI, Vol. 12, No. 10 (1990) 993–1001
2. Korbicz, J., Uciski, D., Obuchowicz, A.: Artificial Neural Networks. Foundamentals and Applications, AOW, Warszawa (1994) (in Polish)
3. Mangasarian, O., Street, N., Wolberg, W.: Breast Cancer Diagnosis and Prognosis via Linear Programming, Mathematical Programming Technical Report 94-10 (1994) (available via WWW from ftp.cs.wisc.edu/math-prog)
4. Rojas, R.: Neural Networks. A Systematic Introduction, Springer-Verlag, Berlin Heidelberg (1996)
5. Sung-Bae, Cho: Neural-network classifiers for recognizing totally unconstrained handwritten numerals , IEEE Trans. on Neural Networks, Vol.8, No.1 (1997) 43–53
6. Xu, L., Krzyzak, A., Suen, Ch.: Methods of Combining Multiple Classifiers and Their Applications to Handwriting Recognition, IEEE Trans. on Systems, Man, Cybernetics, Vol.22,. No. 3 (1992) 418–435

Fuzzy Modeling of Dynamic Non-Linear Processes - Applied to Water Level Measurement

Anke Traichel, Wolfgang Kästner, Rainer Hampel

Institute of Process Technique, Process Automation and Measuring Technique (IPM)
Department Measuring Technique/Process Automation
at the University of Applied Sciences Zittau / Görlitz,
D - 02763 Zittau, Theodor-Körner-Allee 16
r.hampel@htw-zittau.de

Abstract. Rule-based models realized with the help of Fuzzy Logic are more and more applied as an alternative or redundancy to classic analytical models. The algorithms of Fuzzy Logic are suitable for the modeling of the behavior between input and output variables of the process which is characterized by features like high-dimensional, high-dynamic, strong non-linear ones. The paper deals with the application of Fuzzy Logic for the modeling of the collapsed level and mixture level within a pressure vessel with water-steam mixture during accidental depressurizations. The chosen process example is characterized by high dynamics and strong non-linearities. Based on the analysis of experimental data and simulation data the input and output variables as well as the structures of the Fuzzy Models (Mamdani Fuzzy Model) are defined. The quality of the developed Fuzzy Models is validated on the basis of the comparison between experiment and model.

1 Introduction

In opposite to linearized systems for the modeling and for the controller as well as observer synthesis of non-linear processes, a general theory does not exist. As a result of the complexity of problems which occurred in connection with non-linear processes the developed design algorithms are valid for special classes of systems or processes only. Fuzzy Systems are advantageously applicable for the modeling of non-linear processes as a result of their general formulation.

Furthermore Fuzzy Systems allow the description of non-linear systems:

⇒ which cannot be described by axiomatic algorithms

⇒ which are characterized by high complexity and fuzzy parameters.

The paper is focussed on the application of Fuzzy Systems for the modeling of non-linear processes but not for their control. In this sense the Fuzzy Models are used for the identification, estimation and monitoring of the process state.

For the application of relational Fuzzy Models (linguistic formulation of the condition and conclusion [2,3]) and functional Fuzzy Models (crisp formulation of the conclusion in form of analytical function [4]) we find a number of references. Exemplarily, the application of Fuzzy Models for mechatronic systems [5], for a pneumatic conveying system for particles [6], for a fluidized bed combustion plant for sludge incineration [7], and for magnetic bearings [8] are mentioned.

The paper demonstrates the capability and the advantages of Fuzzy Logic with regard to the modeling of non-linear systems on the example of the collapsed and mixture level within a pressure vessel with water-steam mixture. The determination of this process variables in the case of accidental process states, initiated by negative pressure gradients (leaks), is realized by a fuzzy-based parallel model. The presented process example is characterized by high process dynamics and strong non-linearities. The emphasis of the investigations is the application of relational Fuzzy Models.

The paper starts with a description of the water level behavior during negative pressure gradients. For a selected process example the reproduction quality of the developed Fuzzy Models are demonstrated by the comparison between the results of the experiments / simulations and the Fuzzy Models.

The subject of the following sections are the problems:

⇒ Which input and output variables of Fuzzy Model have to be defined?

⇒ Which structure of Fuzzy Model has to be chosen to realize the demonstrated reproduction quality of process state?

Based on the experimental and simulated data the input and output variables of the Fuzzy Models are defined and justified. For the determination of the collapsed and mixture level the different structures of Fuzzy Models (with and without a feedback), the basic rules, and the fuzzy sets are explained. Furthermore the proof of the general validity is furnished.

2 Modeling of Water Level Behavior

The modeling of the process behavior of a water-steam mixture within pressure vessels during negative pressure gradients (e.g., as a result of leaks) is made more difficult by:

⇒ strong non-linearities,

⇒ fast changing of process dynamics,

⇒ fuzziness of parameters,

⇒ limited model quality (observer model) as a result of limited measuring range.

To illustrate the mentioned influences a blow down experiment is presented in Fig. 2.1 by the following process variables:

⇒ pressure p (global parameter)

⇒ pressure gradient dp/dt (global parameter)

⇒ wide range collapsed level lc_w (characterizes the water inventory within the vessel, global parameter)

⇒ narrow range collapsed level lc_n (characterizes the water inventory within a limited measuring range, local parameter)

⇒ mixture level lm (characterizes the volume of water-steam mixture within the vessel, global parameter)

The chosen experiment is characterized by the initial pressure $p_0 = 22$ bar and the initial collapsed level $lc_{w0} = 155$ cm. Before depressurization a constant value of pressure and a similar value of the different water levels can be observed. As a result of the leak (opening of a valve) at $t = 62$ s a swift decrease of pressure occurs (discernible at the response characteristic of the pressure gradient). As a result of the steam content, generated by the evaporation, a rapid increase of the collapsed and mixture level can be indicated (flashing of water-steam mixture). A continuous decrease of all water levels follows as a result of the mass loss beyond the leak. The gradual decrease of the narrow range collapsed level and mixture level is nearly similar to the wide range collapsed level decrease. The narrow range collapsed level and the mixture level are characterized by a different value of flashing in comparison to the wide range collapsed level. It is a result of the different values of steam content within the zones of the pressure vessel. The closing of the valve at $t = 128$ s leads to a collapse of the narrow range collapsed level and the mixture level. The process state within the pressure vessel reaches a new steady state characterized by a constant pressure and constant water levels, which are similar.

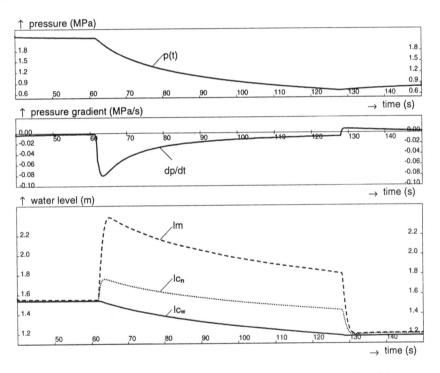

Fig. 2.1. Pressure, pressure gradient, and water levels (wide range collapsed level lc_w, narrow range collapsed level lc_n, mixture level lm) for the blow down experiment BD22H155

The figure illustrates that for the reproduction of the narrow range collapsed level lc_n and the mixture level lm the following effects have to be considered:

1) rapid increase of both levels as a result of the initiated evaporation (opening of leak)

2) gradual decrease of all water levels as a result of the loss of mass beyond the leak

364

3) rapid decrease of both water levels as a result of the collapse of water-steam mixture (closing of leak)

To demonstrate the quality and capability with regard to the reproduction of narrow range collapsed level and mixture level by the developed Fuzzy Models some results are presented. For the above mentioned blow down experiment the reproduced water levels by the Fuzzy Model are compared with the real levels (simulated with the help of the complex thermalhydraulic code ATHLET).

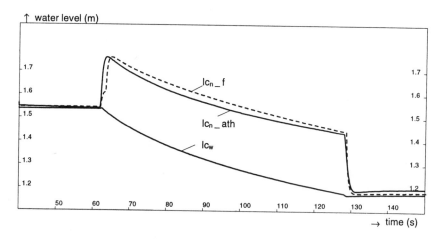

Fig. 2.2. Comparison between the fuzzy-based reproduced (lc_n_f) and the real response characteristic (lc_n_ath) of narrow range collapsed level, BD22H155

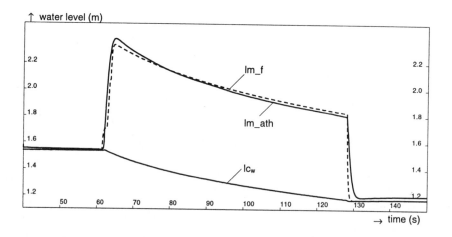

Fig. 2.3. Comparison between the fuzzy-based reproduced (lm_f) and the real response characteristic (lm_ath) of mixture level, BD22H155

The reproduction result for the narrow range collapsed level lc_n is shown in Fig. 2.2. The relative error of the fuzzy-based reproduced level lc_n_f to the measured narrow range collapsed level lc_n_ath is about 5% in maximum, which can be classified as a sufficient result.

Figure 2.3 presents the reproduction result of the mixture level by means of the fuzzy-based parallel model lm_f for the mentioned experiment BD22H155. The dynamics of the process behavior is reproduced with a sufficient quality. In comparison to the real mixture level lm_ath (ATHLET-Code) a good correspondence of the mixture level before, during, and after depressurization could be achieved.

How this reproduction quality for such strong non-linear and high-dynamic process variables can be achieved? Starting from the aim to design rule-based Fuzzy Models (Fig. 2.4) the following problems have to be solved:

⇒   Definition of the input and output variables of the Fuzzy Model!

⇒   Design of the Fuzzy Model structure to realize an exact reproduction of the level!

Fig. 2.4. Fuzzy Model

3   Input and Output Variables of the Fuzzy Models

Data Analysis

A number of blow down experiments were carried out with the help of the pressurizer test facility of the IPM [9]. The experiments serve for the determination of the parameters which have an influence on the narrow range collapsed level and mixture level as well as for the generation of a data base for the Fuzzy Model design. The initial conditions of the experiments characterized by the initial values of pressure p_0 and collapsed level lc_{w0} (wide range) are varied within the technical-related parameter range of the test facility. Table 3.1 gives an overview about the realized experiments. The initial conditions are varied as follows:

⇒   initial pressure: $p_0 = 22 / 18 / 14$ bar

⇒   initial wide range collapsed level: $lc_{w0} = 105 / 115 / 135 / 155 / 170 / 195$ cm

Table 3.1. Blow down (BD) experiments with different initial values of pressure p_0 and collapsed level lc_{w0} at the beginning of depressurization

Experiment	Initial Pressure		
Initial Collapsed Level	$p_0 = 22$ bar	$p_0 = 18$ bar	$p_0 = 14$ bar
$lc_{w0} = 195$ cm	BD22H195 **R**	BD18H195 O	BD14H195 **R**
$lc_{w0} = 170$ cm	BD22H170 O	BD18H170 O	BD14H170 O
$lc_{w0} = 155$ cm	BD22H155 **R**	BD18H155 O	BD14H155 **R**
$lc_{w0} = 135$ cm	BD22H135 O	BD18H135 O	BD14H135 O
$lc_{w0} = 115$ cm	BD22H115 O	BD18H115 O	BD14H115 O
$lc_{w0} = 105$ cm	BD22H105 **R**	BD18H105 O	BD14H105 **R**

R - experiments with the character of reference points
O - experiments for verifying the algorithm

The experiments were post-calculated with the help of the ATHLET-code. The quality of the ATHLET-simulation was verified by experimental data. For the further investigations the ATHLET-data set serves as substitution of the real process. This leads to the advantage that all measurable (pressure, wide range collapsed level, narrow range collapsed level, ...) and all non-measurable variables (mixture level, ...) can be provided by ATHLET [9].

From the number of realized experiments reference experiments were selected which serve as data base to generate the Fuzzy Models. These reference points were chosen in such a way that the technical related parameter range of the test facility was covered. The experiments with the character of a reference point were signed by 'R'. The experiments signed by 'O' were used to check the quality and the general validity of Fuzzy Models.

Output Variables

The wide range collapsed level lc_w is a global, accurate measurable process variable which characterizes the actual process state. Furthermore it is not affected by a limitation of measuring range. Therefore the wide range collapsed level lc_w is favored as basic parameter for the determination of the narrow range collapsed level lc_n and the mixture level lm. This leads to the idea to determine the spread value of the narrow range collapsed level and mixture level in comparison to the wide range collapsed level in form of a ratio factor K (compare Fig. 2.1). The equations (3.1) and (3.2) define the ratio factors K_c for the narrow range collapsed level and K_m for the mixture level.

The ratio factor K_c characterizes the ratio between the local process variable narrow range collapsed level lc_n and the global process variable wide range collapsed level lc_w.

$$K_c = \frac{lc_n}{lc_w} \qquad (3.1)$$

The ratio factor K_m describes the ratio between the global process variables mixture level lm and wide range collapsed level lc_w, which can be observed over the total vessel height.

$$K_m = \frac{lm}{lc_w} \qquad (3.2)$$

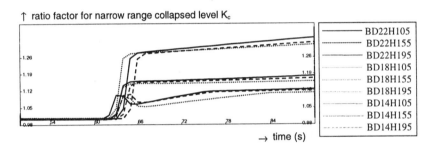

Fig. 3.1. Response characteristics of the ratio factor K_c for the narrow range collapsed level for the transients of the reference points within the simulation period t= 50 ... 90 s

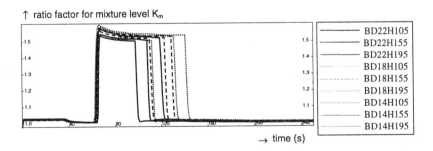

Fig. 3.2. Response characteristics of the ratio factor K_m for the mixture level for the transients of the reference points

Fig. 3.1 shows the response characteristics of the ratio factor K_c for the narrow range collapsed level K_c within the simulation period t= 50 ... 90 s. The ratio factor changes depending on the initial pressure p_0 and depending on the initial collapsed level (wide range) lc_{w0}. The response characteristics during the pressure disturbance are not constant, that means, that the dynamics of the ratio factor K_c must be considered.

Fig. 3.2 shows the response characteristics of the ratio factor K_m for the mixture level. The ratio factor changes also depending on the initial values of pressure and collapsed level. The response characteristics are nearly constant during the depressurization which leads to the conclusion that the dynamics of the ratio factor K_m is negligible.

The ratio factors K_c and K_m are defined as output variables of the Fuzzy Models.

Input Variables

A definite classification and description of the ratio factors can be realized by the following significant process parameters and process signals [9]:

$\Rightarrow$ process signal: pressure disturbance p_{dist} (detection of beginning and end of pressure disturbance)

$\Rightarrow$ process parameter: initial pressure p_0 at the beginning of depressurization

$\Rightarrow$ process parameter: initial collapsed level (wide range) lc_{w0} at the beginning of depressurization

$\Rightarrow$ process parameter: maximal value of time derivation of wide range collapsed level $dlc_w/dt\,|_{max}$

Fig. 3.3 illustrates the relations between the above mentioned input variables and the output variables ratio factors K_c and K_m which have to be described by the Fuzzy Models. In a second step the real process variables narrow range collapsed level lc_n and mixture level lm are calculated based on the measured wide range collapsed level lc_w and the fuzzy-based determined ratio factors.

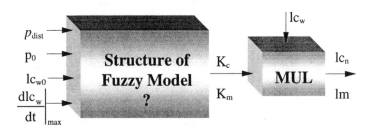

Fig. 3.3. Fuzzy modeling of the ratio factors K_c and K_m and calculation of the water levels lc_n and lm

The defined input variables are shown in Fig. 3.4. The signal pressure disturbance p_{dist} indicates the period of the depressurization. The parameters - initial pressure p_0 and initial collapsed level lc_{w0} (wide range) - serve for the classification of the ratio factors. The maximal value of the time derivation of wide range collapsed level $dlc_w/dt\,|_{max}$ allows a validation of the gradient of the disturbance (analogous to the parameter pressure gradient).

These process signals and process parameters which serve as input signals for the Fuzzy Models are generated by classical algorithms based on the measurable process variables pressure $p(t)$ and collapsed level (wide range) $lc_w(t)$.

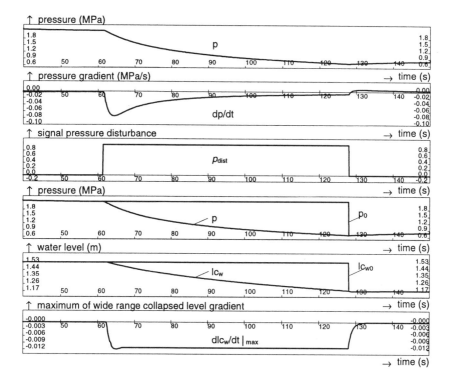

Fig. 3.4. Response characteristics of the process variables pressure $p(t)$, pressure gradient dp/dt, wide range collapsed level $lc_w(t)$ and the generated input variables for the Fuzzy Models:
- process signal: pressure disturbance p_{dist}
- process parameter: initial pressure p_0
- process parameter: initial collapsed level (wide range) lc_{w0}
- process parameter: maximal value of wide range collapsed level gradient $dlc_w/dt\,|_{max}$

Fuzzy Characteristic Features

Based on the experiences of former investigations [9] the following Fuzzy characteristic features are used for the design of the Fuzzy Models:

⇒ value of overlapping: 100% (except the signal pressure disturbance)
⇒ T-Norm (inference):   Algebraic Product
⇒ S-Norm (accumulation):   Algebraic Sum
⇒ defuzzification method: Singleton
⇒ shape of membership function: lambda function
    (except the maximal value of collapsed level gradient – trapeze function)

The design and the simulation of the Fuzzy Models were realized by means of the simulation tool DynStar with Fuzzy Shell [11]. The Fuzzy Models which were developed for an exact reproduction of the relation between the input and output variables are explained in the following sections.

4   Fuzzy Model for Collapsed Level Determination

Structure

As a result of the fact that the ratio factor of the narrow range collapsed level K_c is characterized by a non-negligible dynamics the application of a dynamic structure in form of a Dynamic Fuzzy Model was necessary analogous to the investigations carried out in [10].

Fig. 4.1 illustrates the structure of the Dynamic Fuzzy Model in connection with the defined input and output variables as well as the calculation of the narrow range collapsed level lc_n.

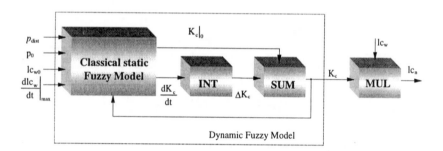

Fig. 4.1. Dynamic Fuzzy Model for the determination of the ratio factor K_c and calculation of the narrow range collapsed level lc_n

The direct output variable of the Fuzzy Model is the time derivation dK_c/dt of the ratio factor. As second output variable the initial value of the ratio factor $K_c|_0$ is determined. The absolute value of the ratio factor K_c which is the output variable of the Dynamic Fuzzy Model structure is fed back as additional input variable of the Fuzzy Model. Furthermore it is used to calculate the actual value of the narrow range collapsed level lc_n.

Basic Rule

The basic rule of the Fuzzy Model to determine the ratio factor K_c for the narrow range collapsed level can be defined as follows:

IF $"p_{dist}"$ AND $"\dfrac{dlc_w}{dt}\bigg|_{max}"$ AND $"p_0"$ AND $"lc_{w0}"$ AND $"K_c"$

$$(4.1)$$

THEN $"\dfrac{dK_c}{dt}"$ AND $"K_c\big|_0"$

Fuzzy Sets

The distribution of the representatives of sets was realized depending on the reference points and adapted on the requirements of the non-linear process behavior. Fig. 4.2 to Fig. 4.7 show the defined Fuzzy Sets. The Fuzzy Sets for the initial values of pressure p_0 and collapsed level lc_{w0} (wide range) are presented in Fig. 4.2 and Fig. 4.3. Their representatives are determined corresponding to the reference point values.

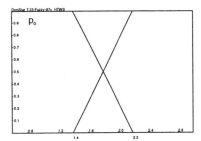
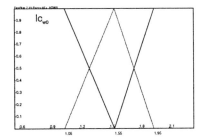

Fig. 4.2. Fuzzy Set for the initial pressure p_0

Fig. 4.3. Fuzzy Set for the initial collapsed level (wide range) lc_{w0}

For the linguistic variable maximum of the collapsed level gradient $dcl_w/dt\,|_{max}$ (wide range) a trapezoid was used as shape of membership function, which leads to a better description of the different parameter ranges (Fig. 4.4). The Fuzzy Set for the signal *pressure disturbance* p_{dist} was defined in form of two separate membership functions of lambda-type. As a consequence the value is definite *0* or *1* (Fig. 4.5).

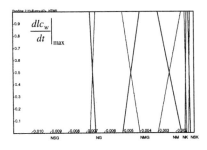
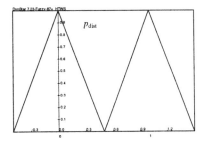

Fig. 4.4. Fuzzy Set for the maximum of the collapsed level gradient $dcl_w/dt\,|_{max}$

Fig. 4.5. Fuzzy Set for the pressure disturbance signal p_{dist}

Fig. 4.6 and Fig.4.7 show the Fuzzy Sets for the time derivation dK_c/dt of the ratio factor and the ratio factor K_c which is the output variable of the Dynamic Fuzzy Model structure.

Investigations [9] lead to the conclusion that for the description of dynamic processes a higher number of Fuzzy Sets of the linguistic output variables are necessary in comparison to the number of Sets of linguistic input variables. It is a supposition for an exact reproduction of the input-output behavior.

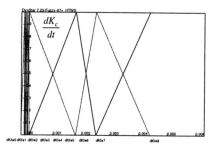

Fig. 4.6. Fuzzy Set for the time derivation of the ratio factor dK_c/dt

Fig 4.7. Fuzzy Set for the ratio factor K_c

The single rules for the Fuzzy Model were generated based on the experimental and simulated data as well as on process experiences. The rules were formulated correspondingly to the basic rule (4.1) and listed within a rule table [9].

General Validity

Fig. 4.8 demonstrates the reproduction results of the narrow range collapsed level lc_n for two additional transients.

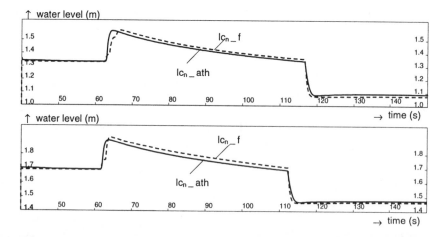

Fig 4.8. Comparison between fuzzy-based reproduced (lc_n_f) and real (lc_n_ath) response characteristics of the narrow range collapsed level for the experiments: BD18H135, BD14H170

The simulations serve for the check of the Fuzzy Model under unknown initial conditions. The results confirm the capability of the Fuzzy Model to realize an interpolation between the reference points. The reproduction quality of the Fuzzy Model can be assessed as sufficient. The relative error is lower than 10 % for all investigated transients.

It can be emphasized that the narrow range collapsed level can be reproduced very well by the Fuzzy Model within the technical related parameter range of the test facility. With the developed Fuzzy Model a rule-based model redundancy for the process variable narrow range collapsed level exists, which reflects the non-linearities with a good quality [9].

5   Fuzzy Model for Mixture Level Determination

Structure

As it was concluded from the data analysis the dynamics of the ratio factor K_m of mixture level during the depressurization is negligible. Therefore a classic static Fuzzy Model can be used for the determination of the ratio factor. Fig. 5.1 shows the Fuzzy Model in connection with the defined input and output variables as well as the calculation of the mixture level lm based on the measured wide range collapsed level lc_w and the determined ratio factor K_m.

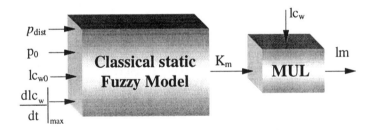

Fig. 5.1. Fuzzy Model for the determination of the ratio factor K_m and calculation of mixture level lm

Basic Rule

The basic rule of the Fuzzy Model for the mixture level ratio factor K_m is defined as:

$$\text{IF } ''p_{dist}'' \text{ AND } ''\left.\frac{dlc_w}{dt}\right|_{max}'' \text{ AND } ''p_0'' \text{ AND } ''lc_{w0}''$$

$$\text{THEN } ''K_m''$$

(5.1)

Fuzzy Sets

Fig. 5.2 and Fig. 5.3 illustrate the number and distribution of Fuzzy Sets for the linguistic input and output variables of the Fuzzy Model. The Fuzzy Sets for the initial collapsed level lc_{w0} (wide range) and the signal *pressure disturbance* p_{dist} are chosen analogous to the Fuzzy Model of Sect. 4. The Fuzzy Set of the initial pressure p_0 was expanded by an additional representative in comparison to Sect. 4 to realize a better differentiation. Considering the results of the data analysis a low number of Fuzzy Sets and a symmetric distribution of representatives for the static model output variable K_m were used. The single rules were formulated based on the basic rule (5.1) and summarized within a rule table. Analogous to Sect. 4 additional single rules were defined to describe the steady state behavior of process.

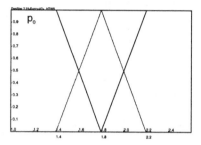 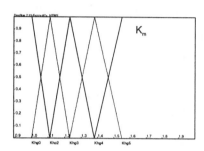

Fig. 5.2. Fuzzy Set for the initial pressure p_0 **Fig. 5.3.** Fuzzy Set for the ratio factor K_m

General Validity

To check the general validity of the developed Fuzzy Model within the defined parameter range a lot of transients with unknown initial conditions were simulated [9]. For all transients the Fuzzy Model realizes a sufficient reproduction quality.

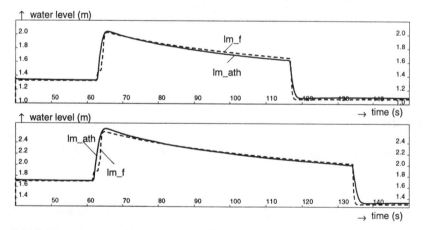

Fig. 5.4. Comparison between fuzzy-based reproduced (*lm_f*) and real mixture level (*lm_ath*) for the experiments BD18H135, BD22H170

Two selected test simulations with unknown initial conditions are presented in Fig. 5.4. The dynamics of the mixture level response characteristics during the pressure disturbance were reproduced well [9].

6 Conclusions

For the process variables - narrow range collapsed level and mixture level - pure fuzzy-based parallel models were developed which can be applied as knowledge-based redundancy for the water level measurement. For the generation of the Fuzzy Models the following conclusions can be summarized:

⇒ using ratio factors on the basis of the wide range collapsed level

⇒ using a Dynamic Fuzzy Model structure for narrow range collapsed level modeling

⇒ using a classic Static Fuzzy Model structure for mixture level modeling

⇒ good reproduction quality for all transients

It can be emphasized that Fuzzy Logic can be applied advantageously for the design of models to describe strong non-linear processes with high dynamics.

References

1. Kaufmann, A., Gupta, M. M.: Fuzzy Mathematical Models in Engineering and Management Science. North-Holland (1991)
2. Tong, R. M.: Synthesis of Fuzzy Models for Industrial Processes - Some Recent Results. Int. Journal General Systems 4 (1978) 143-162
3. Tong, R. M.: The evaluation of Fuzzy Models Derived from Experimental Data. Fuzzy Sets and Systems 4 (1980) 1-12
4. Sugeno, M., Tanaka, K.: Successive Identification of a Fuzzy Model and its Application to Prediction of a Complex System. Fuzzy Sets and Systems 42 (1991) 315-334
5. Bertram, T., Küpper, K., Schwarz, H.: Fuzzy Modelling for the Control of Mechatronic Systems. Proc. of the 2nd Conf. on Mechatronics and Robotics, Duisburg (1993) 89-106
6. Kroll, A., Gerke, W.: Modellierung eines pneumatischen Feststofffördersystems mit relationalen Fuzzy-Modellen. Automatisierungstechnik at 11 (1995) 525-530
7. Kroll, A., Gerke, W., Jordan, B.: Fuzzy-Modellierung und -Regelung eines Wirbelschichtofens zur Klärschlammverbrennung. Automatisierungstechnische Praxis atp 39 (1997) 49-55
8. Xiao, W., Weidemann, B.: Fuzzy Modelling and its Application to Magnetic Bearing Systems. Fuzzy Sets and Systems 73 (1995) 301-312
9. Hampel, R. u. a.: Meß- und Automatisierungstechnik zur Störfallbeherrschung -Methoden der Signalverarbeitung, Simulation und Verifikation. Abschlußbericht BMBF-Projekt 150 10 15, HTWS Zittau/Görlitz (FH) (1999)
10. Fenske, A., Kästner, W., Hampel, R.: Model-Based and Knowledge-Based Measuring Methods for the Observation of Non-linear Processes. Proc. of the 5th Zittau Fuzzy Colloquium, September 4-5, Zittau, Germany (1997) 118-127
11. DynStar: Ein Simulationsprogramm für Automatisierungstechniker. Programmbeschreibung, HTWS Zittau/Görlitz (FH) (1997)

Fuzzy Modelling of Multidimensional Non-linear Processes – Design and Analysis of Structures

Andrzej Pieczynski[1], Wolfgang Kästner[2]

[1]Technical University of Zielona Gora, Department of Robotics and Software Engineering,
Pl - 65 - 246 Zielona Gora, ul. Podgorna 50, Poland
a.pieczynski@adb.pl
[2]Institute of Process Technique, Process Automation and Measuring Technique, (IPM),
Department Measuring Technique / Process Automation,
D - 02763 Zittau, Theodor-Körner-Allee 16, Germany
w.kaestner@htw-zittau.de

Abstract. The response characteristic between input and output variables can be modelled by knowledge-based methods of signal processing like Fuzzy Logic. Based on a low number of data sets Fuzzy Logic can be applied advantageously for non-linear processes, especially. The rule-based description of the non-linear process behaviour can be realised by means of different structures of the fuzzy algorithm. The paper presents and compares three structure variants (complex, parallel and cascaded structure) of the fuzzy model design to reproduce the input-output behaviour. The structure analysis was carried out for the fuzzy-based modelling of parameters which are necessary to describe the process state of pressure vessels with water-steam mixture during accidental depressurizations.

1 Introduction

Fuzzy Logic as modern method of signal processing is applied not only for the control, but also for modelling of non-linear processes. Based on a data base and a knowledge base, Fuzzy Logic allows a rule-based description of the response characteristic between the input and output variables. The advantage in opposite to the application of Neural Networks is the low number of data sets, which are necessary for the reproduction of the non-linear process behaviour. The characteristic field produced by the fuzzy controller represents the correlation between the input and output variables of the fuzzy model and is very suitable for the verification of the fuzzy algorithm. Additional to the degrees of freedom like: number of fuzzy sets, distribution of fuzzy sets, membership functions, operators, the structure of the fuzzy model can be varied to reproduce the input-output behaviour of the non-linear process.

Within the paper different structures of fuzzy models are explained and compared based on a process example. As process example the determination of the collapsed level in a pressure vessel with water-steam mixture during accidental depressurizations was used. The chosen process example is predestined for the methodical investigations and for testing the designed fuzzy algorithms.

2   Description of Process Example

The presented example to apply modern methods of signal processing like Fuzzy Logic is the description of the strong non-linear thermodynamic and thermohydraulic effects within the pressure vessel during accidental depressurizations. The subject pressure vessel with water steam mixture is characterised by the combination of a lot of parameters which have an influence on the process state (pressure, water level, temperature, steam content, mass flow, heat flux). The thermohydraulic and thermodynamic processes in pressure vessels with water-steam mixture are characterised by non-linearity, which can be classified as follows:

♦   *non-linear structure of the analytical model,*
    products of state variables and input variables within the state equations,
♦   *time-dependent coefficients in the model equations,*
    non-linear change of thermodynamic properties (dependence of density and enthalpy on pressure),
♦   *non-linear behaviour as a result of the change of process state,*
    disturbances like feed-in, bleed-off, spray.

The global aim is the calculation of the collapsed level hc (representing the water inventory) within defined zones of the pressure vessel during negative pressure gradients ($dp/dt<0$) which occur as a result of leaks.

In [1] a Hybrid Observer was proposed to solve this problem of process monitoring. The Hybrid Observer combines a classical linear Observer with a fuzzy-based adaptation of the observer model matrices.

The classical observer based on the linearized observer model equation (2-1) which describes the response characteristic between the pressure gradient $d\Delta p/dt$ and the collapsed level Δhc in a number of n zones of the pressure vessel.

$$\Delta \dot{hc}_n = A_b \cdot \Delta hc_n + B_b \cdot \frac{d\Delta p}{dt} \text{ with } A_b = \begin{vmatrix} a_{11} & \cdots & a_{1n} \\ \vdots & \ddots & \vdots \\ a_{n1} & \cdots & a_{nn} \end{vmatrix} \text{ and } B_b = \begin{vmatrix} b_1 \\ \vdots \\ b_n \end{vmatrix} \qquad (2-1)$$

The elements $a_{11} \ldots a_{nn}$ of the system matrix A_b and the elements $b_1 \ldots b_n$ of the input matrix B_b depending on time variable process parameters like pressure, pressure gradient and so on. The fuzzy-based algorithm for the adaptation of the elements of the matrices A_b and B_b enables the compensation of non-linearity and an improvement of the estimation quality. The structure analysis is focused on the design of the fuzzy algorithms for the calculation of the elements $b_1 \ldots b_n$ of the input matrix B_b.

For analyzing the process a number of blow down experiments at the pressurizer test facility of the IPM were carried out characterized by different values of the initial pressure p_0 and initial collapsed level hc_0 at the beginning of depressurization.

In Table 2.1 the experiments and their initial conditions are listed. The defined reference experiments (R) were used for the design of the fuzzy models while the experiments signed as (O) were used to check the capability of interpolation by the fuzzy models.

378

Table 2.1. Blow Down (BD) experiments with different initial values of pressure p_0 and collapsed level hc_0 at the beginning of depressurization

Experiment	Initial Pressure		
Initial Collapsed Level	$p_0 = 22$ bar	$p_0 = 18$ bar	$p_0 = 14$ bar
$hc_0 = 195$ cm	BD22H195 R	BD18H195 O	BD14H195 R
$hc_0 = 170$ cm	BD22H170 O	BD18H170 O	BD14H170 O
$hc_0 = 155$ cm	BD22H155 R	BD18H155 O	BD14H155 R
$hc_0 = 135$ cm	BD22H135 O	BD18H135 O	BD14H135 O
$hc_0 = 115$ cm	BD22H115 O	BD18H115 O	BD14H115 O
$hc_0 = 105$ cm	BD22H105 R	BD18H105 O	BD14H105 R

Fig. 2.1 shows the non-linear dependence of the element b_1 on the actual pressure $p(t)$ calculated by the ATHLET-code and validated by six depressurization experiments (reference points of Table 2.1). The experiments are characterized by the initial pressures $p_0 = 14$ bar and $p_0 = 22$ bar and different initial collapsed levels $hc_0 = 105$ cm, $hc_0 = 155$ cm, $hc_0 = 195$ cm.

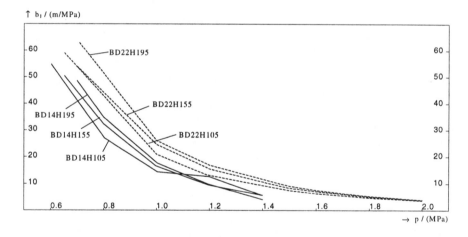

Fig. 2.1. Dependence of the input matrix element b_1 on the actual pressure $p(t)$ for Blow Down (BD) experiments

Within the paper the analysis of the different fuzzy structures for the determination of the matrix element b_1 is demonstrated. For the other input matrix elements qualitatively similar relations can be found. Therefore the common variable b was used to mark the fuzzy-based determined matrix element.

3 Fuzzy Modelling

3.1 Preliminary Remarks

The fuzzy algorithms are applied for the reproduction of the response characteristic between the input variables actual pressure $p(t)$, initial pressure p_0, initial collapsed level hc_0 and the output variable matrix element $b(t)$ which is necessary for the calculation of the collapsed level based on the observer model (2-1).

For the reproduction of the input-output behaviour three variants of fuzzy model structures were developed and compared regarding their quality and design effort: *complex* model structure, *parallel* model structure, *cascaded* model structure.

In the first structure, defined as *complex,* one multi-dimensional fuzzy model was used. The second structure, *parallel,* is characterized by a number of one-dimensional fuzzy models arranged in parallel. The third structure, *cascade,* contains two-cascaded two-dimensional fuzzy models [2].

During the creation of a fuzzy model a few tasks have to be solved: fuzzification, knowledge base and defuzzification [3,4]. By means of the Fuzzification operation the number of fuzzy sets and the shape of membership function are defined. Based on the specific problems of non-linear characteristics approximation in the fuzzification process the triangular and sigmoid membership function are used [5,6].

The fuzzy model can be created on the Mamdani or the Sugeno model base. In the Mamdani model the crisp output was obtained after defuzzification process. In this case the number of sets, membership function shape and defuzzification method have to be defined.

In Sugeno model appears a problem of linear output function definition. In the three dimensional object case the output function can be described by the following equation:

$$f_i\{p(t), p_0, hc_0\} = k_{1i}\, p(t) + k_{2i}\, p_0 + k_{3i}\, hc_0 \qquad (3\text{-}1)$$

where: k_{ij} is a weight of a input signals influence, i is a number of the output Sugeno function and j is a number of the input signal. All weights have to be defined in a model adjusting process.

The validation of the developed fuzzy models was carried out based on the criterion sum-squared error function

$$e_{sq} = \sum_k \left[b_R(k) - b_{FM}(k) \right]^2 \qquad (3\text{-}2)$$

where b_R is a matrix element of reference point, b_{FM} is a matrix element of fuzzy model.

3.2 High-Dimensional Fuzzy Controller (3D-Controller)

The fuzzy model based on a high-dimensional Fuzzy Algorithm with the input variables actual pressure $p(t)$, initial pressure p_0, initial collapsed level hc_0 and the output variable matrix element $b(t)$.

The three input signals were fuzzified on a different number of fuzzy sets. The linguistic variable initial pressure p_0 was represented by two fuzzy sets, *low* and *high*. The linguistic variable initial collapsed level hc_0 was described using three fuzzy sets (*low, normal, and high*). For fuzzification of the pressure $p(t)$ 14 fuzzy sets and an asymmetric fuzzification were used (Fig. 3.1).

In the Mamdani model a sufficient quality was reached using 29 nearly triangular fuzzy sets for the output variable matrix element $b(t)$ (Fig. 3.2).

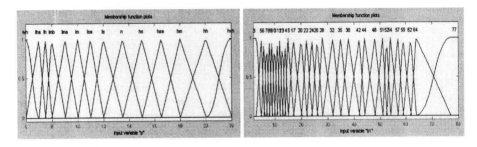

Fig. 3.1. Asymmetric pressure fuzzification   **Fig. 3.2.** Matrix element fuzzification

The fuzzy-based generated surfaces are shown for the Mamdani model in Fig. 3.3a and for the Sugeno model in Fig. 3.3b.

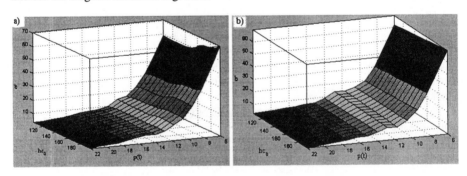

Fig. 3.3. The Mamdani (a) and Sugeno (b) model surface for p_0=14 bar

The Mamdani model is characterized by a better approximation accuracy shown by the error values e_{sp} of Table 3.1. This accuracy was reached by the high number of fuzzy sets for the output variable. For the design of the Sugeno model nine 3D- functions were used which results in a comprehensive design effort. The design of both models was characterized by the asymmetric fuzzification of the linguistic variable $p(t)$ to obtain a good accuracy and a smooth surface (Fig. 3.3) and by a high number of rules.

Table 3.1. Comparison of the error e_{sp} for the applied fuzzy models and for the reference points

BD14H105	BD14H155	BD14H195	BD22H105	BD22H155	BD22H195
Complex Mamdani Model					
2.29	1.16	3.05	1.86	2.70	2.20
Complex Sugeno Model					
25.07	33.01	25.74	61.03	91.58	75.81

3.3 Parallel Structure of 1D-Controllers

The determination of the matrix element $b(t)$ was realized by a parallel structure of 1D-Controllers for the defined reference points and a weighting of the controller outputs by an additional algorithm.

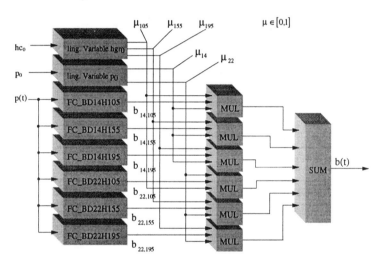

Fig. 3.4. Determination of the matrix element $b(t)$ by means of a parallel structure of 1D-Fuzzy Controllers and an additional weighting algorithm; BD - Blow Down Experiment, initial pressures $p_0 = (14; 22)$ bar, initial collapsed levels $hc_0 = (105; 155; 195)$ cm

The parallel structure (Fig. 3.4) consists of:

♦ fuzzy controllers designed for defined reference points (R in Table 2.1) characterized by defined values for the initial pressure p_0 and the initial collapsed level hc_0 with the input variable actual pressure $p(t)$ and the output variable matrix element $b(p_0, hc_0, t)$ for this reference points

♦ special fuzzy models to calculate a value of membership for the initial pressure p_0 (fuzzy sets *low* and *high*) and for the initial collapsed level hc_0 (fuzzy sets *low*, *normal* and *high*)

♦ an additional algorithm to determine a final value for the matrix element $b(t)$ based on the crisp controller outputs which are weighted in dependence on the real initial values of pressure p_0 and collapsed level hc_0

Equation (3-3) describes the weighting of the crisp fuzzy model output signals b_{p_0,hc_0} with the membership values μ_{p0}, μ_{hc0} of the fuzzified initial values of pressure p_0 and collapsed level hc_0. The weighting algorithm can be interpreted as a fuzzy-based interpolation between the reference point data.

$$b = \sum_{(R)} \mu_{p0} \cdot \mu_{hc0} \cdot b_{p_0,hc_0}$$ (3-3)

In the simple model for fuzzification of pressure $p(t)$ 5 fuzzy sets were used. Fig. 3.5 represents the surfaces of the simple fuzzy models designed for the initial value of pressure p_0 = 14 bar and different values of initial collapsed level hc_0 = 105 cm, hc_0 = 155 cm and hc_0 = 195 cm. It illustrates the splitting of the surface shown in Fig. 3.3a (complex model) in three separate surfaces, which are defined for the special values of the initial collapsed levels.

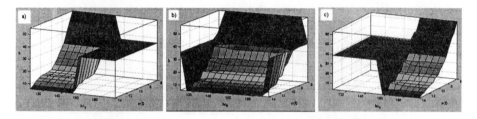

Fig. 3.5. Surface of three simple fuzzy models designed for p_0=14 bar and different initial collapsed levels hc_0 = 105 cm, hc_0 = 155 cm and hc_0 = 195 cm

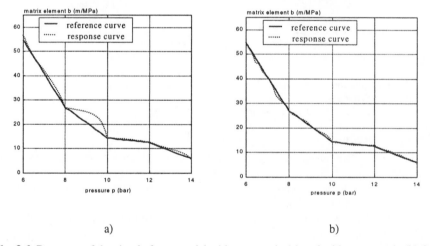

a)   b)

Fig. 3.6. Response of the simple fuzzy model with symmetric (a) and with asymmetric (b) fuzzification

A response of the simple fuzzy model for $p_0 = 14$ bar, $hc_0 = 105$ cm and symmetric fuzzification by means of 5 fuzzy sets is shown in Fig. 3.6a. Fig. 3.6b presents a response of the fuzzy model with asymmetric fuzzification by means of 9 fuzzy sets.

Fig. 3.7 presents the surfaces for the simple fuzzy models ($p_0 = 14$ bar, $hc_0 = 105$ cm) generated by means of symmetric and asymmetric fuzzification of the linguistic variable pressure $p(t)$.

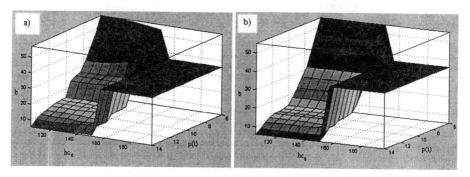

Fig. 3.7. Surface for simple model with symmetric (a) and asymmetric (b) fuzzification

The approximation accuracy e_{sq} for the described simple models is listed in Table 3.2. Different variants of set number as well as symmetric and asymmetric fuzzification of the linguistic variable $p(t)$ are compared. To obtain a good accuracy and a smooth surface an asymmetric fuzzification and a high number (in maximum 14) of sets for the linguistic variable $p(t)$ have to be used.

Table 3.2. Comparison of error e_{sp} for the applied Mamdani Fuzzy Models and for the reference points

BD14H105	BD14H155	BD14H195	BD22H105	BD22H155	BD22H195
7 fuzzy sets and symmetric fuzzification; operation And (min), implication (min).					
33.88	21.25	19.31	30.10	21.55	27.19
7 fuzzy sets and symmetric fuzzification; operation And (prod), implication (prod).					
20.76	10.87	9.87	18.58	12.13	15.98
14 fuzzy sets and asymmetric fuzzification, operation And (prod), implication (prod).					
2.61	1.08	1.94	3.57	2.43	2.88

Furthermore the influence of different operators for conjunction, implication was investigated. The applications of the operator *prod* leads to an improvement of accuracy in comparison to the operator *min*. The change of aggregation and alternatives operators has no influence on the accuracy. An advantage of the parallel structure of simple models is the low number of rules in each simple model.

3.4   Cascaded Structure of 2D-Controllers

The determination of the matrix element $b(t)$ is realized by a cascaded structure of fuzzy models which is designed for the *low* and *high* value of initial pressure (Fig. 3.8):

♦   fuzzy model with the input variable actual pressure $p(t)$ and the output variable $b^*(t)$

♦   fuzzy model with the input variables $b^*(t)$ and initial collapsed level hc_0 and the output variable matrix element $b(hc_0, t)$

♦   algorithm for weighting the output of each cascade based on the initial pressure value p_0 fuzzification

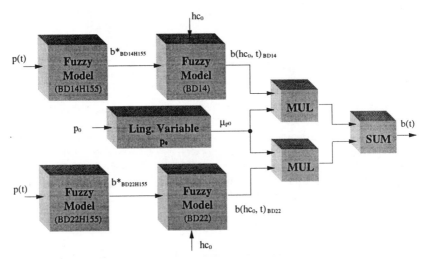

Fig. 3.8. Determination of the matrix element $b(t)$ by means of a cascaded structure of fuzzy models

The first blocks of the cascades are similar like the simple models of the reference points BD14H155 and BD22H155. The fuzzification process of the first block input signal $p(t)$ is similar like in complex fuzzy model (Fig. 3.1).

The second block of each cascade is a two dimensional model to move the output of first block. The direction of the motion depends on the second input variable hc_0. To obtain a smooth surface within the second block a lot of input fuzzy sets are necessary. The fuzzification of the second block input signal $b^*(t)$ is more complicate. The third block is comparable with the weighting algorithm of Sect. 3.3 but only applied for the determination of initial pressure weighting factor. The outputs of both cascades are used to calculate the final value of the matrix element $b(t)$ in dependence of the initial value of pressure. The weighting algorithm can be interpreted as a fuzzy-based interpolation between the values of initial pressure *low* and *high*.

The approximation accuracy e_{sq} for the designed models is listed in Table 3.3. The preferred Mamdani model was investigated, asymmetric fuzzification of linguistic variable $p(t)$ was used as result of the conclusion of Sect. 3.2 and Sect. 3.3.

Table 3.3. Comparison of error e_{sp} for the applied Mamdani Fuzzy Models for the reference points

BD14H105	BD14H155	BD14H195	BD22H105	BD22H155	BD22H195
3.31	1.54	2.93	3.81	2.53	4.14

3.5 Comparison of the Different Structures

For all three structures of fuzzy modelling the capability of interpolation in the case of unknown values of input variables which are not reference points were proofed. All investigated structures lead to good results for the unknown input parameters. The smoothest surface is given by the parallel structure. Based on the error values of the six reference points an average error for the three structures was calculated and compared in Table 3.4.

Table 3.4. Comparison of different features of the fuzzy model structures

Features	Structure of Fuzzy Model		
	Complex	Parallel	Cascade
number of input fuzzy sets	19	10 for each simple model +5	9+2 / 14+2 first block 35 - second block
number of output fuzzy sets	29	8 ÷ 12	30 - first block 40 - second block
number of rules	69	8 ÷ 12	9 / 14 - first block 93 - second block
average approximation error	2.21	2.42	3.04

It can be stated that the average error will be nearly the same value for all structures. The cascaded structure leads to a higher value in opposite to the other structures. The structure of the cascaded model is more complicate as the other ones (higher design effort) especially in the second block. The simple blocks of the parallel structure can be designed very easy and can be duplicated for the reference points in connection with a modification and adapting for the special conditions.

For this special application the parallel structure is the best, it allows furthermore a very easy optimization, and a very simple expansion if additional influence parameters must be considered.

An advantage of the parallel and cascaded structure in opposite to the complex is the possibility to validate the input-output behaviour based on the surfaces (which can be represented in the three dimensional space). In this sense a simple method for verification is given on the base of the smoothness of surface or deformation of surface. Furthermore an easy indication of incorrectness of data or fuzzy algorithm is possible by means of analysis of surface deformation.

386

4 Conclusions

⇒ Fuzzy Logic can be used with success for the modelling of non-linear characteristics.

⇒ Symmetric or asymmetric fuzzification methods have a big influence on the accuracy of modelling. The asymmetric fuzzification leads to a higher deformation of the fuzzy characteristic field and to a better adaptation on strong non-linearity.

⇒ Based on the *complex* fuzzy model a sufficient preciseness of modelling can be achieved but a high number of output fuzzy sets and rules that are necessary.

⇒ The best modelling results have to be obtained using a few one-dimensional fuzzy models in *parallel* arrangement.

⇒ The *cascaded* structure is a compromise between the modelling accuracy and complexity of fuzzy model.

⇒ Low-dimensional *cascaded* or *parallel* structures allow the representation and verification of high-dimensional correlation by means of the fuzzy characteristic field.

References

1. Kästner, W., Fenske, A., Hampel, R.: Improvement of the Robustness of Model-Based Measuring Methods Using Fuzzy Logic. In: World Scientific, Proceedings of the 3rd International FLINS Workshop. Antwerp, Belgium (1998) 129-142
2. Chaker, N., Wagenknecht, M., Hampel, R.: Fuzzy Controller Structure Transformation. World Scientific, Proceedings of the 3rd International FLINS Workshop. Antwerp, Belgium (1998) 99-110
3. Frank, P.M.: Fuzzy supervision – Application of Fuzzy Logic to Process Supervision and Fault Diagnosis. In: Proc. Int. Workshop on Fuzzy Intelligent Systems, Duisburg, Germany (1994) 36-59
4. Patton, R.J.: Fuzzy Observers for Non-linear Dynamic Systems Fault Diagnosis. In: Proc. of the 37th IEEE Conf. On Decision & Control, Tampa, Florida, USA (1998) 84-89
5. Pieczy•ski, A.: An Expert System with Fuzzy Knowledge Base. In: Proc. 3rd. Int. Conf. New Trends in Automation of Energetic Processes'98, Zlin, Czech Republic (1998) 374-378
6. Pieczy•ski, A.: Integrated Neural Network and Fuzzy Knowledge Base System for Fault Detection and Isolation in Steam Power Plant. In: Proc. of the Fifth Int. Symp. on Methods and Models in Automation and Robotics, MMAR'98, Miedzyzdroje, Poland (1998) 707-712

Fuzzy Modeling of Uncertainty in a Decision Support System for Electric Power System Planning

Izebe O. Egwaikhide

Forschungszentrum Karlsruhe GmbH, Institute of Applied Informatics (IAI)
P.O. Box 3640, D-76021, Karlsruhe, Germany
egwaikhide@ieee.org

Abstract. This paper reports on the specification of a fuzzy logic based knowledge modeling concept for the development of a decision support system (*DSS*) christened "Power System Outage-plan Validator" (*PSOV*), which is expected to assist utility engineers in the medium-term outage planning of the Nigerian Electric Power System (*NEPS*). The object-oriented (*OO*) development concept for PSOV is tailored to facilitate internet/web-enabled interactive decision making in the medium-term (yearly) generation outage planning process of an increasingly deregulated power market, with new stakeholders and structural components. The varying types of uncertainties, ambiguities and contradictions in the semi-structured problem domain necessitate a robust modeling system for knowledge analysis and user interface development, which fuzzy logic provides.

1 Introduction

Traditionally, decision making in electric power system (*EPS*) planning is pervaded by lots of uncertainties, caused by the technological complexity, the inevitable interactions with other systems, and the conflicting objectives (e.g.: meeting expected loads, minimization of power generation unit (*PGU*) fuel costs and waste emission).

The cost minimization goal can be formulated as in Equation 1 below; showing the interdependent and conflicting relationships between the goals in the planning process.

$$\underset{\underline{e}}{\text{Min}}\left(\sum_{i=1}^{i=n} C_{gen-PGU_i}(\underline{e})\right) . \tag{1}$$

With:

$\underline{e}$:	Vector of input variables (e.g. incremental cost curves etc.)
$C_{gen-PGU_i}(\underline{e})$:	Generation costs of PGU no. i, *(which itself is a function of $\underline{e}$)*
n:	Number of PGUs in the operation plan in the planning period.

IT solutions, which increasingly incorporate knowledge-based systems (*KBS*) (e.g. [3]), DSSs [28], and lately, other technologies like Genetic Algorithms [30] and Fuzzy Logic [2]), where developed to deal with problems of complexity, uncertainty and performance. The DSSs however mainly dealt with uncertainty using probabilistic methods and contingency analysis (compare with [16,36]), and where based on non-structured development concepts which focused on internal corporate use, and are thus not suitable for use in the emerging iterative cross-corporate decision making processes; mainly due to low portability and inadequate security measures for internet use.

The deregulation of the formerly monopolized power industry worldwide (see [21] and [32]) has however given birth to a varying set of new commercial and structural components, organizational frameworks ("Power Pool" and "Negotiated Third Party Access" [32]), distribution side generation, competition and consequently new uncertainties and risks regarding availability, quality and cost of resources and services.

The traditionally simple structural classification of EPSs into generation, transportation and distribution subsystems has seen an influx of new components and power utility companies (*PUCs*); mainly: Generators (*G*), Power Marketers (*PM*), Scheduling Coordinators (*SC*), Ancillary Service Providers (*AS*), Independent System Operators (*ISO*), Power Exchange (*PE*), Transmission Owners (*TO*), Retail Service Providers (*RSP*) and Distribution Service Providers (*DSP*), as shown in Fig. 1 below [23].

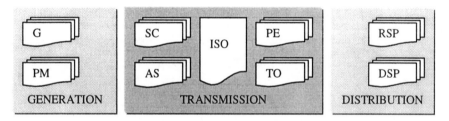

Fig. 1. Emerging structural components of deregulated power markets

Competition has also led to the use of "*calculated risks*" (inherent in other none network-based industries), where, traditionally, crisp values where applied for goals (e.g. security and quality) in EPS planning.

The changes in structure and planning principles further necessitate an efficient and flexible framework for the modeling of uncertainty in resulting cross-corporate DSSs. Besides this, the distributed character of the emerging EPS makes internet solutions for e-business the choice mediums of communication during negotiations which makes portability of developed user interfaces a key goal. Strategies also have to be developed and implemented to raise end-user acceptance, since domain experts (the target end-user group of PSOV), are prone to be suspicious of, and resist the introduction of automated systems, due to fear of loss of jobs and/or unwillingness to replace traditional methods that "work well" (compare with [13]). This social constraint is resolved through end-user integration in the business process reengineering (*BPR*) concept of the PSOV project [5],

The next section provides a very brief insight into the problem domain, through a simplified sub-process of the medium term outage planning procedure in the Nigerian Electric Power System (*NEPS*) [6], whilst Section 3 then briefly (exemplarily) introduces the fuzzy modeling concept and the guiding ideas for choosing the applied theories and methods, before the paper is rounded up with concluding remarks and an outlook on ongoing and future work, in Section 4.

2   PSOV   Concept and Development Strategy

A lot of the DSSs developed for EPSs, till date, used tools and non-structured development processes, which only allow a limited portability, and thus low cross-corporate adaptability, since they where meant for the homogenous centralized planning departments of past monopolies (see [4] and [7]). The user acceptance of these DSSs was consequently also impaired by low reusability, user friendliness and maintainability.

In line with large efforts aimed at bringing structuredness into Knowledge Engineering (*KE*) methodologies (see [26], [24] and [9]), the Combined Object-Oriented (Knowledge-based & Algorithmic) development Life-Cycle (*COOKADLC*) [5], applied in developing PSOV, brings issues of Change Management, BPR, Reusability as well as Project & Software Life-Cycle Management to the forefront.

OO in COOKADLC provides a means of reducing complexity of an EPS system through abstraction, and development of a simplified, "natural" and flexible model (compare with [4]). COOKADLC also integrates the Unified Modeling Language (*UML*) [29] an OO modeling language standardized by the Object-Management Group (*OMG*) [22], which has found its way into major software engineering methodologies like the "Unified Process" from Rational [29], and KE methodologies like CommonKADS[1] [24] for modeling and documentation.

The generation outage planning process, can be summarized as follows (see [5]):

- Submission of outage requests by the PUC planners or power stations (G).
- Verification of the requests based on a set of conflicting objectives by the SC.
- Suggestion of alternatives to PUC planners whose requests violated the goals.
- Preparation of the final maintenance plan after negotiations.

The UML Use-Case Diagram for this process could thus be drawn as in Fig. 2 below.

---

[1] CommonKADS Homepage: URL: http://www.commonkads.uva.nl/

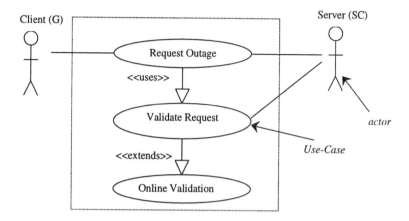

Fig. 2. Simplified UML use-case notation of the outage scheduling process

The Use-Case Diagrams, basically describe the interaction of a group of people (*Actors*) with a system, based on a set of "*Use-Cases*", and enable the documentation and graphical presentation of results in the conceptualization phase to both technical and non-technical stakeholders of the DSS project - e.g. end users, domain experts and managers - since they have been found to possess a relatively high level of comprehensibility compared to other notations (see also UML usability discussions in [1]).

Fig. 3 shows the web-based interaction between generators (*G*) – *the clients* - and the Scheduling Coordinator (*SC*) – *the server* - during the planning process. It also shows the basic components' structure of PSOV, with the fuzzy components highlighted.

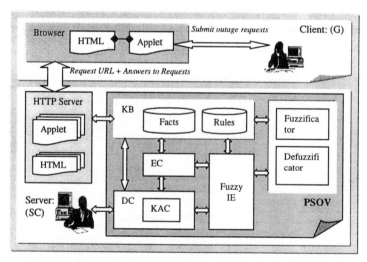

Fig. 3. Basic component and interaction structure of PSOV

The truncated components in Fig. 3 are: the Inference Engine (*IE*): the Knowledge Base (*KB*), the Dialog Component (*DC*), the Explanation Component (*EC*) and the Knowledge Acquisition Component (*KAC*) [28].

The COOKADLC concept integrates fuzzy logic - to model the linguistic uncertainties encountered in the emerging electric power industry - into its evolutionary prototyping reusability-driven and architecture-based development life-cycle framework for hybrid DSSs.

The fuzzy modeling of uncertainty in PSOV is also expected to enhance the ergonomics (naturalness) of the man-machine interaction through the KAC and EC modules, making room for a higher acceptance on the side of the end-users as discussed in the next section.

3 The Object-Oriented (OO) Fuzzy Modeling Concept in PSOV

Fuzzy logic is based on fuzzy-set theory, an innovation from Prof. Lotfi A. Zadeh (see [34] and [14]). Fuzzy Logic is a superset of conventional (Boolean) Logic that has been extended to handle the concept of partial truth - truth values between "completely true" and "completely false". The main area of application of fuzzy logic and fuzzy set theory in EPSs is for the control of non-linear processes and in planning and operation (e.g. for fault diagnosis [2] and for operations planning [8]).

The main advantage of fuzzy logic, as opposed to other mostly probability-based methods for modeling uncertainty, e.g. the Bayesian Extension, the Dempster-Shafer Theory of Evidence and the Certainty Factor Methods [28], is its ability to handle both imprecision and uncertainty in rules and data using the same framework (fuzzy sets, rules and inferences), and in a more natural manner (see [36] and [35]).

The OO fuzzy modeling concept in COOKADLC is explained below, based on the generation reserves optimization objective of the outage planning process which is traditionally "crisply" formulated as in Equation 2 below, [3]:

$$P_{apres\text{-}EPS}(S) \geq P_{npres\text{-}EPS}(S) \tag{2}$$

Whereby $P_{npres\text{-}EPS}$ is the required planned generation reserves to ensure a given security of supply (S), and $P_{apres\text{-}EPS}$ is the available planned generation reserve.

By fuzzifying the reserve objective in Equation 2, a more human-like modeling of the rules applied in the decision making process is reached. For example, the membership functions shown in Fig. 4 could be derived for $P_{npres\text{-}EPS}$ and S.

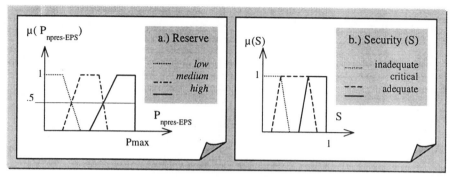

Fig. 4. Membership functions for the fuzzification of the reserve objective

Whereby, *Pmax* denotes the maximal generation capacity of the EPS in question. In the PSOV concept, membership functions which cross at half height, as in Fig. a, are used for fuzzification when crisp outputs are needed, whilst membership functions which cross at full height, as in Fig. b, are used when a linguistic (natural) output expression is needed, to allow a modeling of a high degree of ambiguity [25].

The fuzzy rule to replace Equation 2 in PSOV could be then formulated as follows:

```
IF  Reserve = "low" THEN Security = "inadequate"
```

Whereby "*Reserve*" is the input variable (name for a knowledge value), "*S*" is the output variable (name of a data value due to be computed); *low* and *inadequate* are membership functions (fuzzy subsets) defined on *Reserve* and *S* respectively (Fig.).

A rule object (or better said frame) in PSOV thus has the following structure:

```
RULE(No., Confidence, Antecedent, Consequent, Meta-Tag)
```

The *Meta-Tag*, which could be an object of any form, contains information (mainly strings) for identification and explanation purposes.

The choice of fuzzy sets which cross at full height, as shown in Fig. b, enables simple linguistic (natural) output expressions in the dialog component, a high degree of ambiguity, and consequently reduces the work in the conceptualization and formalization phases, in accordance with the evolutionary prototyping strategy in COOKADLC. Prototypes can thus be developed faster, and presented to stakeholders, leading to a reduction of the traditional bottleneck of knowledge acquisition in KBSs.

The simple example above shows the basic idea behind the incorporation of fuzzy logic into the knowledge modeling process in the PSOV development strategy. Taking other stakeholders in the development of such a cross-corporate tool as PSOV, e.g. non-technical managers, into consideration, the number of applied fuzzy concepts are kept as simple as possible through the choice of just two possibilities for each core function - i.e. for the statement of membership functions (for fuzzification), inference strategies and defuzification strategies - as summarized in Table 1 below.

Table 1. A summary of fuzzy theory concepts used in the PSOV framework

Membership Functions	Inference Strategies	Defuzification Strategies
Triangular	Max-Min	Center of Gravity
Trapezoid	Max-Prod	Average of Maxima

This underlying principle of "as simple as possible", drawn from experience in end-user feedback from practical expert systems (ES) development projects, aimed at avoiding an overly high level of "mathemathisation" of models besides the often associated use of new untested concepts - as ideally done in the field of research -, not only ensures acceptance by end-users, but also enables a future maintenance of the DSSs knowledge content by knowledge managers, without deep knowledge of fuzzy logic theory and concepts.

PSOV also incorporates a truth maintenance system for conflict resolution, based on concepts already tested in the FLOPS[2] shell, presented in [25]. The tools chosen for the PSOV development - based on availability, integrability, and capability to use - are *JESS*[3] and *FuzzyJESS*[4] for KE, *JBuilder 3.0* for conventional software modules and for the Graphic User Interface, and the UML-based CASE tool, *Together/J* for Java, for parallel use in all phases of KE and software engineering, since all tools mentioned above are "pure-Java" based, meaning that their codes can be manipulated in Together/J enabling easy UML-based OO modeling and documentation of all phases and prototypes, as requested by the COOKADLC concept.

The homogenous Java-based strategy also readily accommodates the specification of interface classes; which are a good prerequisite for modeling and designing those dynamic features of PSOV, which can be affected by changes in other elements of the integrated environment. It also makes room for a conversion of the DSS into a community of intelligent agents with varying levels of autonomy, for which interfacing with conventional legacy systems is still an issue of research [15]. See [19], and [31] for an introduction to the ARCHON system, an exemplary application of EPS applications of intelligent agents to EPSs.

4 Conclusion

This paper briefly introduced a DSS for medium-term outage planning in the NEPS, as well as the applied OO structural modeling concept, and fuzzy theory based concepts for modeling uncertainty in its IE and KB. The use of COOKADLC, an OO UML-based (JAVA optimized) methodology, is aimed at avoiding the limitations found in other DSS developments for EPSs, i.e. low reusability of code and low portability (single platform architecture). The PSOV concept is web-enabled (extranet-based for security purposes) to allow easy cross-corporate interaction.

---

[2] Fuzzy LOgic Production System
[3] Java Expert System Shell (JESS) Homepage: URL: http://herzberg.ca.sandia.gov/jess/
[4] FuzzyJESS: URL: http://ai.iit.nrc.ca/IR_public/fuzzy/fuzzyJavaToolkit.html

Despite the increasing availability of different shareware and commercial fuzzy tools (specifically shells, e.g.: FuzzyCLIPS[5] and FuzzyTECH[6]), there is still a need to incorporate Fuzzy Logic, Sets and Numbers into basic programming environments and modeling languages like Java and UML, to further promote the acceptance and development of hybrid DSSs and KBSs in industrial environments - a step forward in promoting knowledge management and conservation (as supported by the ISO[7] 9000 series of norms [12]); both prerequisites for good corporate quality management in the emerging knowledge driven economies. Work on usability of UML notations for visual communication of results in (fuzzy) DSS project groups is also subject of present work on the COOKADLC concept.

A further point of interest is the future role of fuzzy logic principles in the emerging standard communication languages for agents in Multi-Agent based Systems (*MAS*) (see [20] and [31]), which are quite suited for complex cross-corporate decision making and negotiation domains (tasks) in integrated cross-corporate environments like the emerging communication networks of the deregulated EPSs.

Important work needs to be done in development of ontologies (shared message content and concept representation)[8] in this area; one of the pitfalls described in [19] and [33], which presently make an implementation of PSOV as an MAS unrealistic. However, in view of the ongoing MAS standardization work (mainly in FIPA[9] and the OMG - through the MASIF[10] project), concepts are being worked on to enable an integration of MASs at a later date. Present research work on the COOKADLC concept is thus looking into the possibility for a paradigm-neutral conceptualization phase, based on agent-oriented (social level or knowledge level[11]) analysis, to enhance the flexibility of formalization and implementation phases. This should facilitate future up-scaling or downscaling of interaction needs of identified active entities (agents) and passive entities (objects); resulting in MAS, or distributed OO architectures, (e.g. Component-ware[12], CORBA[13] or Distributed Artificial Intelligence) as deemed necessary by the DSS developers in conjunction with other stakeholders.

---

[5] URL: http://ai.iit.ncr.ca/fuzzy/fuzzy.html.

[6] URL: http://www.fuzzytech.com/.

[7] International Standards Organization (ISO).

[8] Standardization of fuzzy theories and methods, e.g. the proposed IEC Standard, IEC 61131-7 [10], and its integration with mainstream standards like the UML - for notation - and emerging MAS standards, discussed above, are prerequisites for this application

[9] Foundation for Intelligent Physical Agents (its prescriptions): URL: http://www.fipa.org/ and the experimental MAS framework FIPA-OS.

[10] Mobile Agent System Interoperability Facility.

[11] Introduced by A. Newell in [18].

[12] JavaBeans is probably the most popular architecture: See URL: http://java.sun.com/beans/

[13] OMG CORBA: See URL: http://www.corba.org/

References

1. Arlow, J.: Literate Modeling: Capturing Business Knowledge with the UML. In: Proc. The Unified Modeling Language, UML'98, Lecture Notes in Computer Science (LNCS), Vol. 1618. Springer-Verlag, Heildelberg (1998)

2. Arroyo-Figueroa, G., et al.: SADEP - A Fuzzy Diagnostic System Shell - an Application to Fossil Power Plant Operation. Expert Systems with Applications 14 (1998) 43-52

3. Bretthauer, G., et al.: RESEDA – A Program Package for Securing a given Security of Supply in Electric Power Generation Systems. Energietechnik 39 (1989) 289-292 (in German)

4. Dillon, T.S., Chang, E.: Solution of Power System Problems through the Use of the Object-Oriented Paradigm. International Journal of Electrical Power & Energy Systems 16 (1994) 157-165

5. Egwaikhide, I.O.: COOKADLC: An Introduction: URL: http://cookadlc.homepage.com/

6. Egwaikhide, I.O.: Specification of the NEPS as a Model for the Simulation and Test of a KBS for Maintenance Scheduling. Research Report, Institute of Automatic Control, TU B.A. Freiberg, Germany (1996)

7. Felice, P.-D.: Why Engineering Software is not Reusable: Empirical Data from an Experiment. Advances in Engineering Software 29 (1998) 151-163

8. Handschin, E., et al.: Coordination of the Long- and Mid-Term Electric Power Generation System Operation Plans with Fuzzy Logic. Elektrizitätswirtschaft 93 (1994) 204-210 (in German)

9. Harmelen, F.V., Fensel, D.: Formal Methods in Knowledge Engineering. The Knowledge Engineering Review 10 (1995) 345-360

10. Hendricks, P.: Envisioning Knowledge-based Systems Impacts: A GroupWare facilitated Simulation Approach. Expert Systems with Applications 15 (1998) 143-154

11. IEC Homepage: URL: http://www.iec.ch/

12. ISO Homepage: URL: http://www.iso.ch/

13. Kunnathur, A.S., et al.: Expert Systems Adoption: An analytical Study of Managerial Issues and Concerns. Information & Management 30 (1996) 15-25

14. Kruse, R., et al.: Foundations of Fuzzy Systems. Teubner, Stuttgart (1994)

15. Lieberman, H.: Integrating User Interface Agents with Conventional Applications: In: Proc. of the Int. Conference on Intelligent User Interfaces, San Francisco, (1990)

16. Merrill, H.M., Wood, A.J.: Risk and Uncertainty in Power System Planning. Electric Power & Energy Systems 13 (1991) 81-90

17. Neumann, U., Handschin, E., Bretthauer, G., et. al.: Computer Aided Planning System in Electric Power Systems. VDI Report, No. 1252. (1996) 213-222 (in German)

18. Newell, A. The Knowledge Level. Artificial Intelligence, Vol. 18. (1982) 87-127

19. Nwana, H. S., Ndumu, D.T.: A Perspective on Software Agents Research. The Knowledge Engineering Review 14 (1999) 1-18

20. Nwana, H. S.: Software Agents: An Overview. Knowledge Engineering Review 11 (1996) 1-40

21. OECD (Ed.): Application of Competition to the Electricity Sector. Report OCDE/GD(97)132, Paris (1997)

22. OMG Task Force on UML, Homepage: URL: http://uml.shl.com/

23. Rahimi, F.A., Vojdani, A.: Meet the Emerging Transmission Market Segments. IEEE Computer Applications in Power 12 (1999)

24. Schreiber, S. Th., et al.: Knowledge Engineering and Management: The CommonKADS Methodology. The MIT Press, Massachusetts (1999)

396

25. Siler, W.: Building Fuzzy Expert Systems: URL: http://members.aol.com/wsiler/
26. Studer, R. et al.: Knowledge Engineering: Survey and Future Directions. In: Puppe, F. (ed.): Proc. of the 5th German Conference on Knowledge-based Systems, Lecture Notes in Artificial Intelligence (LNAI), Vol. 1570. Springer-Verlag, Heidelberg (1999)
27. Sun Microsystems Inc.: JAVA Homepage: URL: http://java.sun.com/
28. Turban, E.: Decision Support and Expert Systems: Management Support Systems. Macmillan Publishing Company, New York (1993)
29. UML Resource Center: Rational Software: URL: http://rational.com/uml/
30. Utesch, M., Egwaikhide, I.O., Bretthauer, G.: Application of Genetic Algorithms for the Query Optimization in the Data Management of a KBS. In: Proc. of the 42. Int. Scientific Colloquium, TU Ilmenau, Ilmenau (1997) 359-364 (in German)
31. Varga, L.Z., Jennings N. R., Cockburn, D.: Integrating Intelligent Systems into a Cooperating Community for Electricity Distribution Management. Expert Systems With Applications 7 (1994) 563-579
32. Verstege, J., et al.: Liberalization of the Power Supply - Effects on Planning and Optimization Tasks, VDI Reports, No. 1352 (1997) 9-24 (in German)
33. Wooldridge, M., Jennings, N. R.: Pitfalls of Agent-Oriented Development. In: Proc. 2nd Int. Conf. On Autonomous Agents (Agents-98), Mineapolis, USA, (1998) 385-391
34. Zadeh, L.A.: Fuzzy Sets. In: Information and Control, 8 (1965) 333-353
35. Zadeh, L.A.: The Role of Fuzzy Logic in the Management of Uncertainty in Expert Systems. Fuzzy Sets and Systems 11 (1983) 199-227
36. Zadeh, L.A.: Is Probability Theory sufficient for dealing with Uncertainty in AI: A negative View. In: Kanal, L.N and Lemmer, J. F. (eds.): Uncertainty in AI. Amsterdam (1986)

Application of the FSOM to Machine Vibration Monitoring

Ingo Jossa [1], Uwe Marschner [1], Wolf-Joachim Fischer [1]

[1] Dresden University of Technology
Semiconductor and Microsystems Technology Laboratory
D-01062 Dresden, Germany
jossa@ihm.et.tu-dresden.de

Abstract. In this paper we are describing the application of the Fuzzy Self-Organizing Map (FSOM) and some feature extraction methods to a intelligent microsystem for online machine vibration monitoring. We have developed such a microsystem with some of the most advanced semiconductor and microsystem technologies. On the basis of these technologies it is possible to perform the complete condition monitoring and diagnosis stage directly in the microsystem which is applied to a observed machine.

1 Introduction

The integration of semiconductor, micro and intelligent (softcomputing) technologies enables the possibility to develope a new class of powerful machine condition monitoring and diagnosis microsystems with exceptable prices. In contrast to traditional systems these new intelligent microsystems integrating all the condition monitoring and diagnosis steps in a small housing which can be directly applied to the observed motor or machine. The housing of our system is 3 cm long and has a diameter of 1,5 cm. In this small volume we have integrated a 16 bit DSP with 100MIPS of computing power, 512Kbyte SRAM, 1Mbyte EEPROM, a field bus interface and the vibration sensor [2]. With this computing power and memory space it is possible to use advanced signal processing and nonlinear methods for feature extraction and classification. The large memory allows a online and autonomous observation of machine vibration signals including long time trend analysis of machine failures which are slowly growing up. Over the field bus interface it is possible that the microsystem communicates with a host computer located in a industrial process control center or in the World Wide Web.

For the classification task we are using the FSOM algorithm which becomes as input a feature vector computed by different feature extraction methods.

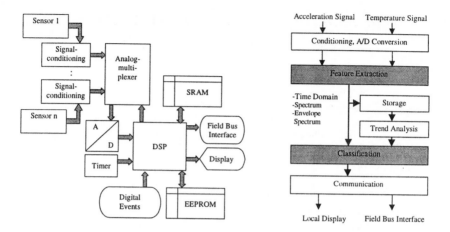

Fig. 1. Schematic and dataflow realized in the microsystem

2 Feature Extraction Methods

In the time domain we are using some statistical methods like median, range, standard deviation, root mean square value (rms), quadratic mean value, kurtosis and percentiles (lower quartile, upper quartile, quartile range). Our best results by using statistical methods we have found with the quartile range and the range of the vibration signals (Fig.2).

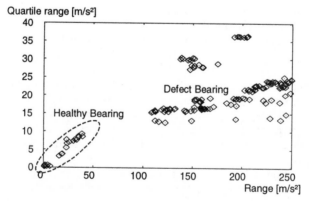

Fig. 2. Quartile range and range of the vibration signal as features

We also using the envelope and the spectrum of the envelope as features. The envelope spectra (Fig.3) shows the frequency of the failure and its higher harmonics. We are using a threshhold value for the detection of components with a magnitude higher than the threshold. This means that we are making a integration over all the spectral components with a magnitude higher than the threshold. The value of this

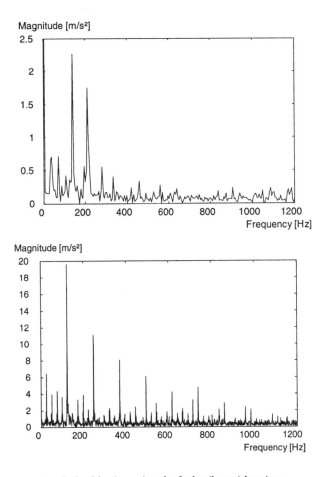

Fig. 3. Envelope spectra of a healthy (upper) and a faulty (lower) bearing

integration we are using as one of the features. In fact of the high computing power and the large memory space available in our microsystem we are actually working on the application of advanced signal processing methods like time frequency distributions and wavelet transforms.

We are computing the time frequency distribution with the so called Wigner-Ville distribution (WVD). A typical fault detection procedure are based on visual observation of the WVD contour plots. Dark zones and curved bands in the contour plot are the main features of an impulse caused by a bearing fault. The progression of a fault can be monitored by observing the changes in the WVD contour plot. Such a visual observation is difficult and needs an great amount of computational power and memory space which is only availlable on workstations. In fact of this our method differs from the above. We are only using a part of the time frequency plane and to this part we are applying a coarse grained pattern recognition method. For this we are subdividing the time frequency plane (Fig. 4) between the frequencies of 0...4 KHz in 32 areas. The energy represented by every of this areas is used as input to a self-organizing neural network. This neural network delivers on its output a value that is used as one of the features. After our first experiences with the WVD we are now testing the so called Choi-Williams distribution (CWD) which shows not so strong cross and inference terms as the WVD.

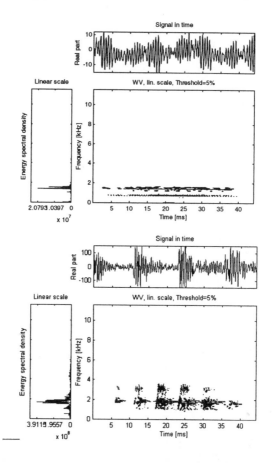

Fig. 4. The Wigner-Ville distribution as feature for healthy (upper) and faulty (lower) bearings.

On nearly the same way as used by the feature extraction with the WVD we are using the wavelet transform as feature extraction method. For this we are subdividing the lower half of the frequency plane (Fig.5) in 32 areas and using the energy represented by every of this areas as input to a self-organizing neural network. The output value of this neural network is also used as one of the features. We also using the binary tree like structure of the wavelet decomposition of the vibration signals as basis for feature extraction. This method is much faster as the analysis of th energie distribution in the time frequency plane.

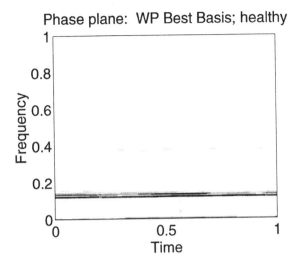

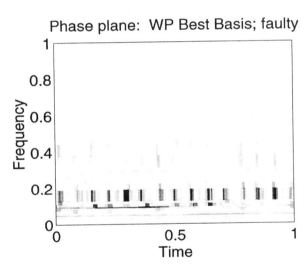

Fig. 5. The wavelet transform as feature for healthy (upper) and faulty (lower) bearings.

3 Fuzzy Self-Organizing Map

Two algorithms are integrated in the Fuzzy Self-Organizing Map (FSOM), the Self-Organizing Map (SOM) and the Fuzzy c-Means algorithm (FCM) [1].

The SOM introduced 1982 by T.Kohonen [3] is one of the most popular neural network algorithms with a wide range of applications. Kohonen describes over 1500 references in his 1995 book [4].

Main principle of the SOM is the vector quantization of a n-dimensional input space $\Re^n$, that means, the infinit number of input data points in $\Re^n$ is represented by a small set of reference vectors $\mathbf{w}_i \in \Re^n$ with $m \leq n$. In the most cases $m = 2$ and the SOM defines a mapping from the input data space $\Re^n$ onto a regular two-dimensional array of nodes. The node i represents with its weight vector the reference vector $\mathbf{w}_i \in \Re^m$. The lattice type of the array is mostly defined as rectangular. An input vector $\mathbf{v} \in \Re^n$ is compared with all the $\mathbf{w}_i \in \Re^m$ and the best match defines the node c with the greatest "response" in the map. Is the Euklidean distance $\|\mathbf{v} - \mathbf{w}_i\|$ used for the comparisation the best matching node $\mathbf{w}_c$ is defined as:

$$\|\mathbf{v} - \mathbf{w}_c\| = \min_i \{\|\mathbf{v} - \mathbf{w}_i\|\} \tag{1}$$

The input vector $\mathbf{v}$ is mapped to the node $\mathbf{w}_c$. This mapping is a kind of nonlinear projection of a propability density function of high-dimensional input data onto a two-dimensional array. During the learning process of the SOM, those nodes which are topologically close in the array will learn from identical inputs. Useful values of the $\mathbf{w}_i$ are given by the convergence limits of the following learning process:

$$\mathbf{w}_i(t+1) = \mathbf{w}_i(t) + h_{ci}(t) \cdot [\mathbf{v}(t) - \mathbf{w}_i(t)] \tag{2}$$

where t is the discrete-time coordinate and $h_{ci}(t)$ is the so-called neighborhood kernel used on the lattice points. Usually $h_{ci}(t) = h(\|\mathbf{r}_c - \mathbf{r}_i\|, t)$, where $\mathbf{r}_c \in \Re^2$ and $\mathbf{r}_i \in \Re^2$ are the location vectors of nodes c and i. A widely applied neighborhood kernel is the bell-shaped function:

$$h_{ci}(t) = \alpha(t) \cdot e^{\left(-\frac{\|\mathbf{r}_c - \mathbf{r}_i\|}{\sigma^2(t)}\right)} \tag{3}$$

where $\alpha(t)$ is some monotonical decreasing function of time $0 < \alpha(t) < 1$, and $\sigma(t)$ is also a monotonical decreasing function of time that defines the width of the neighborhoodkernel.

Tsao et. al [5] has introduced the FSOM which delivers some advantages over the both integrated algorithms. The sequential presentation of input pattern and also the sequential adaption of the weight vectors like the SOM are replaced by the parallel method of the FCM. This parallel strategy makes the algorithm independent from the presentation order of the input pattern. The parameter λ of the SOM is replaced by the membership values of the FCM.

The learning process of the FSOM starts with the choice of the number n of neurons or reference vectors, the termination criterion ε and the distance measure (at most the Euclidian distance) followed by the initialisation of the weight vectors $\mathbf{W}_i$ with random numbers. The next step is the choice of the values for the exponent $m_0 > 1$ and the stepsize for the exponent. At a given time step t the membership value $u_{ik,t}$ is computet by the use of all the weight vectors $\mathbf{W}_{i,t}$ and the input vectors $\mathbf{V}_{i,t}$

$$u_{ik,t} = \frac{1}{\sum_{j=1}^{n}\left(\frac{\left\|\mathbf{v}_k - \mathbf{w}_{i,t}\right\|}{\left\|\mathbf{v}_k - \mathbf{w}_{j,t}\right\|}\right)^{\frac{2}{m_t-1}}} , \qquad \begin{matrix} i = 1,\ldots,n \\[2mm] k = 1,\ldots,K \end{matrix} \qquad (4)$$

Using the membership values and the actual exponent weight m_t the learning rate $\lambda_{ik,t}$ for every neuron is

$$\lambda_{ik,t} = \left(u_{ik,t}\right)^{m_t} \qquad (5)$$

this means that every neuron have for every input vector a special learning rate. The schedule of m_t is realized with

$$m_t = m_0 - t \cdot \Delta m \qquad (6)$$

This algorithm realizing an equivalent computation to the self-organizing process in the SOM. The neighborhood kernel that is used for the weight adaption in the SOM is now replaced by the exponent weight m_t which can also seen as a parameter for the fuzziness.

With the special learning rate $\lambda_{ik,t}$ for every neuron and the decreasing exponent weight $m_t \to 1$ the lateral distribution of learning rates is a function of time, which is „sharpens" at the winner neuron as $m_t \to^+ 1$. Is $m_t \approx 1$ the adaption is applied only to the winner neuron . The adaption step at a given time t is

$$\mathbf{w}_{i,t} = \mathbf{w}_{i,t-1} + \frac{\sum_{k=1}^{K} \alpha_{ik,t} \cdot \left(\mathbf{v}_k - \mathbf{w}_{i,t-1}\right)}{\sum_{s=1}^{K} \alpha_{is,t}}, \qquad i = 1,\ldots,n. \tag{7}$$

For the control of the termination the changes of $u_{ik,t}$

$$E_t = \sum_{i=1}^{n} \sum_{k=1}^{K} \left\| u_{ik,t} - u_{ik,t-1} \right\|^2 \tag{8}$$

are used. Is the error E_t smaller than the termination criterion ε the learning process is finished.

After the training the FSOM is used for the observation of the state space of a machine. The state space is defined by the computed features. Every subspace defined by the nodes of the

FSOM represents faulty or healthy machine states. All the subspaces that representing healthy states are concluded to a one large common healthy subspace. The same is possible with the faulty subspaces. After this the classification of the faulty and the healthy machine state can be realized by the FSOM. Advantages of the FSOM over the SOM is the better modelling of the floating and drifting nature of machine state spaces and also the faster convergence of the algorithm. The better modelling of the floating and drifting nature of machine states means that for a healthy machine state located nearly the border to a faulty subspace a significant membership value for the faulty machine state is computed. These membership value makes a better observation of the machine state changes possible.

3 Summary

In this paper the integration of the FSOM and some feature extraction methods in a intelligent microsystem for online machine vibration monitoring is reported. With some of the most advanced semiconductor, micro and intelligent technologies we have developed a powerful microsystem which has enough computing power and memory space to use statistical and advanced signal processing methods for feature extraction and the FSOM for classification.

The prototype of our system is now realized and the real world application to motors of our industrial partner is planned for the end of this year.

Actually we are working on the improvement of our classificator. For these we are realizing a great amount of measurements on different motors.

References

1. Bezdek, J.C.: Pattern Recognition with Fuzzy Objective Function Algorithmen. Plenum, New York (1981)
2. Jossa, I., Marschner, U., Fischer, W.-J.: A Intelligent Microsystem for Machine Vibration Monitoring. Proceedings KES'99, Adelaide, Australia (1999)
3. Kohonen, T.: Self-Organizing Formation of Topologically Correct Feature Maps. Biol. Cyb. 43 (1982) 59-69
4. Kohonen, T.: Self-Organizing Maps. Springer, Berlin Heidelberg (1995)
5. Tsao, E.C.-K., Bezdek, J.C., Pal, N.R.: Fuzzy Kohonen Clustering Networks. Pattern Recognition 27 (1994) 757-764

Near-Term Forecasting of Drilling Cost of Borehole by Fuzzy Logic

G. Sementsov, I. Chigur, I. Fadeeva

Ivano-Frankivsk State Technical University of Oil and Gas
Karpatstka str. 15 Ivano-Frankivsk,
76019, Ukraine

Abstract. Attempts to provide self-control of process of long holing on oil and gas have not given due effect owing to complication of object, it fuzzy and equivocation of the information. In this connection it is offered to use for management of drilling expert systems, which one use fuzzy models and methods of the theory of fuzzy control systems.

Introduction

The chisel enterprises constantly aspire to augment profitability of drilling. One of the relevant paths in reaching this purpose is use of the optimized management of the chisel enterprise. It means, that each trip of a chisel should ensure maximum heading on a chisel, minimum costs of time and minimum cost of meter heading of a well.

The second problem - limitation on buyings of tubes, chisels, reactants from other enterprises. In different time they can be various. Therefore modern control systems are indispensable for the chisel enterprises. They allow to optimize process of drilling of borehole. As cost of meter heading of a well can be determined only at the end of trip of a chisel, for the chief the large concern introduces forecasting cost of meter heading of a well for one or two days - daily budgeting, forecasting till some hours, that is operative budgeting. The purpose of such forecasting of cost of meter heading of wells to be near-term forecasting of the cost price of drilling of borehole.

1 Feature of Management of Drilling on Oil and Gas

Management of drilling on oil and the gas has the following features:

- From the stanpoint of managing a problem we are, among others, faced with multiplicity that cannot be formalized completely.
- From the ecological point of view if not the special measures of urging are envisioned, the costs of increased quality of borehole drilling are caused by the enterprise, whereas the effect of quality increase is of advantage for the oil or gas mining enterprise.

- From the point of view of comprehensive management, in a system should be aggregated budgeting of drilling of borehole, organization of management and management of know-how.
- From the point of view of production, some groups of problems are separated: management of technological processes, quality control of drilling, quality control of work of performers.

For constructing algorithms of management it is necessary to decide problems of statistical check and regulations, optimization in conditions of equivocation of the information both variations of pattern of model and yardsticks with magnification of borehole depth.

The process of management of drilling of borehole can be introduced as process, which one is characterized by a gang of functions of the purpose, which one will be realised at all stages of drilling and encompass all levels of management.

The special significance there is a nonseparable connection of problems.

2 Statement of a Problem of Optimization of Drilling of Borehole

For increase of performance of drilling of borehole it is necessary to decide problems of optimization of know-how, operating control and functioning of each subsystem by yardsticks, which one influence the cost price of drilling of borehole at all stages.

At a solution of such problems the different analytical and statistical methods, methods of simulation on the computer are used. The new possibilities for an integrated solution of this problem give methods of a simulation modelling, which one are developed in a real time, and use the fuzzy logicians.

The problems of optimization of drilling are formulated in this case as follows:

It is considered that during drilling of borehole the set of technological variables is inspected:

$$x_1, x_2, ..., x_n$$

which one influence economical indexes of process

$$y_1, y_2, ..., y_k,$$

and on losses on each lease:

$$q_1, q_2, ..., q_p.$$

The limitations, which are preselected by the technological operating instructions, sound as

$$\begin{cases} x_n^0 \leq x_n \leq \tilde{x}_n, \\ k_i^o \leq \sum Q_i x_i \leq \tilde{k}_i, \end{cases}$$

where $k_i^0, \tilde{k}_i, Q_i$ - numerical constants.

By optimization of drilling follows by methods of analytical simulation to select limitations:

$$x_n^0, \tilde{x}_n, k_i^0, \tilde{k}_i.$$

So that all indexes are guaranteed were stacked in boundaries $y_k^0, ..., \tilde{y}_k$, and all indexes of losses did not exceed some boundary significance $\tilde{q}_p$.

In turn, $y_k^0, \tilde{y}_k, \tilde{q}_p$ are set so that in these conditions to receive a well with higher consumer qualities, and also to lower the costs.

In the integrated intellectual control system of drilling of borehole intercoupling of these problems of management of technological processes and building of a well with problems of supply of the minimum cost price of a well to the greatest degree can be taken into account.

3 Cost Model of Drilling Boreholes

From mathematical model [1] the forecast significance of endurance of drilling $t_{...}$ for a case is instituted, when the wear of arms advances wear of bearings of a chisel

$$t_d = \frac{\left[(1+b)^2 - 1\right]}{k_2 p^{\alpha_2} n^{\beta_2}}. \tag{1}$$

And also forecast significance of endurance of drilling $t_{...}$, when the wear of bearings advances wear of arms of a chisel.

$$t_{cn} = \frac{1}{k_3 p^{\alpha_3} n^{\beta_3}}. \tag{2}$$

After substituting (1) and (2) into the expression for the objective function we get

$$B = \frac{\left[B_2(t_d + t_{cn}) + B_g\right]}{h_p}, \tag{3}$$

where: B - cost of a business hour of drill unit in DM;

   t_d- drilling-time, which one is instituted by stability of bearings and arms of a chisel at the preselected significances of an axial load of bit speeds, hour;

   B_g- cost of a chisel, DM.;

   t_{cn}- time, which one is spent for implementation of trippings, hour;

   h_p - heading for trip of a chisel at the preselected significances of an axial load on a chisel and rotating speeds that corresponds to drilling-time $t_{..}$.

$$B_{on} = \left[B_d(k_3 p^{\alpha_3} n^{\beta_3})^{-1} + B_{cn}t_{cn} + B_g\right] \cdot \left[kp^{\alpha} n^{\beta} \ln(k_0 p^{\alpha_0} n^{\beta_0} + 1)\right]^{-1}. \tag{4}$$

For a case, when the wear of arms advances wear of bearings:

$$B_{oz} = \left[B_d(1+m)^2(k_2 p^{\alpha_2} n^{\beta_2})^{-1} + B_{cn}t_{cn} + B_g\right] \cdot \left[Qkp^{\alpha} n^{\beta} \ln(1+m)\right]^{-1}, \tag{5}$$

where

$$k_0 = k_2 k_3^{-1},$$
$$\alpha_0 = \alpha_2 - \alpha_3,$$
$$\beta_0 = \beta_2 - \beta_3.$$

Basic data for account of a cost of drilling of one meter of a well B_{on} and B_{oz} are the significances of coefficients k_i, α_i, β_i, which one can be received as a result of a solution of a problem of identifying of arguments of model, and also cost of a chisel B_g and time of trippings $t_{..}$.

Equations (4) and (5) enable to determine a cost of drilling of one meter of a well in a real time and to use this information for forecasting a cost of drilling of a well B_c, which one is equal:

$$B_c = \sum_{i=1}^{n} B_{on(oz)} . \tag{6}$$

Here, n is the amount of trips of chisels in a well.

At the same time it is necessary to allow, that during drilling of borehole the natural factors - property of formations, formation pressure, temperature, and also property of a chisel are inflected. In this connection the coefficients of the equations (4) and (5) become obscure and are subject to repetitive identifying. The equivocation of the information about process of drilling results in necessity of use of the rules of a fuzzy system.

The principle of constructing of model based on fuzzy logic is shown in a Fig. 1. It is based on analysis of technological arguments $x_1, x_2, ..., x_n$, influencing the natural factors and their proportions, and also economical indexes $y_1, y_2, ..., y_k$.

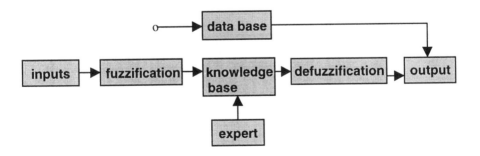

Fig. 1. The principle of constructing a model based on fuzzy logic

The proposed method for a solution of a problem of near-term forecasting is better than idle time of regression analysis. This method uses fuzzy logic rules [2].

The proportion between accessible by inputs and outputs is introduced the fuzzy rules *P: IF...THEN...ELSE....* It has allowed to describe •omplicated situations by a simple method and to construct an expert system ECON-DRILL.

The inputs will be converted to a fuzzy kind by fuzzification and create the base of knowledge. The outputs receive the information from a knowledge base after transforming to a precise kind by defuzzification.

The complementary knowledge of the experts of process can be included in a knowledge base, that is the knowledge base can be reamed in time.

Besides it is necessary to allow, that the designed proportions at drilling are inflected in time. On the other hand, to become clearly that the well customized fuzzy system of the rules promotes occurrence of diverse sight on process of drilling of borehole.

Conclusion

The designed model of near-term forecasting of cost of meter heading of a well and cost price of drilling of borehole allows together with the fuzzy attitude of technological situations in conditions of information equivocation to provide process control of drilling and essentially to diminish the cost price of a well and to increase its profitability.

References

1. Sementsov, G.: Automatization of Drilling. An Introduction. Ivano-Frankivsk, Ukraine (1999)
2. Sementsov, G., I. Chigur, I.: The Method of Technical Condition Monitoring of Rock Bit During Hole Drilling Basing on Fuzzy Logic. In: Proc. 7th Zittau Fuzzy Colloquium, Zittau, Germany (1999) 108-118